Biostratigraphy

Microfossils and Geological Time

Using fossils to tell geological time, biostratigraphy balances biology with geology. In modern geochronology – meaning timescale-building and making correlations between oceans, continents and hemispheres – the microfossil record of speciations and extinctions is integrated with numerical dates from radioactive decay, geomagnetic reversals through time, and the cyclical wobbles of the Earth-Sun-Moon system. This important modern synthesis follows the development of biostratigraphy from classical origins into petroleum exploration and deep-ocean drilling. It explores the three-way relationship between species of micro-organism, their environments and their evolution through time as expressed in skeletons preserved as fossils. This book is essential reading for advanced students and researchers working in basin analysis, sequence stratigraphy, palaeoceanography, palaeobiology and related fields.

BRIAN MCGOWRAN worked as a governmental and consultant geologist-palaeontologist in Australia, New Guinea and the southwest Pacific for several years. For three decades he was an academic, teaching and supervising in a wide range of the Earth sciences, and he has held visiting positions at the Universities of Vienna and Princeton and at the Geological Survey of Austria. His previous publications include technical books on geological excursions through the famous and touristic Cenozic outcrops of southern Australia, as well as over 200 journal articles. Oceanic drilling in the Indian Ocean in the early days of the Deep Sea Drilling Project anchored his long-time interest in the history of the Cenozoic Era and especially on the interplay through geological time of the three environmental realms – the terrestrial and oceanic realms, and the shallow seas spilling across the continental margins, the neritic realm, in between. His main focus in recent years has been the geological record of southern Australia, a locale that has had the front-row seat witnessing the birth and development of the Southern Ocean, engine room of the cooling planet during the Cenozoic Era.

Biostratigraphy

Microfossils and Geological Time

BRIAN MCGOWRAN
The University of Adelaide

CAMBRIDGE
UNIVERSITY PRESS

CAMBRIDGE UNIVERSITY PRESS
Cambridge, New York, Melbourne, Madrid, Cape Town, Singapore, São Paulo

CAMBRIDGE UNIVERSITY PRESS
The Edinburgh Building, Cambridge CB2 2RU, UK
Published in the United States of America by Cambridge University Press, New York

www.cambridge.org
Information on this title: www.cambridge.org/9780521837507

First published 2005

Printed in the United Kingdom at the University Press, Cambridge

A record for this publication is available from the British Library

ISBN-13 978-0-521-83750-7 hardback
ISBN-10 0-521-83750-2 hardback

For the six women in my inner universe:

Fiona, Heidi, Lisi, Rosi, Jasmine

and

Susi

Contents

Preface

Some geologists and palaeontologists recount their awakenings as occurring while collecting minerals and rocks, or when suddenly seeing – really seeing – fossils in rocks for the first time. Others chose geology to fill in the fourth subject in first-year science, as not much more than a happy accident. Many of the teachers at my high school were of the Depression generation for whom teaching was the main career option before the war, and some were indeed excellent in the mathematics-science area. But becoming a successful junior cricketer was more important to me than becoming a scientist. In those days science for the A-stream comprised physics, chemistry and maths, all of which held some interest for me, but that was all. Botany or physiology were for girls, and evolutionary biology and geological time remained as much a shrouded mystery for my generation as it was for our otherwise well-educated antecedents. I came upon W. W. Norton's *Physical Geology* (1915) by chance at the age of 16 – it had belonged to an aunt who did Geology I with Douglas Mawson – and I became entranced with theories of the landscape and its history which, together with other natural history and human history, were quite absent from my formal education. When I should have been studying for the public exams, I was digging around in the geology section of the Adelaide public library.

Without being very self-aware about it, I was groping towards the realization that Earth and life evolution and history were what really mattered. There were unsettling experiences on the way – my first-ever Geology I practical was on wooden models of crystals and the first zoology lectures were on the chemistry of carbohydrates. This was the old inductionist philosophy at work – begin with the atoms and molecules and work up to grand notions such as Earth history or organic evolution, no matter that it is they that make their sciences autonomous (I encountered that word only years later). I learned a lot from rather few people and more from chatting with some of them than sitting through lectures.

Another life-bending experience was a summer in the Bureau of Mineral Resources in Canberra after second year – the field veterans advised me that palaeontology was women's work and tended not remunerate as well as the managing of field-mapping parties did (because camp leaders managed cooks and drivers and palaeontologists did not). Constructive encouragement from numerous geologists and palaeontologists when Australian Geological Surveys and academic departments were strong was the best education possible for a young, naïve and brash enthusiast. Education was a bit patchy, but discovering the soul of geology in rock relationships and Earth history surely compensated for the gaps. I owe Martin Glaessner and Mary Wade incalculably for encouraging voraciously wide, curiosity-led reading. Pettijohn's sediments, Marshall Kay's geosynclines and Kuenen's marine geology were exciting enough in the mid 1950s, but pale in retrospect compared to the Big Three of the second Darwinian revolution in evolutionary biology, the modern synthesis – Dobzhansky's *Genetics and the Origin of Species* (1937), Mayr's *Systematics and the Origin of Species* (1942), and Simpson's *Tempo and Mode in Evolution* (1944).

My first research was in Cambrian stratigraphy and field mapping before entering micropalaeontology, not answering any great calling or foreseeing a powerful research programme but for the (perceived) money and a confused dream of singing calypsos on Trinidadian beaches one day. Instead, I became very interested in the microanatomy, taxonomy and stratigraphy of the foraminifera. With no scanning electron microscope and no computer drafting, we did our own thin sections, washing and picking, drawing and photography of specimens and plate preparation for publication (my amateurish drawings became embedded in the main reference works of the field). Writing a thesis on the Mesozoic–Cenozoic boundary in Australia, publishing too little of it, and becoming a consultant to exploration was followed by an appointment to supervise the Palaeontology Section in the Geological Survey of South Australia – all of which gave me an abiding respect for how applied palaeontology interacted with mapping and drilling and for the workings of organizations outside the ivory tower. And here we are.

About this book

George Gaylord Simpson prefaced his *Principles of Animal Taxonomy* with this quote from A. J. Cain: 'Is it not extraordinary that young taxonomists are trained like performing monkeys, almost wholly by imitation, and that in only the rarest cases are they given any instruction in taxonomic theory?' So, too in biostratigraphy – how do we expound our subject? There are those with horrible memories of having to memorize lists of index fossils, for example of the British

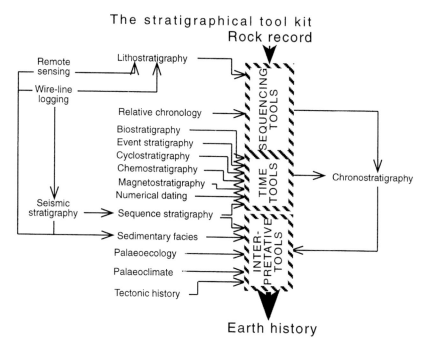

Figure 0.1 From the rock record to Earth history: the context of modern
biostratigraphy in the stratigraphic tool kit with its three boxes of sequencing,
time and interpretive tools (Doyle and Bennett, 1997, with permission). The modern
kit is more complex in its available technologies than were the kits of earlier times.
Biostratigraphy is not the only biology feeding into this kit of sequencing tools, time
tools and interpretative tools transforming the rock record into earth history. Nor is it
desirable to constrain the scope of the term 'biostratigraphy' to fossil ranges and
zones. For example: event stratigraphy contains many bioevents as well as physical
events; and chemostratigraphy relies heavily on an understanding of the biology of
the organisms that record chemical signals in their fossilizable skeletons.

or the German Mesozoic; and many will have been lost to geology forever, for
that feat of memory. Is there any more to biostratigraphy than that – than
determining fossils and testing their constant ranges, over and over again, and
doing an inventory of the various combinations prefixing '-zone'? What could be
more empirical than the building of a fossil succession in the absence of a
coherent theory of that succession, as in the early nineteenth century? Fossils
are where you find them – a guide fossil remains a guide fossil only until it has
been discovered in the wrong place. All very inductive, all very solemn and, well,
excitement-challenged. Modern stratigraphy and biostratigraphy reside in a
more complex and integrated 'tool kit' employed more ambitiously (Fig. 0.1).

What is the meaning of 'biostratigraphy', the word itself? The best, most
comprehensive still modern exposition is Teichert's (1958). According to

Teichert, the word itself seems to have been introduced by Louis Dollo in 1904 to cover 'the entire research field in which paleontology exercises a significant influence upon historical geology', and biostratigraphy thus defined would now embrace several other palaeodisciplines – all of those concerned with the reasons why fossils are found where they are – Jukes's (1862) factors of environment, geography and time (Chapter 1). Diener's *Grundzüge der Biostratigraphie* (1925) covered all of this. Biostratigraphy was defined later as palaeontological stratigraphy, or as stratigraphy with palaeontological methods. (A corollary was: sedimentology is stratigraphy without the fossils.) Teichert outlined the three aspects of the study of fossiliferous stratified rocks thus: (i) their division into locally mappable units, (ii) the local sequence of fossil assemblages in the rock units, and (iii) the correlation of the rocks through the evidence of their contained fossil assemblages. We find that little had changed by the 1970s in the *International Stratigraphic Guide*, in which biostratigraphy was defined as 'the element of stratigraphy that deals with the remains or evidences of former life in strata and with the organization of strata into units based on their fossil content' (Hedberg, 1976). It is clear enough, and reasonable, that the geology of fossils receives more emphasis in biostratigraphy than does the biology of fossils, for biostratigraphers work in a geological environment. Even so, a recent and forward-looking definition is interdisciplinary (Simmons *et al.*, 2000): 'Biostratigraphy is the study of the temporal and spatial distribution of fossil organisms.'

Biotaxonomy and biostratigraphy are two of the quintessentially historical-scientific disciplines, linked through their bonds in the fact and theories of organic evolution at geological timescales; and linked too in their common concerns with historical contingencies – the emergence and the extinction of countless species, for a start. An enquiry into biostratigraphy should include our correlations and age determinations. What are we actually employing under the appellation 'fossils' – species or higher taxa, or biocharacters? This question confronts the evolutionary dualism of taxic evolution and transformational evolution (Eldredge, 1979), which (I find) is a useful way to approach the question of what we actually correlate with (Chapter 4). Why, for example, have biostratigraphers had a cavalier attitude toward fossil population samples and what was once known as the 'species problem'? (Known, that is, in palaeontology; the species problem still thrives in philosophy (Chapter 8).) Can we have finely split, index-fossil 'species' simultaneously with taxa sufficiently robust to contribute to the rejuvenating field of the fossil record in macroevolution? Or will we make a clean break with the clutter of the Linnaean system of classification and identification and go down the road of some kind of 'operational taxonomic unit' (in the charmless neologism of the 1960s: Hull, 1988) more amenable to the accountancy of the computer age? Also surviving is the ancient question of why some fossils or groups empirically seem to be more useful than others – the opposite of the 'living fossil'. There is yet another hoary problem of interest and (I think) importance: are there natural biostratigraphic units? I refer to the perceptions of fossil assemblages as remains of ancient communities and to the search for punctuations of the fossil succession, both subsumed at times under 'ecostratigraphy', which implies something beyond the dry empiricism of selecting and baptizing zone fossils. This is drawn together in Chapters 4 and 6 – biostratigraphy and biohistorical theory.

But the results of Phanerozoic organic evolution comprise a vast and sprawling panorama of body fossils and trace fossils and their assemblages. There has to be a selection of material. Easily the best approach is via the record revealed by the Deep Sea Drilling Project (DSDP) and the Ocean Drilling Project (ODP) and their forerunners in subsurface geology and basin analysis driven by petroleum exploration. I am referring to marine micropalaeontology in the later Phanerozoic eon and mostly in the Cenozoic Era. Certainly the planktonic foraminifera and other phyto- and zoo-protists have no monopoly over our changing insights into biostratigraphy in recent decades. But they do have a dominating physical presence in the literature at the cutting edge of the disciplines, just as they are retrieved in their thousands in cores from the depths of continental and oceanic basins, in the latter often actually constituting their own sediment. And so I justify Chapter 2 on the biostratigraphy of fossil foraminiferal microplankton, from its beginnings in the zonation and correlation of the deepwater sediments of the Alpine mountain belts to the modern worries about succession – homotaxy – and diachrony. Modern biostratigraphy sits in more complex conceptual and technological contexts than did its antecedents (Fig. 0.1) and zonations have to be integrated into modern geochronology, as in Chapter 3.

Likewise, the polemics over fossils and time have continued into the microfossil domains, although there are signs – more than signs – of a mounting impatience with the old arguments over the synonymy or otherwise of biostratigraphy with chronostratigraphy. Just as some would sidestep species on their way into automated methods of correlation, so too is the zone being subverted in the 'age models' of modern palaeoceanography. Meanwhile, Chapter 7 on biostratigraphic and chronostratigraphic classification is an opportunity to parade case histories from the Cenozoic. The topics here have many connections with the topics of Chapter 3 and the division is somewhat arbitrary. Nor need we adhere to the cases from the marine record. The successions of macro- and micro-mammals not only have an intrinsic worth and interest, but their interactions with the marine successions are an essential part of the enquiry. Ages and stages are used more freely in the pre-Cenozoic, and probably increasingly in the Cenozoic, not least due to the rise of sequence stratigraphy.

As Sir Karl Popper advocated tirelessly, the inductivist model of science, whereby we collect data and draw conclusions from it, is not very useful or stimulating or realistic. No: we always have theories and biases influencing our choice of observations and our perception of 'facts', and biostratigraphy is no different. We do not simply identify, range, zone and correlate in a tedious induction – for we have stubborn, crazy, powerful and pet ideas about tectonism, climatic change, or evolutionary relationship whose triumph or tragedy can depend utterly on the fragility or robustness of the chronology or correlation. Be it in the mode of 'time's arrow' or 'time's cycle' (Gould, 1987), there is a feedback between biostratigraphy and the reasons why it is being carried out. As the historical sciences evolve, the critical and interesting questions change, and our *Weltanschauung* changes. Much of that shift in world view demands renewed scrutiny of the chronology, including greater chronological refinement, and it is now commonplace that fossils alone have not nearly the information for correlation that is available synergistically from their integration with various physical signals. And there is more: there is the choice of an agnostic approach or a thoroughly committed approach towards the nature of the stratigraphic record. Hence *systemic stratigraphy*, a timely resurrection of an earlier merging of history with its controlling chronology, is the central topic of Chapter 5, where we consider the use of Quaternary-style reversible events in the pre-Quaternary, of the integration of fossil events with physical events in event stratigraphy, and of a stratigraphic record consisting of cycles.

We can heal the split between the so-called index fossils and the so-called facies fossils (Fig. 0.2). Fossils carry signals of genealogy and ecology on the one hand, and age and environment of the other. The former ('sequence biostratigraphy') is strengthened in the context of the latter ('evolutionary palaeoecology'). It is difficult or impossible to draw a line anywhere in this diagram to exclude something as non-biostratigraphic. Among the various kinds of microfossil comprising the contents of micropalaeontology (e.g., Glaessner, 1945; Pokorny, 1963; Haq and Boersma, 1978; Brasier, 1980; Bignot, 1985; Lipps, 1981, 1993), the foraminifera are particularly broad in the range of their signals (Fig. 0.3).

This book has had a prolonged gestation and I have accumulated numerous debts of gratitude more wide-ranging than its subject matter and production. I have learned the hard way what many authors learn that way – assistance in all kinds of tangible ways is fine and indispensible, and acknowledgments feel inadequate as one struggles to articulate them. But deep approval of what one is trying to do is on a higher plane. I have benefited from uplifting of that kind ever since I was at school, and the following short list omits some wise and kind people. For early and strong encouragement in Earth and life history and in thinking about biostratigraphy: Martin Glaessner, Mary Wade, David Taylor,

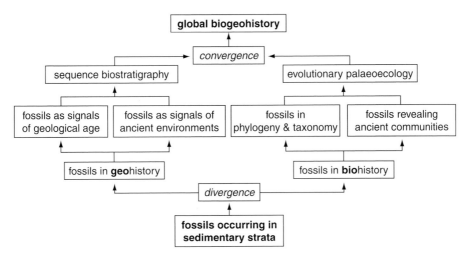

Figure 0.2 From fossils to biogeohistory: pathways diverging and subsequently reunited. The divergence near the base is between fossils in geology and fossils in biology. Within the former, or 'applied palaeontology,' the age/facies split arose from the eighteenth-century question: is that fossil informing us about the age or the depositional environment of this stratum? Hence the two streams, classical biostratigraphic zones progressing (left stream) and progressing concepts in eco- and sequence-stratigraphic events and units (right stream). Sequence biostratigraphy unites the two streams in the modern synthesis of sequence stratigraphy and exogenic biogeohistory (Chapter 5). Evolutionary palaeoecology unites morphology and systematics with community reconstruction at timescales beyond the reach of ecology (Chapter 6). The whole diagram is pervaded with the need for age-control, hence for biostratigraphy.

Roye Rutland, Al Fischer and Bill Berggren. For the cross-disciplinary excitement in the early days of deep-ocean drilling: John Sclater and Steve Gartner. Colleagues and students: Nell Ludbrook, Murray Lindsay, Wayne Harris, Clinton Foster, Rob Heath, Mike Hannah, Geoff Wood, Stephen Gallagher, Guy Holdgate. Collaborations keeping me going in the day-to-day stuff: Amanda Beecroft, Graham Moss and – now well into our second decade of genially apportioned research and writing – Qianyu Li. (Among their scientific efforts, the good people in the room next door helped me into the computer age.) Drafts were read by Graham Moss, Qianyu Li, Marie-Pierre Aubry, Bill Berggren and Al Fischer. I deeply appreciate not only their constructive criticisms but especially their enthusiasm and encouragement. Qianyu Li's help with the array of figures was immense. Sally Thomas at Cambridge University Press was enthusiastic and decisive at all times in getting this project accepted, patiently advising its author and transmuting the manuscript into a book. It has been a pleasure and an education to work with Sally, Carol Miller, and their colleagues.

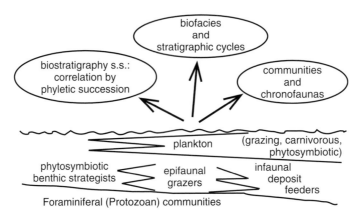

Foraminiferal (Protozoan) communities

Figure 0.3 Reasons for studying the foraminifera. Benthics comprising a mix of epifaunal, infaunal and photosymbiotic strategists fossilize together with planktonics, themselves possessing a range of strategies in lifestyle (often overlapping). Their understanding was driven by classical biostratigraphy s.s. (i.e., correlation and age determination based on the reconstructed ordination of speciations and extinctions), and biofacies (fossil assemblages signalling environments and environmental change and cyclicity). These streams have been united recently in sequence stratigraphy and the search for cycles in the rock record. The flipside of these 'geological' drives is the 'palaeobiological' question of ancient communities and their long-timescale equivalents, chronofaunas – which question unites the three balloons. Not shown: foraminiferal shells carry isotopic signals (mainly δ^{18}O, δ^{13}C, ^{87}Sr/^{86}Sr, but there are others). (McGowran and Li, 2002, with permission.)

Pervading all this was the powerfully supportive presence of Susi McGowran – believing shrewdly in the book and its author in the dark moments and slothful patches, being patient and stinging as necessary, listening to the gripes and doubts in the lows and the enthusiasms and exultations in the highs, and believing in life after biostratigraphy.

Acknowledgments

Figures 1.11, 1.12, 1.13, 1.14, 1.15, 6.30, 8.1, AAPG ©1949, ©1978, ©2001, reprinted by permission of the American Association of Petroleum Geologists whose permission is required for further use.

Figures 4.21, 5.29, 5.30, 6.3, 6.4, 6.5, 6.9, 6.10, 8.4 from *Historical Biology*, ©1996, ©2000, published by permission of Taylor & Francis Ltd (http:// www.tandf.co.uk).

Figures 4.12, 4.13, 5.13, 5.14, 5.15, 5.16, and extracts on pp. 87, 358, from *Paleobiology*, ©1980, ©1981, ©1991, ©1996, ©1998, ©2000, published with permission of The Paleontological Society.

Figures 4.5, 4.6, 4.7, 4.8, 4.9, 4.31, 4.32, 5.2, 5.18, 5.21, 6.20, 6.21, 7.23, and the extracts on pp. 144, 148, 251, ©1977, ©1986, ©1989, ©1991, ©1993, ©1999, published by permission of Springer-Verlag Gmbh & Co. KG.

Figures 4.11, 5.9, and extract on p. 121, reprinted, with permission, from *The Annual Review of Earth and Planetary Sciences*, Volume 16 ©1988 and Volume 22 ©1994 by Annual Reviews (www.annualreviews.org)

Figures 4.3, 5.3, and extracts on pp. 21, 101, from *The University of Kansas Paleontological Contributions*, courtesy of and ©1975, The University of Kansas, Paleontological Institute.

Figure 4.14, reprinted with permission from *Nature*, 336 ©1988, Macmillan Magazines Limited.

Figures 0.1, 4.35, 4.36, 4.38, ©1998, published by permission of John Wiley & Sons Limited.

Figure 7.2, ©1955, published by permission of W. H. Freeman and Company.

Figures 3.10, 3.11, 4.18, 4.23, 4.28, 6.6, 6.7, 6.8, 6.24, 7.26, Table 7.1, ©1976, ©1983, ©1989, ©1992, ©1993, ©1999, published by permission of Elsevier.

Table 1.1, and extract on p. 11, ©1933, published by permission of Oxford University Press.

Figures 4.2, 4.34, 6.14, 7.4, 7.5, 7.6, and extract on p. 370, ©1962, ©1971, ©1983, ©1985, ©1989, ©2000, published with permission of Cambridge University Press.

Figure 2.4, and extract on p. 92, ©1957, U. S. National Museum.

Figures 4.25, 5.17, ©1959, ©1996, published with permission of The Geological Society.

Figures 8.8, ©1977; 3.1, 3.3, 3.4, 3.6, 3.12, 3.13, 3.16, 3.20, 6.19, 6.21, 6.29, 6.31 from *Paleoceanography*, ©1986, ©1987, ©1988, ©1989, ©1994, ©2001, published with permission of the American Geophysical Union.

Figures 0.3, 2.2, 5.7, 6.11, 6.12, 6.13, 6.33, 7.21, Table 6.2, *Memoirs of the Association of Australasian Palaeontologists*, 23, 27, ©2000, ©2002, published with permission of the Association of Australasian Palaeontologists.

Figures 2.5, 2.8, 2.11, 2.12, 3.2, 3.5, 3.9, 3.14, 4.22, 5.6, 5.11, 5.12, 5.22, 5.23, 5.24, 5.25, 5.26, 5.27, 5.28, 6.1, 6.2, 6.15, 7.13, 7.17, 7.24 from *SEPM Special Publications*, ©1981, ©1988, ©1995, ©1998, published with permission of SEPM (Society for Sedimentary Geology).

Figures 2.6, 4.18, 4.19, 4.20, 7.20, and extract on p. 89, *Micropaleontology*, ©1955, ©1968, ©1972, ©1976, ©1988, with permission of Micropaleontology Press

Figures 1.9, 1.10, 3.17, 3.18, 3.19, 5.1, published with permission of Geological Society of America

Figures 3.23, 3.24, 3.25, 3.26, 3.27, 3.28 published in *Proceedings of The Ocean Drilling Program, Scientific Results*.

Figures 2.9, 3.8, 7.19, ©1984, University of Tokyo Press.

Figure 3.7, from *Palaeontology*, ©1985, The Palaeontological Association.

Figures 3.21, 3.22, from *Terra Nova*, ©1999, Blackwell Publishing.

Figure 4.1, published with permission of the Geological Society of India.

Figure 4.30, ©1981, The University of Barcelona.

Figure 5.19, published with permission of the Australian Petroleum Production and Exploration Association Ltd.

Figure 8.9, published with permission of the Geological Society of Australia.

Extracts on pp. 30, 98, ©1969, ©1979, E. J. Brill.

Extract on p. 100, ©171, Edizioni Technoscienza.

Biogeohistory and the development of classical biostratigraphy

Summary

The science of geology emerged from eighteenth-century tensions between the notion of Earth-as-machine and the notion of a recoverable Earth history. Fossils had a central role in identifying formations for mapping, in building and testing a succession of life, in reconstructing ancient environments, and most of all in developing the perception that similarity among assemblages of fossils indicates similarity in geological age. There followed the ecological facies concept and the chronological zone concept, both pre-evolutionary. This chapter takes these themes up to the mid twentieth century when the stratigraphic *Guide* was in preparation and planktonic microfossils were about to dominate the biostratigraphy of the Cenozoic Erathem.

Introduction

Fossils record the fleeting tenure of species as members of the Earthly biosphere. This nagging fact made more sense of the rock relationships in the exposed parts of the Earth's crust, extracting more order from an apparently chaotic jumble, than did any other observation or speculation on rocks, or any exploration and development of mineral resources. The presence of fossils in sedimentary strata could reveal a succession of ancient faunas and floras. Simultaneously, the same observations could be used to define and recognize groups of strata: thus we have both biohistory and geohistory. Sedimentary strata containing trilobites seemed to occur above strata lacking fossils (themselves sitting on the deformed crystallines), and below other strata containing ammonites. Then there was yet another group of strata lacking ammonites but

TERTIARY   or CENOZOIC

Nummulites lævigata

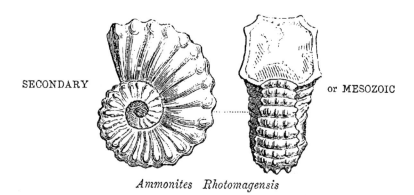

SECONDARY   or MESOZOIC

Ammonites Rhotomagensis

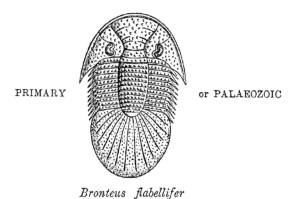

PRIMARY   or PALAEOZOIC

Bronteus flabellifer

Figure 1.1 The fossil-based geological time scale: frontispiece of Lyell's *Student's Elements of Geology* (Lyell, 1871). Trilobites, ammonites, and the large rock-forming foraminifer *Nummulites* characterize the Palaeozoic, Mesozoic and Cenozoic Eras, respectively.

containing nummulites, the 'petrified lentils' observed by the travelling chronicler Herodotus in the blocks comprising the Egyptian pyramids. The three kinds of fossil symbolized the three divisions of the fossil record for Sir Charles Lyell, as shown here (Fig. 1.1) in the frontispiece of *The Student's Elements of Geology* (1871): the Primary or Palaeozoic, Secondary or Mesozoic, and

Tertiary or Cenozoic. That edition of the *Elements* was published about a quarter of a century after the three eras of Earth history were secured on the evidence of their fossil record and no longer on their mineralogy or lithology, and by then Lyell had accepted, ever so tardily, organic evolution as the explanation of fossil succession and its pre-eminent utility in the correlation and classification of strata.

An account of the origins of biostratigraphy, of the science and the arts of using fossils for chronological correlation and geological age-determination, can begin at one of the truly natural turning points in the story. Towards the end of the eighteenth century, James Hutton was discovering deep time, Georges Cuvier was demonstrating once and for all the fact of organic extinction, and geology was rapidly being established as an empirical discipline which would include the systematic mapping of the rocks exposed at the surface of the Earth. That was also the time that the ideas of *pre*history, *bio*history and *geo*history took hold in the collective Judaeo-Christian intellect. Although all of these notions had forerunners and precursors – 'precursoritis' usually leads us back to classical antiquity – Hancock (1977) deemed it necessary to reassert one of the great mainstays of the textbooks on historical geology, at least in the English-speaking world – that the science of biostratigraphy was founded by William Smith, that he owed nothing much of significance to earlier writers, and that the importance of his work is greater than that of any subsequent contributor to the theory of our science.

Significance of fossils

Why does a fossil occur where it does in a sedimentary stratum? Beyond the taphonomic questions of the preservation or destruction of organic remains – fossilization itself – there are the three factors of environment, geography and time. That the three factors have long been known is exemplified in this summary from the textbook by J. Beete Jukes (1862):

1)   First of all, within the same biological province there may have been differences in the 'stations', to use the naturalists' phrase, that is, the place where the fossil was buried may have been at the time either sea or fresh-water, deep or shallow water, near shore or far from it, having a muddy or a sandy bottom, or being a sea clear of sediment, and the fossils entombed at these different stations of the province may have varied accordingly.

2)   Secondly, we may pass from one 'province' to another, the two provinces having been inhabited by different but contemporaneous groups of species.

3)   Thirdly, there may have been a difference in 'time', during which a general change had taken place in the species, those formerly existing having become extinct, and others having come into existence that had not previously appeared on the globe.

It was the first of these, the ecological factor, that was appreciated the earliest, by the Greeks and the men of the Renaissance (Rudwick, 1972; Mayr, 1982). For Leonardo da Vinci and others, a sedimentary rock containing fossil shells like modern shells signified the former presence of the sea, no matter that the modern sea was many leagues' distant. Indeed, James Hutton, the discoverer of deep time (Gould, 1987), was well aware of the significance of fossils – but not as signals of time and history. There is 'not a shred of suggestion that fossils might record a vector of historical change, or even distinctness of moments in time. Fossils, to Hutton, are immanent properties of time's cycle' (Gould, 1987). Instead, the incorporation of fossils into subsequently lithified sediments indicated the operation of heat; and their presence in rocks in continents well above sea level indicated uplifting. Thus we have crucial evidence for the existence of the restorative force necessary for completing each geological cycle. Last, petrified wood was eroded from continents in earlier cycles and hence are clues to the former existence of plants (Gould, 1987). All of these inferences had their basis in ecology and environment, not in history and surely not in any perception of distinctive biological changes during geological time. And Gould probed further, suggesting that our antecedents' awareness of fossil forms not found in the living state merely revealed their ignorance of the modern biota and that this was not just an ahistorical stance but an active denial of history by Hutton.

For Teichert (1958) the science of stratigraphy developed in a logical way. First, there was the recognition and interpretation of physical characteristics of sedimentary rocks, with emphasis on *lithostratigraphy* from Steno to Werner, in the seventeenth and eighteenth centuries. Then there was recognized the orderly and meaningful succession of fossil floras and faunas in sequences of sedimentary strata, and the development of *biostratigraphy* since William Smith. The third step was the recognition of the contemporaneity of dissimilar rocks and fossil assemblages and the subsequent development of the *facies* concept from Gressly in 1838 to Mojsisovics in 1879. *Lithostratigraphy, biostratigraphy, facies*: 'modern stratigraphy rests securely on these three basic achievements of the human mind' (Teichert, 1958). Figure 1.2 exemplifies the complication and apparent falsification of the fossil record in that the primacy of the first or the third of those factors is not always clear. The related fossil species *a* and *b* are confined to different environments reflected by two sedimentary facies. At any one locality *a* is always below *b* and will be considered to be older, but in fact *a* and *b* are contemporaneous species.

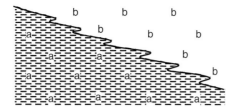

Figure 1.2 Fossils and lithology: time or environment? Two facies-bound species are consistently superposed *a* below *b*, but actually are contemporaneous – an unscaled pattern of diachrony cited by Simpson (1951) as an 'example of complication and apparent falsification of the fossil record'. Some would restrict this diachrony to within a third-order sequence (Chapter 5).

Biostratigraphy itself developed as a discipline essential to the growth of historical science on a three-part foundation (McGowran, 1986a). The three pedestals were: (i) the recognition of successional assemblages of fossils in successional strata; (ii) the successful testing and confirmation of that succession in other localities and other regions; and (iii) the perception that similarity among assemblages of fossils indicates similarity in geological age.

Laudan (1976, 1987, 1989) reassessed what the first of those points means. Is it the succession of faunas in successional sedimentary formations – on the grounds of superposition – that is important, or is it the identification of each formation by its fossil content – the sorting out, the reliable identifying of otherwise confusingly similar but separate and distinct clay strata, say, which are always exposed as discontinuous outcrops and excavations? The two aspects of fossil content are not so much contradictory as differing in emphasis. Where does the *identification* of individual formations of strata end and the *correlation* of formations begin? In the standard accounts, William Smith's use of fossils in stratigraphy may have begun in the former endeavour but established the latter. His subsequent celebrants, beginning with his canonization by Adam Sedgwick in 1831, identified Smith as the person most of all responsible for the overthrow of the neptunist stratigraphies of the eighteenth century, based as they were on a perceived, consistent succession of lithology and mineralogy. This preeminence of fossils in correlation, linked to the independence of fossils from sedimentary facies in what came later to be called the Phanerozoic Eon, was stated most clearly by John Phillips in 1829, and 'this conception can scarcely can have been foreign to William Smith ten years earlier, though we seldom find it formulated' (Arkell, 1933). Arkell continued, interestingly, 'It is only occasionally that a gleam of light reveals the inner working of men's minds about this time, for the output of a great mass of important descriptive matter was engaging most of their attention' – they knew about the temporal significance of fossils but they were too busy exploiting it to write in general terms about it.

But Laudan claimed that Smith's actual work was based instead on the following convictions – the constant *order* of strata and the constant individual *properties* of strata including mineral content, fossil assemblage and, most importantly, topographic expression. Smith's real contribution (in this view) was in tracing and mapping the course of strata from outcrop to outcrop in England rather than in establishing the use of fossils in identifying the strata. In the Paris Basin, Cuvier and Brongniart showed that the Alluvial of the neptunists was a complex succession of formations that could be traced over 120 km and more by means of the consistent succession of their fossils. In both of these programmes credited with establishing historical geology and history biology based in sound biostratigraphy, then, successional assemblages were established as a fact of biohistory that could be confirmed in different sections of sedimentary strata.

Now contemplate Fig. 1.3 and Fig. 1.4, highly idealized and simplified versions of transgression-regression cycles, and quite anachronistic in being cartoons more at home in the twentieth century than in the early nineteenth, being based on Israelsky's (1949) oscillation chart which has some basis in reality (e.g. Poag, 1977, Fig. 4) (although clearly pre-sequence stratigraphy; see Chapter 5). In Figure 1.3 three distinct biotic realms produce fossil assemblages namely plant (non-marine), mollusc (neritic) and foraminifer (offshore). They can be utilized in two distinct ways – to identify and to discriminate those strata in distant locales, along with lithological and mineralogical criteria; and to demonstrate faunal and floral succession in which the higher respective assemblages must be younger by superposition. Note too that within each assemblage there are waxing and waning distributions producing 'time-transgressive' or *diachronous* configurations. The dualism of identification and age demands some consideration of the meaning of *correlation*. Broadly, in stratigraphy, to correlate is to show correspondence in character and in stratigraphic position. That includes the tracing of stratigraphic units between discontinuous outcrops, or through the subsurface from one control section to another using lithological, physical and/or palaeontological criteria. Several authors have advocated that broad use of the term (e.g. Shaw, 1964; Hedberg, 1976), but it refers rather to the *identification* of sedimentary formations, their boundaries, and included members and horizons. More restrictively and more appropriately, according to some (e.g. Rodgers, 1959; Raup and Stanley, 1978), correlation means *chrono*correlation – establishing the time-equivalence of two spatially separate stratigraphic units (McGowran, 1986a).

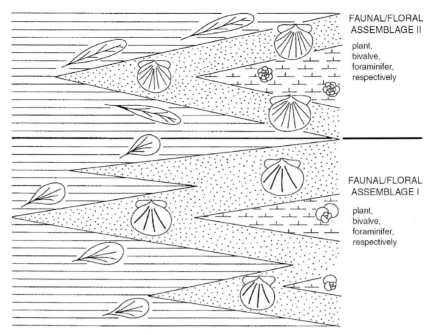

FAUNAL/FLORAL
ASSEMBLAGE II

plant,
bivalve,
foraminifer,
respectively

FAUNAL/FLORAL
ASSEMBLAGE I

plant,
bivalve,
foraminifer,
respectively

Figure 1.3 Fossil succession in three biofacies in a pattern of transgression-regression (McGowran, 1986a). This sketch was contrived to demonstrate two things – lateral movement of non-marine, neritic and marine biofacies in response to environmental shifts but also a change in time, allowing recognition of two successional assemblages within each biofacies. Concurrence of the three ensuing boundaries at the heavy line might be a kind of coordinated stasis (Chapter 6).

We can follow this matter of fossil assemblages and their chronological significance a little further in Figure 1.4, where there are fossil assemblages that follow shifting lithologies (thus shifting environments in life) as in Figure 1.3, in contrast to assemblages that do not so shift. The latter category is illustrated by three successional assemblages of pollen grains whose mutual boundaries cut across lithological boundaries because pollens are blown out to sea (we ignore here such complications as subsequent destruction by oxidation); it is illustrated too by assemblages of planktonic foraminifera whose mutual boundaries likewise cut across lithologies where elements of the living communites come inshore. There are two concepts here. First, there is the concept of *facies* which appeared in the 1820s, on lateral intergradations in lithology (Young and Bird in England; Amos Eaton in New York) and on the observation that the same fossils can occur in different lithologies (Brongniart in France). Brongniart realized the tremendous possibilities afforded by this independence of some fossil distributions from lithological facies (Hancock, 1977) – the

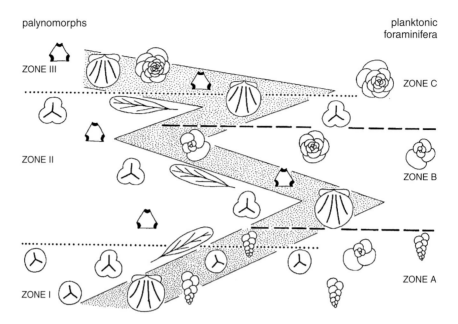

palynomorphs

planktonic
foraminifera

ZONE III

ZONE C

ZONE II

ZONE B

ZONE I

ZONE A

Figure 1.4 Biofacies migrations (non-marine, inshore, offshore) as in Fig. 1.3, with
two sets of biozones based on the fossils of mobile and relatively facies-independent
organisms (McGowran, 1986a). Two sets of three biozones can be recognized on the
highest occurrences respectively of pollens (dotted lines) and planktonic
foraminifera (dashed lines). Nothing in this diagram proves that biozone
boundaries are 'time-parallel' but it is a reasonable and testable working
assumption that they come close to that situation.

possibilities of (in subsequent jargon) long-distance *chronological biostratigraphic
correlation*. This is the second concept (McGowran, 1986a).

Zones and zonation through a century

Laudan (1982) identified a turning point in the 1820s in the advent of
Smith's nephew and protégé, John Phillips:

> In deciding to use fossils as the key to the succession, Phillips was altering
> the whole basis of mapping. On Smith's map, a band of uniform color
> represented strata with particular geographical positions and similar
> surface features, and in addition, Smith assumed without question, a
> similar place in the succession, similar lithology and similar fossils. On
> Phillips' map, however, the bands of uniform colour represented strata
> containing the same fossils, and therefore, he assumed, *occupying the same
> place in the succession whatever their lithology* (emphasis added).

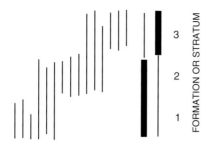

"That a formation or stratum may differ from all those above it, by the presence or absence of certain species, and from all those below it, by the presence or absence of other species:
"That it may contain some particular species, unknown either above or below. We may add, that formations and strata differ by the relative abundance or paucity of their imbedded fossils."
John Phillips, 1829

Figure 1.5 This hypothetical range chart is a reasonable rendering of Phillips's (1829) verbal summary (McGowran, 1986a).

By 1829 Phillips himself could state bluntly:

> for since it thus appears, that a few shells brought from a quarry, are data sufficient to determine the geological relations of the rock, we are entitled to conclude, that in a given district the age and position of certain strata, or groups of strata, are infallibly indicated by their organic contents. These researches, commenced by Mr Smith in England, have been extended with the same results over all parts of Europe, and a large portion of America, and therefore it is concluded that strata, or groups of strata, are to be discriminated in local regions, and identified in different countries, by their imbedded organic remains

Figure 1.5 shows visually Phillips's (verbal) conclusions as quoted therein (McGowran, 1986a). The 'formation or stratum' would appear to be a biostratigraphic zone except for the anachronism – such formalizing of fossil successions simply did not happen yet. It is instructive to consider an authoritative textbook account twenty-odd years later. As quoted already, J. Beete Jukes outlined the constraints on fossil distribution; he used a sketch (Fig. 1.6 herein) to discuss them. 'Let there be', wrote Jukes, 'a great series of rocks divisible into three groups *A*, *B*, and *C*, each with alternations of argillaceous, arenaceous, and calcareous strata. Each lithology in *A* will contain characteristic fossil assemblages *a*, *b*, and *c*, respectively, which also will recur so that the assemblage

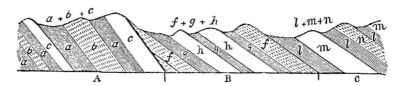

Figure 1.6 Hypothetical fossil succession in a 'great series of rocks' (Jukes, 1862, Fig. 105). Strata are grouped into *A*, *B* and *C*. Each group contains recurring lithologies characterized by (also recurring) fossil assemblages (*a*, *b* and *c* in group *A*). In the higher groups of strata, the still-recurring lithologies contain new fossil assemblages which recur for a time in their turn ($f + g + h$ in *B*; $l + m + n$ in *C*).

overall for group *A* will be $a + b + c$. But as we pass up into group *B* we will encounter a different set of assemblages, $f + g + h$ in their respective lithologies, even though those lithologies may be indistinguishable from their counterparts in group *A*. And likewise for assemblages $l + m + n$ in group *C*.' Jukes's point was that there are two reasons for differences among fossil assemblages – environmental contrasts and the lapse of geological time: what he called *the law of the distribution of fossils*. Interestingly, Jukes began this discussion with three groups of strata but he does not end it with any zonation, or any other classification of fossil distribution, even though the detailed collecting with reference to stratal position, and that careful biotaxonomy on which progress depends, had been proceeding in various parts since the 1820s.

For Jukes did not refer to the work of Albert Oppel, published in 1856–58 and identified in due course as the 'birth of biostratigraphy as a separate discipline' (Hancock, 1977). What was special about the work of this man 'who was to place the whole science of stratigraphical geology on a new footing and to breathe new life into it' (Arkell, 1933) and then died, even younger than Mozart? Adapted from a figure by Berry (1977, Fig. 1), Figure 1.7 is intended to illustrate Oppel's principle of biostratigraphic zonation. There are two noteworthy points. First, the column is composite, representing a district in which several exposed sections of strata contribute to the succession – the process of piecing a succession together is there right at the beginning; and likewise with the ranges of carefully collected and identified fossils. Second, there are two ways in which the zones labelled I to IV are distinguished. The zone I/II boundary, for example, is in the vicinity of three last appearances and two first appearances of species. Each of those species can contribute to the recognition of that boundary in some other district if so required. Also, however, the association of species characterizes each zone. That is, we have here both *assemblage* criteria and *boundary* criteria. It is the first point that is the more important – Oppel emphasized that whilst the correlation of groups of strata had been achieved, 'it has not been

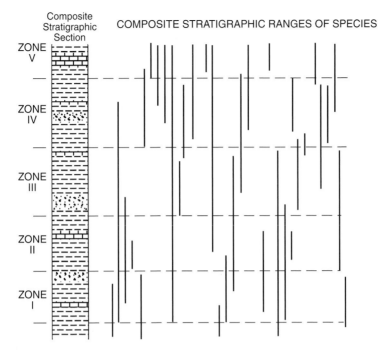

Figure 1.7 Oppel's principle of zonation (McGowran, 1986a, based on Berry, 1977). The section and the range chart are both composite for the district or region. The divisions labelled zones are clear enough – but were Oppel's zones in the rocks (the 'British' view) or were they idealized or abstracted, temporal terms (the 'German' view)?

shown that each horizon, identifiable in any place by a number of peculiar and constant species, is to be recognized with the same degree of certainty in distant regions. This task is admittedly a hard one ...' (from Arkell, 1933). And from these regional profiles one may develop the ideal profile, of which '... the component parts of the same age in the various districts are characterized always by the same species' (from Arkell, 1933). Oppel did not tease out the somewhat pedantic classification of zones, as happened later; and, indeed, he neither invented nor anywhere defined what he meant by a zone (Arkell, 1933). There was already a respectable list of forerunners to Oppel in the study of fossils in strata (e.g. Arkell, 1933; Moore, 1941, 1948; Conkin and Conkin, 1984; among many), but to him

> ... is due not the credit for the inception of the zonal idea, but for a very great refinement in its use, and, most important of all, for emancipating the zones from the thralls both of local facies, lithological and palaeontological, and of cataclysmic annihilations, thus giving

them an enormous extension and transferring them from mere local records of succession to correlation-planes of much wider (theoretically universal) application (Arkell, 1933).

'For we have here the beginnings of a detailed and generally applicable time scale, abstracted from local lithological and paleontological considerations', said Arkell, who made a striking comparison with the affairs of men:

> Before it geological history had been as confused as the history of Assyria and Babylonia at the time of the city-kingdoms, each with its own local chronology, overlapping those of its neighbours. Since Oppel, historians have been provided with an orderly system of dynasties, subdivided into reigns, and even in countries as distant as the Himalayas it has been possible to discern marks appropriate to the periods when the more important of the dynasties held sway, although the influence of the individual reigns was not always felt outside North-Western and Central Europe.

For Schindewolf (1950, 1993) palaeontological zonation is chronology – a 'purely temporal' system and not actually stratigraphy; he was certain that both d'Orbigny and Oppel assigned a temporal, abstract meaning to 'zone', and he rejected the spatial concept of a zone comprising the actual rocks with their fossils.

Since the times of Oppel and with one major exception remarkably little has happened in the field of zonation, *sensu stricto*. Consider Figure 1.8, which summarizes various kinds of biozone defined and discussed by the International Subcommission of Stratigraphic Classification (ISSC) (Hedberg, 1976) – a century and more later (McGowran, 1986a). The zones fall largely into three general types: (i) there is the 'distinctive natural assemblage' which allows grouping of strata into an *assemblage zone*; (ii) the range or ranges of selected taxa give us *range zones* including the various kinds of interval zones, whose distinction is rather pedantic; (iii) fluctuations in the abundance of a taxon give the *acme-zone* (of 'lesser importance'). There is little here that was unknown to Oppel. Arkell's (1933) superb discussion of the topic devoted most space to the changes in abundance on which Buckman based the hemera – the first unit of geologic time using the acme of a taxon. Probably the major advances in the late nineteenth century were Charles Lapworth's on Ordovician–Silurian graptolites (Fortey, 1993) – but these were applied in unpacking structural complexity. Indeed, Fortey emphasized the durability of biostratigraphic data in contrast to the contingencies of structural and palaeo-geographic inference.

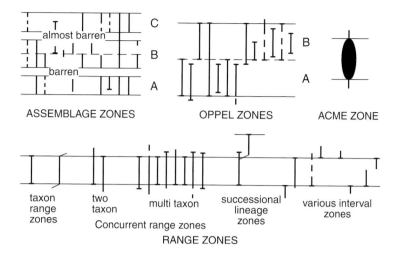

ASSEMBLAGE ZONES   OPPEL ZONES   ACME ZONE

RANGE ZONES

taxon range zones

two taxon

multi taxon

successional lineage zones

various interval zones

Concurrent range zones

Figure 1.8 Kinds of biostratigraphic zones, redrawn from ISSC figures (Hedberg, 1976; McGowran, 1986a).

But lurking in Figure 1.8 are two examples of a notion not available in any cogent way to Oppel, even though it was to erupt towards the end of the same decade – the notion of organic evolution and consequently of ancestor-descendant relationships among species and the shape of their genealogy, or *phylogeny*. There is quite a difference between a range zone, of whatever stripe, based on rigorous, comprehensive collecting, identification of species and compilation of species' ranges, and a range zone based on the phyletic emergence of a species from its ancestor and its subsequent extinction. Likewise, it is one thing to to define a zonal boundary on the top of the range of a species in the local rocks, and quite another to define it on that species' extinction (although the acceptance of the fact of extinction preceded acceptance of the fact of phylogenetic origin by half a century) (McGowran, 1986a). Although in both cases one might reasonably expect the field observations cumulatively to approach the evolutionary interpretation asymptotically, there is a major conceptual shift involved.

But 74 years after the publication of *On the Origin of Species*, Arkell spent very little time on lineage zones. Although studies such as the lineage zonation based on the evolution of the Late Cretaceous echinoid *Micraster* date back to the 1890s, they do not seem to have loomed large in Jurassic biostratigraphy by the 1930s. On the other hand, Arkell did focus on the difference between the 'total' range of a taxon on which the biozone is based, and the 'local' range in the rocks, which gives the teilzone. If the time-equivalent of the biozone is the biochron (Table 1.1), then 'The ideal biochron is as elusive as the ideal hemera';

Table 1.1 *Zones (assemblage and single species) and their chronological equivalents*

Basis	Stratal term	Chronological term
Zones based on assemblages		
acme or duration	faunizone	secule or moment
	(German Faunenzone)	(Zeitmoment or Zonenmoment)
Zones based on single species		
acme	epibole	hemera (Blützeit einer Art)
absolute duration	biozone	species-biochron (Absolute Lebensdauer einer Art)
local duration	teilzone	teilchron (Locale Existenzdauer einer Art)

From Arkell (1933), with permission.

and, we can dispense with the local range-zone, the teilzone, 'only when we are able to deal with lineages' – but, 'Unfortunately, opportunities for making use of lineages in zonal work are extremely rare'.

We find a closely similar outlook and assessment in Moore's review of stratigraphical palaeontology (1948). 'The concept of biozones seems to have little practical value, inasmuch as the total range of the guide fossil controls definition; the observed vertical distribution of most fossil organisms varies from place to place, and total range always is difficult to determine with certainty.' Moore gave us a comprehensive if fictitious sketch of taxa and ranges to show the relations of time divisions and their equivalents based on fossil invertebrates (Fig. 1.9). It is revealing to one inured in the use of microfossil zones and datums (Chapter 2) to see how Moore's chart treats biozones–biochrons and especially teilzones–teilchrons. Thus, the total local or 'absolute' range is treated for each taxon. There is no discussion here of the notion of lining up – *ordinating* – events from different taxa in succession – first and last appearances; tops and bottoms of the teilzones – so that that succession can be subjected to test elsewhere and the nagging problem of incomplete ranges can be resolved. Perhaps that is the most telling illustration of the difference between the essentially neritic fossil record, including the remains of mobile and often highly mobile organisms, and the mostly bathyal and oceanic fossil successions to be considered in Chapter 2.

Even so, there is no clear caesura from the noble traditions of invertebrate fossil biostratigraphy to the newer notions of micropalaeontology. We shall see that there is more in common between the classical times of the discipline and the present than we proselytes tend to remember (see also Kleinpell, 1979).

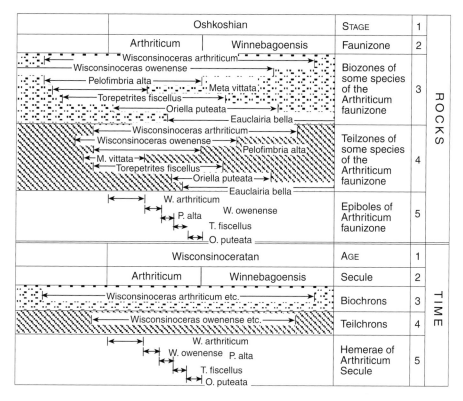

Figure 1.9 Relationships of time-rock and time divisions defined mainly or wholly on fossil invertebrates (Moore, 1948, Fig. 5, with permission). Moore's caption continues: 'The divisions of varying rank are designated by fictitious stratigraphic and paleontologic names, which are nonexistent in literature. They are intended to illustrate concepts in zonation and corresponding segmentation of geological time.'

Schindewolf (1950, 1993) acknowledged the applied and economic impact of micropalaeontology (Croneis, 1941) but stoutly rejected any claims of a new and revolutionary methodology, of epistemological autonomy, or of it being 'the paleontology of the future'.

Whilst Moore (1941) was presenting a splendid, still relatively early example of the power of microfossil (foraminiferal) zones dipping seawards in the Tertiary of the US Gulf Coast region (Fig. 1.10), problems were accumulating in perceiving distinctions between facies fossils and chronologically significant fossils. Among several examples appreciating this divergence, the paper by Lowman (1949) is outstanding in its imaginative use of the dense subsurface sampling of the US Gulf Coast and its exploiting the actualistic link between modern and ancient patterns in foraminiferal distribution and biofacies. The appreciation of such patterns was not new but the sheer accumulation of both

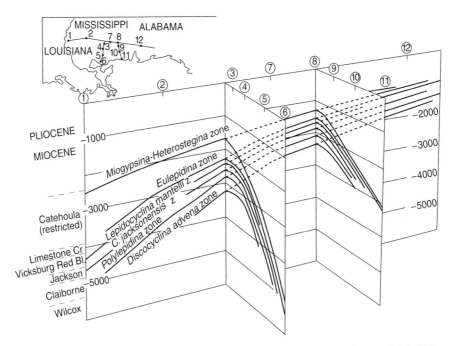

Figure 1.10 Foraminiferal zones in the subsurface, Eocene–Miocene, US Gulf Coast (Moore, 1941, Fig. 12, with permission). The genera are large, photosymbiotic, warm-water, benthic forms. The succession is consistent, along strike and downdip, and could be used in rotary cuttings, not just cores.

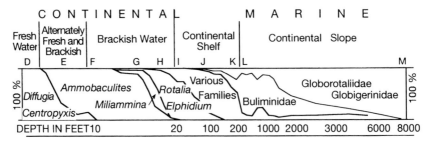

Figure 1.11 Percentage abundances of foraminifera delineating modern biofacies in a composite profile, Mississippi delta and Gulf of Mexico (from Lowman, 1949, Fig. 12, with permission). The pattern was built by connecting bar graphs at each station D to M.

samples and specimens was overwhelming. Lowman demonstrated the biofacies belts from freshwater environments to the slope (Fig. 1.11); if sea level rose or fell, not too fast for the communities to keep up, then biofacies must be diachronous (Fig. 1.12). Thus an updip-downdip section (Fig. 1.13) will display a 'climb' across bedding planes in the downdip or seaward direction by the

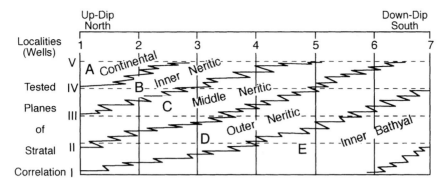

Figure 1.12 Diachronous biofacies 'climbing' in downdip direction as they cross 'tested planes of stratal correlation' I–V during sustained regression (Lowman, 1949, Fig. 28, with permission). The facies could be rapidly determined in the Oligocene and Neogene using the broad modern pattern shown in Figure 1.11.

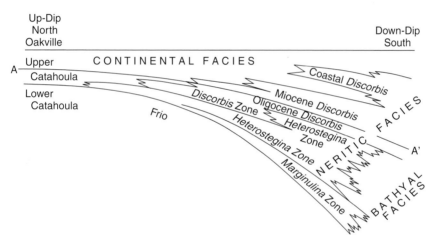

Figure 1.13 'Climbing' downdip (seaward direction) by the (benthic foraminiferal) *Heterostegina* and *Discorbis* zones is illustrated in an Oligo-Miocene depositional unit (Lowman, 1949, Fig. 3, with permission).

Discorbis and *Heterostegina* zones, based on two prominent genera of neritic benthic foraminifera. A generalized sketch demonstrated a perceived distinction between the environmentally more robust species, longer-ranging and distributed more widely, and the narrowly constrained guide species (Fig. 1.14). Another, emphasizing the penetration of neritic facies by richly fossiliferous spikes from the bathyal realm (Fig. 1.15), foreshadowed the notion of the maximum flooding surface, forty years later (as pointed out by Loutit *et al.*, 1988). Most of the intervening years were devoted to the development of the

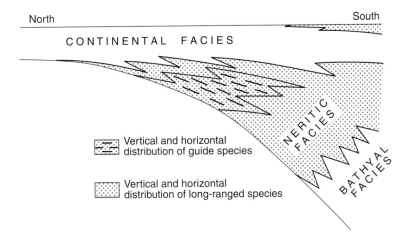

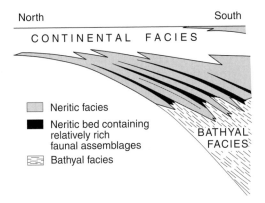

Figure 1.14 'Diagrammatic cross-section of a cyclical sedimentary unit, showing the distribution of long-ranging species and guide species' (Lowman, 1949, Fig. 26, with permission).

Figure 1.15 'Diagrammatic cross-section of cyclical sedimentary unit, showing distribution of richly fossiliferous streaks (black) in neritic facies' (Lowman, 1949, Fig. 27, with permission). This figure was used by Loutit *et al.* (1988) to illustrate the notion of condensed sections and flooding surfaces (Chapter 5).

'true' index fossils – the microplankton – and so this split between 'facies fossils' and 'index fossils' was perpetuated. We deal with the plankton beginning in Chapter 2, but we return to this dichotomy in Chapter 5. Lowman's superb demonstration of foraminiferal biofacies in space and time is a natural point to conclude this selective outline of 'classical' biostratigraphy.

2

The biostratigraphy of fossil microplankton

Summary

Although micropalaeontology is almost as old as biostratigraphy itself, it actually flourished as a tool in petroleum exploration from the 1920s (benthic foraminifera), and 1930s–1950s (planktonic foraminifera). Other microfossil groups flourished during the postwar resurgence in marine geology which brought forth the DSDP and ODP. During this progression there was a shift in biostratigraphic emphasis from assemblage zones, where the emphasis was on the fossil contents of the designated stratal section, to species range zones defined on boundary events, i.e., first and last appearances, which led in turn to the ultimate 'events', namely speciations and extinctions which could be used in defining rock-free phylozones. We come thereby to (bio)chronozones and datums, which contribute an irreversible succession of events (because evolutionary events are unique) to geochronology.

Introduction

Planktonic micropalaeontology as an active biostratigraphic discipline spans only seven decades. It played no part in the building and embellishment of the geological timescale during the nineteenth century. Adolph Brongniart used the benthic *Nummulites* in his stratigraphic studies as long ago as the early 1820s and the foundations of micropalaeontology were laid by the great visionary Alcide d'Orbigny – who, however, in this instance concentrated not so much on succession and correlation but rather on the description, general classification and distribution of foraminifera (Glaessner, 1945). Several decades more elapsed before foraminifera were employed in the analysis of strata in the

Table 2.1 *A whiggish scenario of stimulus and progress in foraminiferology. Advances in technology included microscopy (optical, revealing a whole world of biodiversity; much later, scanning-electron, revealing new detail in shell architecture) and deep-ocean drilling; demand came from geological mapping and petroleum and mineral exploration; modern integrating disciplines include especially palaeoceanography and evolutionary palaeoecology. As well as preserving information as fossilized organisms and assemblages (and perhaps communities), the shells preserve geochemical signals in their calcite (O, C, Sr).*

A whiggish ascent in foraminiferology	
8	Evolutionary palaeoecology
7	Cyclostratigraphic calibration & chronology
6	Palaeoceanography
5	Ocean Drilling Project
	Deep Sea Drilling Project
4	Geological Survey mapping
	e.g. the mapped Stages of the NZ Cenozoic
3	Petroleum exploration
	e.g. deep water facies, Caucasus & Alps
	deepwater facies, Caribbean
	neritic: Gulf Coast wedges
	neritic carbonates: East Indies Letter Classification
2	Global oceanic expeditions
	HMS *Challenger*, 1870s: flourishing extant diversities and diverse environments
1	Microscopists' curiosity
	late 1700s and 1800s; Darwin's discouragement

search for water and petroleum. Stratigraphic correlations between boreholes on the basis of microfossils found to occur in common developed in several countries in Europe and in North America in the 1870s–1890s (e.g. Stainforth *et al.*, 1975). The first effort in petroleum geology, by Grzybowski in 1897 in the flysch facies of the Polish Carpathians, did not stimulate more such research, presumably because of the language barrier and because of the material, consisting of poorly preserved agglutinated foraminifera (Glaessner, 1945). At any rate, credit for the first publication to bring applied micropalaeontology to the forefront in petroleum exploration and development is given to Applin *et al.* (1925) even as Vaughan (1924) and Diener (1925) were still denying that the smaller foraminifera had any stratigraphic importance. When A. M. Davies published his *Tertiary Faunas* in 1934, he ranked the larger benthic foraminifera as taking 'a good second place' to the Mammalia in the Tertiary era in approaching the ideal zone fossil – 'a species that can spread over the whole earth in a time which is negligible compared with its duration as a species, though that in turn is very short on the geological time scale'. There was no mention of planktonic foraminifera or of other microplankton by Davies but it was also in 1934 that M. F. Glaessner and N. N. Subbotina first realized the value of the planktonics in regional correlation in the Soviet Union or, for that matter, in the world (Berggren, 1960). The year 1934 was of good vintage: Thalmann urged the value of the planktonic genus *Globotruncana* in achieving far-reaching Cretaceous correlations.

An ascending conceptual series in planktonic zones

What are the ideas that have driven the development of this discipline through its brief history? We begin with two. The first is the assemblage zone and oppelzone, essentially as outlined in Chapter 1. The second is the critical distinction between species that lived on the mud or in it and species that lived in the pelagic realm – the plankton (Fig. 2.1). R. M. Stainforth expressed the distinction very well:

> A basic principle since the earliest days of stratigraphy is that similar fossils express similar ages for the rocks which contain them. As studies spread around the world, certain fossil groups were recognized gradually as providing especially precise and widespread indices (e.g., Paleozoic graptolites and Mesozoic ammonites). The dual reason underlying the utility of such fossils is that, in life, the organisms drifted near the surface of the sea and thus were dispersed widely by ocean currents; then, on death, they sank randomly to become a unifying element among sea-floor biotas (communities) which, because of susceptibility to local ecologic factors, differ widely and in extreme cases have no common species other than the extraneous planktonic forms. The original axiom, therefore, has been modified to recognize that similarity of planktonic (surface-and near-surface-living) fossils is a reliable criterion of similar age, whereas resemblances of benthonic (bottom-living) fossils may reflect identity of environment as much as (or even more than) identity of age. The tendency to differentiate taxa more narrowly among planktonic than among some benthonic organisms also implements stratigraphic utilization of the former (Stainforth *et al.*, 1975, pp. 15–16)

Figure 2.2 begins with those notions and it charts the concepts that guided the subsequent development of the discipline. It is a guide to what follows here, which traces the changing concepts rather than giving a detailed history; and concept-tracing is a messy business. Take the kinds of zones, for example. Because

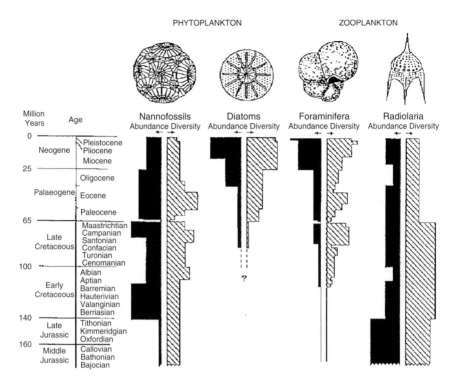

Figure 2.1 Abundance and diversity patterns of pelagic microfossils (Seibold and Berger, 1993, after H. R. Thierstein *et al.*, in G. B. Munch, Ed., Report of the Second Conference on Scientific Ocean Drilling Cosod II, European Science Foundation, Strasbourg). Nannofossils and foraminifera produce calcareous skeletons, radiolarians and diatoms, siliceous, producing pelagic–biogenic sediments (see Fig. 2.15). Abundances and diversities are not scaled, with permission.

oppelzones are a mix of assemblages and ranges, they do not separate cleanly from the clutch of zones that rely on tops and bottoms – the taxon-range zone, interval zone, and concurrent-range zone; nor can the latter, in turn, be completely separated from the zones founded in lineages – the phylozones. However, we can see a gradual change along that succession from the pioneers in the 1930s to the 1960s, and yet an acute awareness of organic evolution was there from the outset and Glaessner anticipated the trend – in part, at any rate – in 1945.

Planktonic foraminifera in the Caucasus

Glaessner's chart (Fig. 2.3) displayed 39 species and varieties distributed through sections in the Caucasus spanning time from the Cenomanian to the early Oligocene. He recognized divisions I to XV which, although labelled

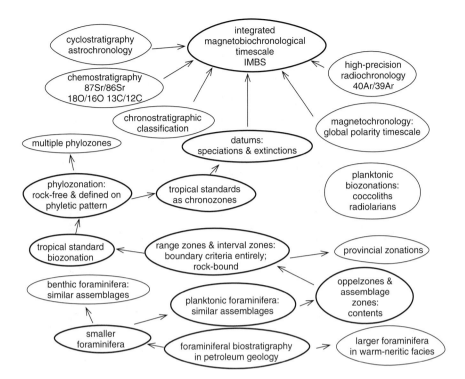

Figure 2.2 An ascending conceptual series in planktonic microfossil zones, to be read from the bottom up. Several ovals appear to be deadends but this is misleading: every oval marks active research. There are numerous inputs towards the top of the diagram, signifying increasing integration of disciplines into modern geochronology. The initial division is between the large and small foraminifera; the latter divide between benthic and planktonic. Planktonic foraminiferal zones have evolved from assemblage zones to oppelzones and range zones, to phylozones, to chronozones, to datums along with datums from other microfossil groups, all of which are integrated with geophysical and geochemical signals in the IMBS. From McGowran and Li (2002, Fig. 2) with permission.

chronostratigraphically, are assemblage zones. (Note that the chart is composite – it is aggregated from sampled measured sections which are not shown; note too that stratigraphic ranges are shown extended where, in Glaessner's judgement, the Caucasus ranges are not the 'full' species ranges.) Although doubtful records are shown, the boundaries of the 15 divisions are the tops and bottoms of all the ranges of the taxa – a characteristic way of displaying the contents of what are assemblage zones.

Glaessner's thinking on micropalaeontological biostratigraphy is revealed more fully in his book (1945). On correlation, under the heading *Assemblages*:

24

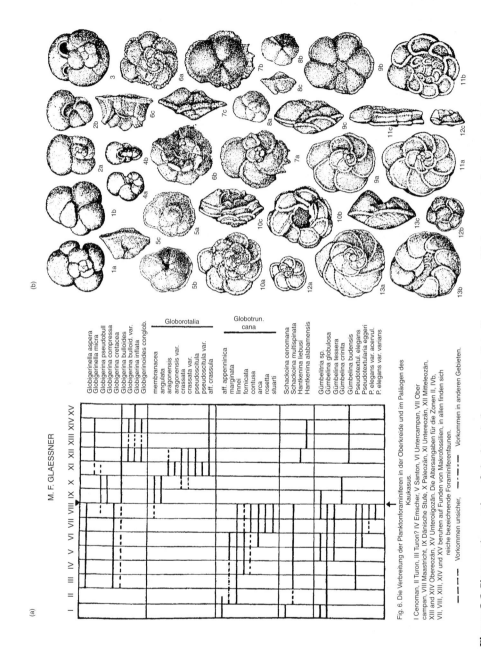

Figure 2.3 Glaessner's zonation of planktonic foraminifera: Late Cretaceous and Palaeogene in the Caucasus (Glaessner, 1937). The boundary between zones VIII and IX is the Maastrichtian/Danian boundary, now also the Cretaceous/Palaeogene boundary (arrow added).

Neither the identity of commonly occurring species alone, nor a high ratio of identical species prove that two assemblages are correlated. The fundamental criteria of inter-regional correlation are the known stratigraphic ranges of species. Species which are frequently found and which always occur in the same limited stratigraphic interval, combined with those defining a similar interval by overlapping of their known stratigraphic ranges, indicate the age of the faunal assemblage containing them (p. 224).

Although Glaessner did not specify type sections or localities in his 1937 paper, it is very clear in his 1945 exposition that zones are the rocks in which the pertinent fossils occur. There are two kinds, faunal and biozones: 'Rapid changes of environmental conditions create *faunal zones* characterized by distinctive assemblages of fossils. Evolutionary changes create *biozones*. This term comprises either the sediments formed during the entire range of time in which a species, genus, or higher taxonomic group existed, or the sediments formed only during its 'acme' or time of maximum development' (1945, p. 214). His discussion of *Index fossils and biozones* concerned mostly the importance of phylogeny and especially the morphogenetic sequences of Tan Sin Hok, to which I return in Chapter 4. A phylogenetic model of species comprising the Cretaceous genus *Globotruncana* notwithstanding, it would seem that the biostratigraphic units recognized by Glaessner in his seminal study of the mid 1930s were assemblage zones.

Parallel work in the Caucasus by Subbotina was more sustained (Berggren, 1960) culminating in her monograph (Subbotina, 1953). Her foraminiferal biostratigraphy of the Danian to Upper Eocene is outlined here in Table 2.2. Subbotina correlated the composite successions in four regions (her Table 2) and presented (her Table 3) the distribution of planktonics according to zones as composite abundances, which play an important part in the characterization of those zones. The Paleocene and Eocene genera *Globigerina*, *Acarinina* and *Globorotalia* are shown as phylogenetic trees, which seem at first sight to be an advance on Glaessner's work. However, virtually all the levels of interpreted speciation and extinction fall at zonal or subzonal boundaries, so that the differences from Glaessner's chart are not conceptual so much as in the systematics of the fossils and in greater detail.

The Trinidad connection

Independently of the work in the Caucasus, planktonic foraminiferal biostratigraphy began in the 1930s in Trinidad (Cushman and Stainforth, 1945). It was needed for unravelling the complexities of local geology in petroleum

Table 2.2 *Subbotina's biostratigraphy of the planktonic foraminifera in the Palaeogene of the Caucasus (Subbotina, 1953)*

Age determination	Foraminiferal zone	Foraminiferal subzone
Upper Eocene	*Bolivina* zone	—
Upper Eocene	Zone of *Globigerinoides conglobatus* and large globigerinids	Subzone of large globigerinids *Globigerinoides conglobatus* subzone
Upper Eocene	Zone of thin-walled pelagic foraminifers	—
Upper Eocene to Middle Eocene	Zone of acarininids	*Acarinina rotundimarginata* subzone *Acarinina crassaformis* subzone
Lower Eocene	Zone of conical globorotaliids	—
Paleocene–Lower Eocene	Zone of compressed globorotaliids	*Globorotalia marginodentata* subzone *Globorotalia crassata* and *Acarinina intermedia* subzone
(?) Danian Stage	Zone of rotaliiform globorotaliids	*Globigerina inconstans* subzone *Globigerina trivialis* subzone

exploration, which expanded rapidly in the Caribbean after World War II, and several authors have paid tribute to H. G. Kugler for stimulating the necessary research and encouraging its subsequent publication. The first zones were published in the 1940s (Bolli and Saunders, 1985) and the 'Trinidad zonation' reached its apotheosis in the 1950s as part of intensive development in the Caribbean region, especially Venezuela (Blow, 1959).

Systematics and biostratigraphy produced and partly published over almost twenty years was synthesized and summarized by Bolli (1957a, b, c). The Paleocene–Lower Eocene zonation based on the distribution of species of *Globigerina* and *Globorotalia* is displayed here (Fig. 2.4). It differs from Subbotina's and Glaessner's patterns in some ways. Based on many more person-hours, it is more detailed – nine zones compared to three (Glaessner) and five (Subbotina) for about the same stratigraphic span, and 38 named taxa against 15 and 31 respectively. Whilst Bolli stressed the composite nature of the range charts and zonations in that they are pieced together from tectonic slices and slip masses, he cited a type locality or drilled interval for each zone; the zones are thoroughly grounded in the rocks. Bolli did not, however, define those zones or repeat previous definitions but, instead, characterized the contents. With exceptions that are obvious in Figure 2.4, most of

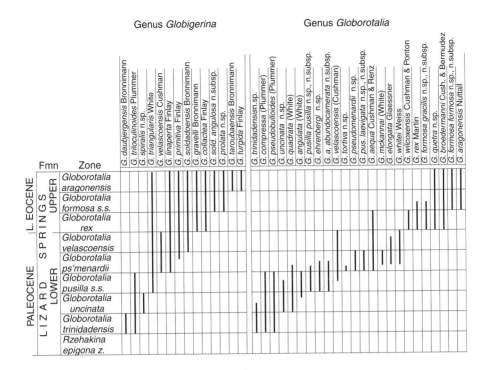

Figure 2.4 Paleocene and Lower Eocene in Trinidad, the Lizard Springs Formation: composite ranges and zonation on the planktonic foraminifera *Globigerina* and *Globorotalia* (highly polyphyletic genera at that time) (Bolli, 1957a), with permission.

the ranges begin and end at zonal interfaces and that was apparent too in the suggested phylogenetic pattern (1957a, Fig. 12). Thus, the Trinidad zonation fits comfortably within Glaessner's characterization of assemblage zones. Even so, it can hardly be doubted from Bolli's text that e.g. the Upper Paleocene *Globorotalia pseudomenardii* zone is a total range zone, or that that the evolution of the *Globorotalia fohsi* lineage provided successional phylozones in the Miocene.

The papers that developed the Caucasus and Caribbean biostratigraphies (see also Blow, 1959) are among the prime documents of our discipline. But was there any major conceptual advance over the situation in biostratigraphy as it was in 1934 – as in Davies' *Tertiary faunas* without the benefit of the planktonic foraminifera? I think not. There was developed a means of biochronological resolution, correlation and age determination that simply was not previously available for analysing marine facies of Cretaceous and Cenozoic age, often lacking macrofossils and often available only as rotary cuttings or cable-tool sludges. The criteria for the ideal group of index fossils as stated by Davies were the same for the planktonic foraminifera, as

Table 2.3 *Biostratigraphy of the Paleocene and lower Eocene, US Gulf and Atlantic coastal plain, based on planktonic foraminifera*

Age	Assemblage	Zone	Subzone
Ypresian	*Globigerina-Globorotalia-Truncorotaloides* assemblage	*Globorotalia rex* zone	
Landenian	*Globigerina*-keeled *Globorotalia* assemblage	*Globorotalia angulata* zone	*Globorotalia velascoensis-G. acuta-Globigerina spiralis* subzone *Globorotalia pseudobulloides* subzone
Danian	*Globigerina* assemblage	*Globorotalia compressa-Globigerinoides daubjergensis* zone	
Maastrichtian	*Globotruncana* assemblage	*Globotruncana* zone	

(Loeblich and Tappan, 1957a, b).

proselytized in the 1950s most notably by Loeblich and Tappan (1957a, b). Table 2.3 summarizes their biostratigraphy as pieced together for the US Gulf and Atlantic coastal plains. Much more than the studies of deeper-water facies already discussed here, this project had to adopt the piecemeal strategy of constructing successive assemblages of taxa by correlations among short sections in neritic facies, rather than by plotting the ranges of taxa in longer sections – to the extent that Young (1960) complained of excessive 'catastrophism' as the main outcome. The essential point here is that the spectacular rise of these fossils as stratigraphic tools did not constitute a conceptual breakthrough. Planktonic foraminifera improved very substantially upon previously studied fossil groups but in a quantitative, not qualitative way. They allowed increased distance and precision in correlation.

From assemblage zones to range zones

In the mid 1960s two brief but quite unusually comprehensive papers were published. One was a summary and modification of the Trinidad zonation which also took account of a rapidly growing literature elsewhere (Bolli, 1966); the other was the definition of the N-zones of low and sometimes mid-latitudes (Banner and Blow, 1965). These papers, both on planktonic foraminifera, had two other and critically important points in common. The first point was that they defined their zones only by bounding events – by first and last appearances. Bolli's paper took account of suggestions from elsewhere of stratigraphic distributions that either could be seen and adopted in the Trinidad succession or could fill the gaps in it. Otherwise, changes in the extent and names of the zones between 1957 and 1966 were not very great. The difference was to be found in the change from zones that were not formally defined to zones that are so defined – by bounding events that are first occurrences and last occurrences. The second important point was that these zones were no longer just Caribbean zones – they were zones that could be applied wherever the fossils could be found. Similarly, Banner and Blow cited recognition of their zones from the Caribbean to the Alps, from the US Gulf coast to Japan and New Zealand. Thus, those zones were lifted out of the rocks, as it were. In the subsequent full explication of the Neogene N-zones, Blow (1969) specified holotype and paratype localities, but the actual definitions were not constrained by such specifications.

Figure 2.5 compares the two schemes of zonation for the Neogene.

From range zones to phylozones

Just as the geological timescale was built without the benefit or hindrance of a cogent theory of organic evolution, so too can we imagine quite easily the construction of systems of assemblage – and range zones without any biohistorical theory in the vicinity. Of course, evolution was here to stay and all the authors from Glaessner onwards presented theories of phylogeny. The most famous analysis of a planktonic foraminiferal lineage concerns the evolution of *Orbulina* from Miocene *Globigerinoides* (Blow, 1956) (Chapter 4). By the early 1960s the reconstruction of foraminiferal lineages was an important component of planktonic foraminiferal research programmes and this development was celebrated in a 1964 symposium on the use of lineages in Neogene biostratigraphy (Drooger *et al.*, 1966).

The boundary events of the N-zones were more than first and last appearances: they were phylogenetic events – speciations and extinctions. Thus, Blow (1969, p. 203), in an extended quotation which was the fullest statement of its time on this matter of evolution and biozonation:

> Wherever possible, the first stratigraphical occurrence of a particular
> taxon has been used, especially where that taxon can be referred
> to a particular stage in the evolutionary development of a known
> phylogenetic lineage. Thus, wherever possible, the base of a zone is
> defined upon the positive evidence of the observable presence of a
> taxon (at its lowest stratigraphical level) where it occurs in association

EPOCH		BOLLI (1957b, 1966, 1970) BOLLI and P. SILVA (1973)	A. Common Planktonic Foraminiferal Events used by both BOLLI and BLOW to define their zones. B.* Foraminiferal Events used by BOLLI, and not used by BLOW.	BANNER and BLOW (1965) BLOW (1969)	Foraminiferal Events used by BLOW and not by BOLLI.
PLEISTOCENE		*Globorotalia truncatulinoides* — *G. bermudezi* / *G. calida* / *G. hessi* / *G. viola*	*G. fimbriata* F.A.* / *G. tumida flexuosa* L.A.* / *G. calida calida* F.A.D. / *G. hessi* F.A.* / *G. truncatulinoides* F.A.D.	N 23 / N 22	
PLIOCENE	LATE	*Globorotalia truncatulinoides cf. tosaensis*		N 21	*G. tosaensis tenuitheca* F.A.
	MIDDLE	*Globorotalia miocenica* — *G. exilis* / *G. trilobus fistulosus*	*G. miocenica* L.A.* / *G. trilobus fistulosus* L.A.*	N 20	
	EARLY	*Globorotalia margaritae* — *G. margaritae evoluta* / *G. margaritae margaritae*	*G. margaritae* L.A.* / *G. margaritae evoluta* F.A.* / *G. margaritae* F.A.*	N 19	*G. pseudopima* F.A. / *S. dehiscens* F.A.
MIOCENE (N)	LATE	*Neogloboquadrina dutertrei*	*N. dutertrei* F.A.*	N 18	*G. tumida tumida* F.A.
		Globorotalia acostaensis		N 17	
		Globorotalia menardii	*N. acostaensis* F.A.	N 16	*G. plesiotumida* F.A.
		Globorotalia mayeri	*G. siakensis* L.A.	N 15	
		Globigerinoides ruber	*G. ruber* L.A.*	N 14	
MIOCENE (M)	MIDDLE	*Globorotalia fohsi robusta*	*G. fohsi robusta* L.A.* / *G. fohsi robusta* F.A.*	N 13	*G. nepenthes* F.A.
		Globorotalia fohsi lobata	*G. fohsi lobata* F.A.*	N 12	*S. subdehiscens* F.A.
		Globorotalia fohsi fohsi	*G. fohsi fohsi* F.A.*	N 11	*G. fohsi (s.l.)* F.A.
		Globorotalia fohsi peripheroronda	*G. insueta* L.A.*	N 10	*G. praefohsi* F.A.
		Praeorbulina glomerosa	*P. glomerosa* F.A.*	N 9	*G. peripheroacuta* F.A.
	EARLY	*Globigerinatella insueta*	*C. dissimilis* L.A.	N 8	*O. suturalis* F.A.
		Catapsydrax stainforthi	*G. insueta* L.A.	N 7	*G. bisphericus* F.A.
		Catapsydrax dissimilis		N 6	
		Globigerinoides primordius	*G. kugleri* L.A.	N 5	
LATE OLIGOCENE		*Globorotalia kugleri*	*G. primordius* F.A.	N 4	
				N 3	

Figure 2.5 Comparison of Neogene zonations by Bolli (left) and Blow (right) (Srinivasan and Kennett, 1981a, with permission). F.A., first appearance; L.A., last appearance.

with other forms, morphologically intermediate between it and its immediately ancestral taxon. Such phylogenetic events have been compared and related, stratigraphically, to stages within the phylogenetic series of other taxa, to the extent that confidence may be placed in their correlative value. For a general example, if A is found to evolve into B at a horizon which is consistently the horizen of evolution of M into N, and also is consistently above the horizon of X into Y, regardless of

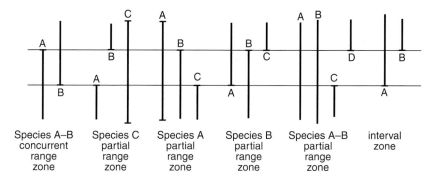

Figure 2.6 Definitions of various kinds of range zone, after Berggren and Miller (1988), with permission.

the geographical area being studied, then the first evolutionary appearances of Y, B and N, in that stratigraphical order, have been adopted as stratigraphical indices; B may be taken, in such a case, as the nominate taxon for a particular biostratigraphic interval, and the distribution of Y and N may be used as controls. For a particular example, *Globorotalia (Turborotalia) acostaensis acostaensis* makes its first evolutionary appearance at the base of, and has been adopted as the nominate taxon for, Zone N.16; this is biostratigraphically controlled by the first evolutionary appearance of *Candeina nitida praenitida* in the uppermost part of Zone N.15 and the first evolutionary appearance of *Globorotalia (G.) merotumida* in the lowermost part (but above the base) of Zone N.16.

Blow went on to express a general distrust of horizons of extinction as isochronous surfaces, and he enlarged on the problem of diachronous extinctions in a subsequent paper (Blow, 1970). As events defining zones, his preference was for evolutionary first appearances, for successional speciations in well-worked-out lineages with, as above, other speciations close in time that could act as constraints. Already by 1969, Blow was suggesting that 'The series of zones proposed in this work seem to represent *a probable maximum subdivision of the planktonic foraminiferal record* in tropical areas, taking into account such factors as the stratigraphic persistence of various easily recognisable taxa, their geographical extent and the evolutionary rates of development of the post-Eocene Globigerinacea' (p. 278; emphasis added).

An important change was occurring here. There is quite a difference between the range and interval zonal boundaries sketched in Figures 1.8 and 2.6 based on rigorous, comprehensive collecting, identification of taxa, and compilation, on the one hand, and zonal boundaries based on speciations or extinctions on the other. It is one thing to to define a boundary on the top of a range of a species in

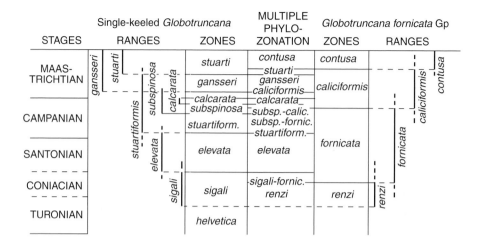

Figure 2.7 Multiple phylozones based on *Globotruncana* bioseries (van Hinte, 1969).

a measured section; it is quite another to define it on the basis of that species' extinction on the other. Certainly, the accumulated observations of samples 'may asymptotically approach theoretical totality' in Blow's (1979) distinctive prose, but there is involved also a major conceptual shift.

In 1957, Bolli cited type localities or drilled sections of strata for the Trinidad zones but they were not repeated in 1966 for the zones, now defined clearly by bounding biostratigraphic events. In *Plankton Stratigraphy* (Bolli *et al.* 1985) the *only* definition deemed necessary in most cases is exemplified by: 'FO of *Discoaster druggi* to LO of *Triquetrorhabdulus carinatus*' (Miocene calcareous nannofossils); or, 'Interval from first occurrence of *Rotalipora appennica* to first occurrence of *Rotalipora brotzeni*' (Albian planktonic foraminifera). The FO ultimately must be a speciation, even though these examples do not actually specify the ancestral-descendent change, as did Blow's definitions. In accordance with the Stratigraphical Code, Blow himself cited holotype and paratype localities for the N-zones whilst at the same time emphasizing the phyletic configurations. The ISSC in the 1970s (Hedberg, 1976) recommended that the definition of a zone include specifically identified strato-type or reference sections, comment on thickness, lateral extent, relation to litho-stratigraphic units. The North American Code (1983) sets out similar requirements for the formal establishment of a zone.

A step beyond phylozones brings us to multiple phylozones. Also in the 1960s, van Hinte began with the fact that we had a reasonable understanding of the phylogenetic pattern of some lineages and showed how lineages of *Globotruncana* in the later Cretaceous could be the basis for combined and mutually supportive phylozones. Figure 2.7 is adapted from van Hinte (1969). Berggren (1971a) developed very similar notions for the Cenozoic succession.

Coping with provinciality

As Drooger (1966) and Bolli (1966), among numerous writers, have pointed out, the establishment of the Trinidad zonation was followed swiftly by its extension and testing around the world. That was the essence of the shift discussed above. But there was another effect: already by 1964 (1966, p. 48) Bolli could generalize that:

> The planktonic Foraminifera populations of the Upper Cretaceous, Paleocene, and Eocene show no or very limited variations in their species within latitudes ranging to at least 45–50° on each side of the equator. As a result, they allow for fine-cut, worldwide stratigraphic correlation. During the past few years it became more and more evident that such populations of the Oligocene and Miocene vary much more geographically. Consequently, worldwide stratigraphic correlations are often difficult or impossible. These differences in populations are probably the result of a more pronounced climatic variation since the beginning of the Oligocene. Reduced temperature tolerance of certain species could be at least partly another reason.

Now that global climatic deterioration was hardly a new discovery, for example, the Edinburgh anatomist Robert Grant developed a theory in the 1820s for life's progressive change in response to the loss of uniformly warm global conditions and the onset of climatic zoning (Desmond, 1989). The striking translatitudinal changes in diversity and composition among modern planktonic foraminiferal faunas were known (Chapter 4). The question was: how to deal with this particular problem of planktonic (open-ocean) foraminiferal biostratigraphy – a problem exacerbated by the fact that stratotypes and other classical European sections accumulated in temperate as well as in neritic seas. What frameworks do we need when tackling the situations of provincial distributions?

One way was to erect local zonal successions, as in the southern (Austral) temperate region (Carter, 1958a, b; Jenkins, 1960, 1966a, 1967, 1971, *inter alia*). It is no accident that chronostratigraphic systems – stages or their equivalents – were particularly strong in isolated bioprovinces (Chapter 7). Jenkins (1985) stigmatized as parochial behaviour the development of local stages and the avoiding of planktonic foraminiferal zonations in research preceding his 1960 study (Chapter 7); however, it was also normal procedure in the history of Late Phanerozoic stratigraphy. As oceanic sections were recovered in deep-sea drilling, pronounced microfaunal changes across major watermass boundaries led to the establishment of quite elaborate zonations in some cases. The prime examples were the tropical, warm subtropical, and transitional systems of the

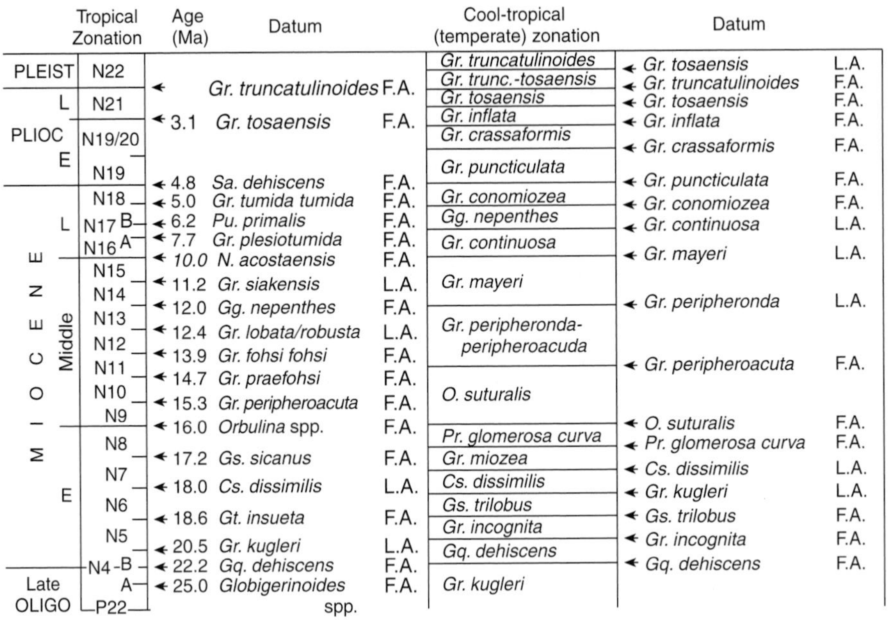

Tropical Zonation	Age (Ma)	Datum	Cool-tropical (temperate) zonation	Datum
PLEIST — N22		◄ Gr. truncatulinoides F.A.	Gr. truncatulinoides	◄ Gr. tosaensis — L.A.
L — N21			Gr. trunc.-tosaensis	◄ Gr. truncatulinoides — F.A.
	◄ 3.1	Gr. tosaensis F.A.	Gr. tosaensis	◄ Gr. tosaensis — F.A.
PLIOC E — N19/20			Gr. inflata	◄ Gr. inflata — F.A.
			Gr. crassaformis	◄ Gr. crassaformis — F.A.
N19	◄ 4.8	Sa. dehiscens F.A.	Gr. puncticulata	◄ Gr. puncticulata — F.A.
N18	◄ 5.0	Gr. tumida tumida F.A.	Gr. conomiozea	◄ Gr. conomiozea — F.A.
L — N17 B	◄ 6.2	Pu. primalis F.A.	Gg. nepenthes	◄ Gr. continuosa — L.A.
N16 A	◄ 7.7	Gr. plesiotumida F.A.	Gr. continuosa	
	◄ 10.0	N. acostaensis F.A.		◄ Gr. mayeri — L.A.
N15			Gr. mayeri	
N14	◄ 11.2	Gr. siakensis L.A.		
N13	◄ 12.0	Gg. nepenthes F.A.		◄ Gr. peripheronda — L.A.
Middle — N12	◄ 12.4	Gr. lobata/robusta L.A.	Gr. peripheronda-peripheroacuda	
N11	◄ 13.9	Gr. fohsi fohsi F.A.		
N10	◄ 14.7	Gr. praefohsi F.A.		◄ Gr. peripheroacuta — F.A.
N9	◄ 15.3	Gr. peripheroacuta F.A.	O. suturalis	
	◄ 16.0	Orbulina spp. F.A.		◄ O. suturalis — F.A.
N8	◄ 17.2	Gs. sicanus F.A.	Pr. glomerosa curva	◄ Pr. glomerosa curva — F.A.
N7	◄ 18.0	Cs. dissimilis L.A.	Gr. miozea	◄ Cs. dissimilis — L.A.
E — N6	◄ 18.6	Gt. insueta F.A.	Cs. dissimilis	◄ Gr. kugleri — L.A.
			Gs. trilobus	◄ Gs. trilobus — F.A.
N5	◄ 20.5	Gr. kugleri L.A.	Gr. incognita	◄ Gr. incognita — F.A.
N4 B	◄ 22.2	Gq. dehiscens F.A.	Gq. dehiscens	◄ Gq. dehiscens — F.A.
Late OLIGO — A — P22	◄ 25.0	Globigerinoides spp.	Gr. kugleri	

(Tropical Zonation left margin: PLEIST, PLIOC (L, E), MIOCENE — L(ate), Middle, E(arly), Late OLIGO)

Figure 2.8 Parallel zonations between strata deposited beneath tropical, warm-sub-tropical, and transitional watermasses, southwest Pacific (Srinivasan and Kennett, 1981b, with permission). F.A., first appearance; L.A., last appearance.

southwest Pacific (Srinivasan and Kennett, 1981a, b) (Fig. 2.8). Berggren (1984) illustrated provincial biostratigraphy in more general terms for the translati-tudinally extensive Indo-Pacific region (Fig. 2.9).

Wade (1964, 1966), less trusting of the local stratigraphic ranges in southern Australia than was Jenkins in 1960, proposed a deliberately looser way of zoning. Wade's main theme was the value of 'tolerant' evolutionary lineages found to develop in lockstep in temperate and tropical regions; and her biostratigraphic strategy was to interleave recommended tropic-temperate zones with less distinctive local (southern Australian) zones. The prime example of a very widespread but evolving group was the evolution of *Orbulina* from *Globigerinoides* in the Miocene (Chapter 4).

Another way was Blow's (1969). Blow was remarkably sanguine about our chances of extending his N-zonation into cooler-water regions. It was not that he could follow his essentially tropical assemblages to higher latitudes but rather that he believed that *some* of the zone-defining phylogenetic events could be recognized there, so that combinations of the N-zones were feasible. Also, Blow advised that hitherto-neglected small species would strengthen tropical-extra-tropical correlations. The essential point here is that separate provincial

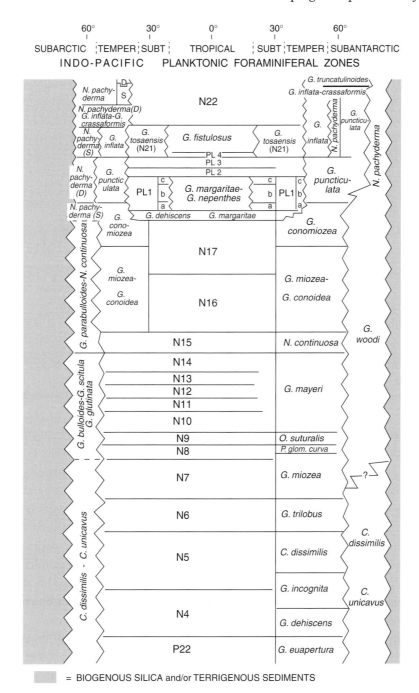

Figure 2.9 Provincial zonation from the equator to high northern and southern latitudes (Berggren, 1984): Neogene planktonic foraminiferal successions in the Indo-Pacific region.

planktonic foraminiferal zonations were not considered necessary even for the Neogene, when the world was tumbling into its present steep oceanic gradients and climatically zoned, icehouse state.

Yet another strategy, advocated by me (McGowran, 1978a, 1986b), is discussed below, under datums and chronozones. Finally, we return to the classical regions of stratigraphy, in this case, to central Paratethys. The centrepiece of modern work in this province is a strengthened provincial chronostratigraphy (Rögl, 1985) (Chapter 7).

Towards the chronozone

The zonal successions established, confirmed and refined in the tropics and the subtropics have become known as the 'standard' zonation, or the 'tropical standard'. This is entirely acceptable – the zones are 'better' for various reasons which are interesting if not entirely understood (Chapter 6). There was not a single standard; the zonations developed in two streams out of their Caribbean origins, one presented comprehensively in Bolli *et al.* (1985) and the other in Blow (1979). But the question that I address here is: do the so-called standard zones change conceptually or in their status as a result of acquiring that aura of importance?

Berggren (1971a) contrasted the 'biostratigraphic nature' of assemblages of fossils in general with the 'chronostratigraphic significance' of those zones based on rapidly evolving lineages, such as the phylozones in the record of the tropical planktonic foraminifera. Figure 2.10 attempts to clarify the business of zones and time – of biostratigraphy and some aspects of chronostratigraphy – in a way contrived to illustrate the main recurring problems of our discipline. Three tropical pelagic lineages give us three speciations on which to draw three zonal boundaries. The boundaries can be drawn to the limits of the actual records of the species concerned: those limits are encountered inshore and at higher latitudes – respectively 'facies' and 'biogeography', as some might prefer. (A third limit is encountered where the skeletons of index species are destroyed by deep-sea dissolution, but that is ignored here.) However, the tropical pelagic zones can be cross-correlated opportunistically with others, because the oscillations of climatically controlled bioprovinces and of sea-level changes and marine transgressions/regressions across continental margins produce an interfingering pattern in the stratigraphic record. With some luck and alertness to opportunity, there develops an interlocking scheme of biozones representing different global provinces and contrasting, mutually exclusive environments (McGowran, 1986a).

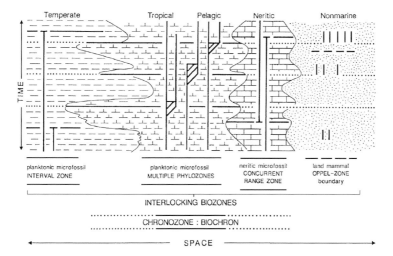

Figure 2.10 Zones and time; chronozones and environments (McGowran, 1986a). A time-space-facies diagram with four facies symbols interfingering to indicate climatic and sea-level fluctuations. Heavy vertical lines represent species' ranges. Thus, in favourable circumstances one can cross-correlate between the facies of different environments. The tropical/pelagic record of microfossils is central, in that (i) those zones are based on *speciations* (shown as not entirely instantaneous), and (ii) phylozones (which are biostratigraphic) are the *basis* for chronozones (which are chronostratigraphic, as the ISSC would have it) from which time-parallel surfaces are extended beyond the facies in which the species are found.

Among the various bio-events displayed in Figure 2.10, the first evolutionary advent of species in the tropical lineages is the 'hardest' data, and the figure suggests that one can extend time lines beyond the environmental and geographic limits of the evolving lineages. The *Guide* characterizes a chronozone as comprising all rocks formed anywhere during the time-range delimited by some geological feature or specified interval of rock strata (Hedberg, 1976). The essential point is that the strata so included formed in all sorts of environments – way beyond the lateral extent of the defining bioevents. Thus the various facies respectively with mammal teeth, reefal large foraminifera, and subpolar planktonics could be included in a chronozone based on a tropical pelagic phylozone.

The *Guide* displays the schematic relationship between biozone and chronozone (Hedberg, 1976, Fig. 12; Salvador, 1994, Fig. 13) without an *x*-axis. Loutit *et al.* (1988, Fig. 3) adapted the concept to the distribution of planktonic microfossils by adding latitude for a species whose true range – the chronozonal limits – is to be sought in the tropics (Fig. 2.11). Loutit *et al.* also illustrated the influence of watermass on producing disjunct distributions (Fig. 2.12).

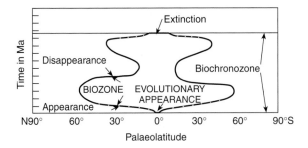

Figure 2.11 Biozone, biochronozone and watermass (Loutit *et al.*, 1988, with permission). The time-latitude envelope describing the biozone is broadly symmetrical because the climatic/watermass fluctuations are broadly bipolar.

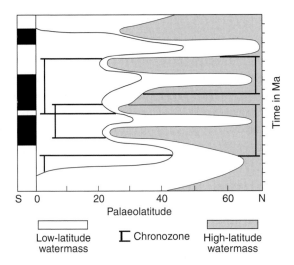

Figure 2.12 Schematic illustration by Loutit *et al.* (1988, with permission) of deep-sea stratigraphy integrating biostratigraphy and magnetostratigraphy. Biochronozones defined and correlated within high- and low-latitude watermasses, respectively, can be cross-correlated because watermasses migrated latitudinally through time. The defining bio-events can be calibrated magnetostratigraphically in favourably preserved and recovered drilled sections.

Datums

To develop this discussion further requires the introduction of the notion of horizons marked by bio-events, namely *datums*. The biosphere has invaded virtually every environment and has left a fossil record in many of them. Figure 2.10 barely touches upon the complexities of the biostratigraphic record of a given slice of time. How far can we reasonably go in capturing the subtleties of the fossil record in space and in time in the unsubtle nets of formally defined zonal systems? An increasingly popular attempt to avoid the clutter and confusion of multiple zonations based on the same major taxon (e.g., planktonic foraminifera) is simply to list the defining events, or datum levels. That is the crispest way to define modern phylozones, as we have seen. But the extant codes of stratigraphy specify formalities associated with the establishment of zones, as we have seen also. As one response to the complexities of the fossil record, I suggested that we abandon provincial zones altogether (McGowran, 1978a, 1986b). That apparent heresy needs explication.

Hornibrook (1969) defined the *datum level* as 'a correlation plane joining levels in rock sequences which on palaeontological or other grounds appear to be isochronous'. He observed that a single world system of Cenozoic planktonic foraminiferal zones is too simplistic an objective and that parallel tropical and temperate successions is a more realistic aim. To focus on the persistent problem of cross-correlating, Hornibrook suggested that greater effort be directed toward the establishment of several '*correlation levels or reference horizons or datum planes*'. The notion of datums was taken up by Jenkins (1966b) and by Berggren (1969b) who listed 23 levels of well-established biohorizons in the Cenozoic; the biostratigraphic approximation of the main divisions of the Cenozoic (series and subseries) is abstracted from that list (Table 2.4). But Blow, having pronounced against parallel provincial zonations (1969), now pronounced magisterially against datum planes (1970): 'these are considered as largely misleading and should not replace the concept of formally defined zones'. Notwithstanding, the need for a simple chronological sequence of biostratigraphic events did not disappear and Hornibrook and Edwards (1971) and Saito (1977) produced tabulations of events. Effective interregional correlation required addressing the following (Saito, 1984): (i) identification of the events in different provinces and some evaluation of their degree of synchroneity; (ii) do they maintain the same order of succession? (iii) which of these events are evolutionary? Hornibrook's proposals originated in New Zealand where the strongest southern extratropical zonation had been developed (Jenkins, 1966a, 1967). Wade's (1964, 1966) interesting strategy of composite zonation was not developed further in southern Australia; instead, Jenkins' system was brought across the Tasman Sea and found to be applicable, with some modification (Lindsay, 1967; Ludbrook and Lindsay, 1969; McGowran *et al.* 1971). But, there had to be some modification by amending zonal definitions – that was the critical point. Could we go on modifying zones as an entirely necessary and legitimate scientific endeavour whilst staying free of the trammels of proliferating nomenclature? Jenkins himself (1974) added subzones that were not simple divisions of

Table 2.4 *An example from the late 1960s of planktonic foraminiferal datums which could be used as fair approximations in recognizing major boundaries within the Cenozoic*

Time/Time–rock division	Planktonic foraminiferal datum points
Pleistocene	FAD *Globorotalia truncatulinoides*
Upper Pliocene	LAD *Globoquadrina*; LAD *Sph'opsis*
Lower Pliocene	FAD *Sphaeroidinella*
Upper Miocene	LAD *Cassigerinella*
Middle Miocene	FAD *Orbulina*
Lower Miocene	FAD *Globigerinoides*
Upper Oligocene	LAD *Pseudohastigerina*
Lower Oligocene	LAD *Hantkenina*
Upper Eocene	LAD *Morozovella*
Middle Eocene	FAD *Hantkenina*
Lower Eocene	FAD *Pseudohastigerina*
Upper Paleocene	FAD *Morozovella angulata*
Lower Paleocene	LAD *Globotruncana*
Maastrichtian	LAD *Rugoglobigerina*

Adapted from Berggren (1971a).
FAD first appearance datum.
LAD last appearance datum.

the zones, i.e. were not nested and hierarchical, thereby adding to the confusion. In southern Australia a difficult local problem in correlation was anatomized as follows (McGowran, 1986b, 1989a): (i) compilation by correlation and ordination of a composite regional succession of biostratigraphic events from scattered, discontinuous, neritic assemblages; (ii) testing the composite succession against an oceanic section in the region; and (iii) continuing the struggle to correlate the composite succession against the tropical standard. I have found over several years that the New Zealand corpus was more assimilable across the Tasman in its datum form than as zones (Fig. 2.13). Where Saito (1977) found that it was 'obvious by now that any attempt to establish interregional correlation by planktonic foraminiferal zonal schemes requires a thorough review of the definition of zones proposed by each author and an understanding of the reference biohorizons on which the zonal scheme is based', we had encountered these problems *within* a small part of the austral province. There was indeed 'inherant instability' in this biozonation (Hornibrook, 1971).

From there, it was but a short step to abandon austral zones altogether. Two things remain to be stated, one of them emphatically. First, it need only be

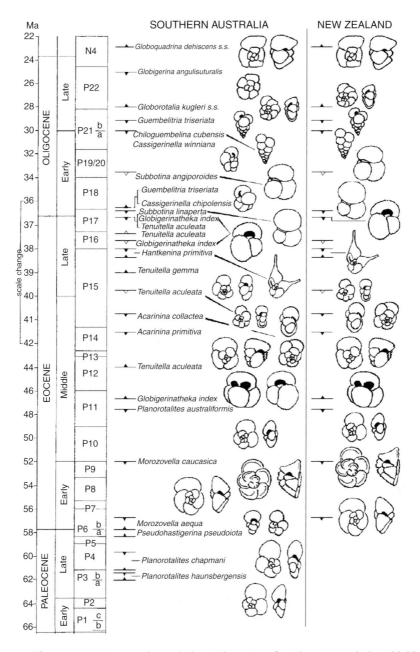

Figure 2.13 Datums and correlation, Palaeogene of southern Australasia. A highly composite regional succession of planktonic foraminiferal events is correlated with the lower-latitude P-zones, which are being employed as chronozones (McGowran and Beecroft, 1985) and the homotaxial succession of events in the southern temperate realm is demonstrated by parallel events in New Zealand (see Fig. 2.15 and discussion).

recognized clearly that these differences have to do with the transmission – the communication – of conclusions as to correlation and *not* with the substance of those conclusions; there is no implication whatsoever that the tropical P- and N-zones are actually being recognized in southern Australasia.

Second, the tropical pelagic foraminiferal zones are being used accordingly as *chronozones*. This development was anticipated in part by Riedel in 1973 when he noted the fact that refined correlations were often more satisfactory than were less refined age determinations – because the former involved comparisons between similar pelagic microfaunas and microfloras but the latter required reference to bereft stratotypes (Chapter 7). Accordingly, 'for many geological purposes involving integration of observations in one or a few biogeographic regions but not over the entire surface of the earth, it may be desirable to express correlations in terms of biostratigraphic zonations (or intervals of time defined by sets of paleontologic events), rather than in epochal terms with their additional uncertainties' (Riedel, 1973, p. 251). I have taken this suggestion a step further; it is a big step, admittedly, because it takes zones right out of the realms inhabited by their organisms.

Cross-correlation, calibration and quality control

If the development of planktonic foraminiferal biostratigraphic zonation was rapid, then the rise of the other groups of pelagic protists has been spectacular. At the time of Glaessner's (1945) summary of the groups, variously they were virtually untapped (the coccoliths), or had the attributes that should make them valuable index-fossils for long-range correlation (the diatoms), or were valuable not in long-range correlation and age determination but as comprising diverse biofacies that are significant in local correlation (the radiolarians). All of that changed in an accelerating way from the 1950s onwards within the framework provided by the planktonic foraminifera and stimulated in turn by (i) the development of marine geology, (ii) the advent of the scanning electron microscope and (iii) the launching of the Deep Sea Drilling Project. In his review in 1973, Riedel contrasted the relatively complete state of the art in Cenozoic planktonic foraminiferal biostratigraphy with the situation in the other groups of pelagic protists – the diatoms, radiolarians, calcareous nannofossils and silicoflagellates (adding the dinocysts would have enhanced the point). From a bibliography for the years 1969 and 1970 (the first volumes on the Deep Sea Drilling Project appeared in 1970), Riedel gave the numbers of published titles: about 500 papers on the foraminifera; 115 on all the rest. In not much more than a decade later, we had Cretaceous and Cenozoic zonations for all the major groups of pelagic, skeletonized micro-organisms (Bolli *et al.*, 1985) –

Table 2.5 *Average durations of the zones correlated and tabulated by Bolli et al. (1985, Ch. 2, Figs. 1 and 2). Subzones are counted as zones.*

Zones and subzones	Durations of pelagic protistan zones average (m.y.)
I. Cenozoic zones	
planktonic foraminifera	
'Trinidad stream'	1.4
P- and N-zone	1.5
P-zones[a]	1.5
calcareous nannofossils	
CP- and CN-zones	1.1
NP- and NN-zones	1.5
radiolarians	1.9
diatoms	1.9
silicoflagellates	3.3
dinoflagellates	5.9
II. Mesozoic zones	
planktonic foraminifera	2.3
calcareous nannofossils	3.0
radiolarians	8.8
dinoflagellates	6.6

[a] Palaeogene zones from Berggren and Miller (1988).

planktonic foraminifera, calcareous nannofossils, radiolarians, diatoms, silico-flagellates and dinoflagellates. A crude but not uninformative comparison can be made on the basis of average zonal durations (Table 2.5).

Glaessner (1967) warned that 'It is most important that we should remain stratigraphers, not strictly specialists in planktonic foraminifera or palynologists or vertebrate palaeontologists, when we discuss time scales and correlations'. Certainly we – most of us – must be specialists: the requirements of modern micropalaeontology are too demanding for anything else. Thus, ocean drilling expeditions might include not only a specialist for each of the main protistan groups but one for the expected Cretaceous foraminifera as well as for the Cenozoic, or for the benthics as well as for the planktonics. And yet the most powerful impact of two decades of drilling the ocean basins is of synergistic cooperation – two specialists will progress more rapidly than two generalists.

Of course, the fossils themselves are specialists, in a sense. Oceanic sedimentary facies boil down to two classes of biogenic facies, plus the terrigenous

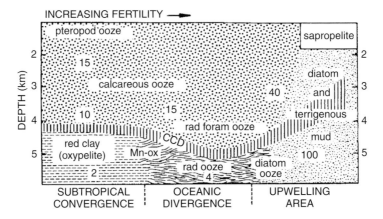

Figure 2.14 Distribution of major oceanic facies in a frame of water depth (km) and fertility, not quantified; however, numbers give typical sedimentation rates in mm/1000 years (= m/million years). Berger (1974; Seibold and Berger, 1993, Fig. 8.2) based this diagram on sedimentary patterns in the eastern central Pacific. Convergence: low fertility thence low productivity, low numbers of siliceous skeletons do not sediment at depth, calcareous ooze above CCD (calcite compensation depth), clay and oxides, etc., by default below CCD. Divergence: increased fertility and productivity, more biogenic sediment (both calcite and opal), CCD depressed and rad (radiolarian) ooze at depth. Upwelling: maximum fertility and productivity, opal becomes dominant (as diatoms); in extreme case organics accumulate as sapropels, with permission.

materials which include turbidites and glacigenes, plus the brown clays which concentrate where it is too deep for calcite to accumulate, where the surface waters are too infertile for opaline silica to fall below the upper kilometer or so before resorption, and where oceanic areas are beyond the reach of terrigenous sedimentary delivery systems. Calcareous ooze contains coccoliths and foraminifera; and zones are, unsurprisingly, developed in that facies at low and midlatitudes (extending to higher latitudes in the Cretaceous and the Palaeogene). The siliceous oozes contain diatoms, radiolarians and silicoflagellates under the zones of high productivity (Fig. 2.14).

The plotting of datums for species' first and last appearances, in sections where the different taxonomic groups are preserved together, has been proceeding since deep ocean drilling began. Hay and Mohler (1969) pointed out that the parallelism of two sets of zones based on two groups of planktonic microorganisms may increase the biostratigraphic *resolution* twofold in parts of the column. Based on somewhat different assumptions and with non-identical strengths and weaknesses, the respective biostratigraphies should keep each

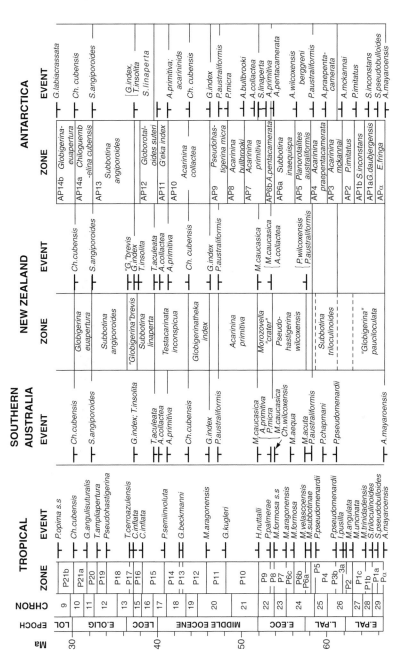

Figure 2.15 Southern datums: planktonic foraminiferal events (first and last appearances) in the Austral Palaeogene. At third-order scales there is not significant diachrony, such as might be imagined in response to global cooling and the retreat of provincial oceanic belts towards the equator. From various sources; Antarctic AP zones from Stott and Kennett (1991).

45

other honest, as it were, and reciprocally illuminate. They should, in harness, add *accuracy* to resolution. And yet, three decades later, rigorous cross-correlations between calcareous and siliceous planktonic zonations and datums have quite some way to go, and Berggren and Miller (1988) could warn that 'There are still surprising uncertainties in the calibrations of calcareous nannoplankton with planktonic foraminifera … and we encourage continued close cooperation between workers in these fields'. That comment still holds.

Correlations in the Austral Palaeogene

As mentioned already and to be mentioned again (Chapter 7), important developments originated in New Zealand. Jenkins (1993) in due course broadened the schema. Here, I want to consider some correlations that extend a long way out of the tropical region using calcareous plankton. The framework is summarized in Figure 2.15. First, biostratigraphic events pieced together in southern Australia are correlated with the tropical standard P-zones without the intermediate step of using Austral zones. Thus, the P-zones are being employed here as chronozones. The correlations are made without the benefit of magnetochronology. Next, the New Zealand zones, their defining events and some other selected events are added. Finally, we have the biostratigraphy of sections drilled on the Maud Rise in the Weddell Sea at about 65 °S (Stott and Kennett, 1990). In this case Stott and Kennett correlated the biostratigraphy with the geomagnetic polarity stratigraphy of the two sites – the magnetochronology is the go-between more than are the biostratigraphic events in common.

Note that there are indeed few events in common between the tropics and the rest. Again, there are several events in common among the three southern regions. I am impressed that we have been able to identify a succession of initial appearances that are in the right positions – *Planorotalites australiformis*, *Acarinina primitiva*, *Globigerinatheka index*, *Chiloguembelina cubensis* – as are several last appearances – *P. australiformis*, *A. primitiva*, *Subbotina angiporoides*, *Ch. cubensis*. The variations in position among last *G. index* and *Tenuitella insolita* at about the Eocene–Oligocene boundary are trivial, pending more work. Of the well-known events in the Austral region, the last occurrence of *Subbotina linaperta* is considerably earlier in Antarctica than elsewhere. However, the independent correlations of southern Australia, New Zealand and Antarctica with the tropical standard reveals an impressive parallelism in the biostratigraphic succession. Contrast that situation with the situation in zonal nomenclature: there are no zones in common between Antarctica and New Zealand and there are different intervals (and intervals overlapping only in part) bearing the same name – potential for more confusion in communication – than if we stick to datums.

3

Biostratigraphy: its integration into modern geochronology

Summary

We come thereby to (bio)chronozones and datums, which contribute an irreversible succession of events (because evolutionary events are unique) to geochronology. This record has been synthesized with magnetochronology (in sediments, volcanoclastics and oceanic crust) and radiochronology to produce the integrated magnetobiostratigraphic scale (IMBS), unique to the Cenozoic Erathem. Homotaxy, the consistent succession of bio-events in space and time, might harbor diachrony, a possibility requiring the disentangling of local or regional biozones from biochronozones. Cyclostratigraphy, explained by Milankovitch astrochronology, contributes not only an independent check on correlation and possible diachrony, but an unprecedented degree of chronological resolution.

Arcane initials aplenty: the CTS, the GPTS and the IMBS

Geological time is extracted from the stratigraphic record comprising the succession of sedimentary strata. Thus we get chronostratigraphy and the construction of the *classical timescale* (CTS) using radioisotopic dates related opportunistically to the stratal succession. Neither biostratigraphic resolution nor accuracy, although highly desirable outcomes in their own right, bears greatly upon the eternal problems of extending correlations into all environments and situations of interest to Earth history.

Funnell (1964) used biostratigraphically controlled radiometric dates to prepare the first relatively precise Cenozoic timescale. Berggren *et al.* (1985c) distinguished several approaches to subsequent Cenozoic geochronology.

Odin *et al.* (1982) emphasized radiometric dates to determine the numerical ages of geological boundaries (*radiochronology*). Berggren (1978) emphasized the role of evolutionary events in organizing the timescale (*biochronology*). The third strand, *magnetochronology*, was exemplified by the radiometrically dated reversal scale for the past few million years (Cox *et al.*, 1964; McDougall and Tarling, 1963) and the first timescale based on seafloor anomalies recorded over the past 80 myr (Heirtzler *et al.*, 1968). With the first versions of the Berggren integrated timescale 'the transition from a purely descriptive to a more quantitative study of Earth history is thus well on the way to becoming established' (Berggren, 1969; the Berggren (1971b) version actually was antecedent).

Magnetochronology is based on the record of the normal and reversed polarity of the geomagnetic field. It is a binary or flip-flop signal, iterative, not unique, and preserved in volcanic and sedimentary sections as well as in the oceanfloor. It is a phenomenon preeminently able to surmount the barriers between environments and provinces – to cross-relate events in the oceans and on the continents. If a local record of polarity changes can be slotted into the global polarity timescale, then it can act as an anchor for all the other components of a geohistorical timescale – especially the fossil succession and the evaluation of isotopic dates. The polarity record can do that because of two properties: reversals are instantaneous, geologically speaking, and they are not restricted environmentally, i.e. they can tie together strata of mutually exclusive facies and fossil content. A crucial example involving the terrestrial succession in the Eocene–Oligocene of North America is discussed in Chapter 7.

Hailwood (1989) gave us a clear statement of the value of the *global polarity timescale* (GPTS). Figure 3.1 develops the problem of relating radiometric dates to a geological timescale. There are actually two problems: firstly, it is very rare to have a good high-temperature date at a geological boundary; secondly, uncertainties in the isotopic dates until recently could be greater than the duration of the interval requiring boundary dates. Therefore, there has to be interpolation between, or extrapolation beyond, the dated horizons, which in turn demands some *linearizing operator*, such as a hopefully uniform sedimentation rate. A powerful alternative linearizing operator is founded in the assumption of uniform seafloor spreading (SFS) at certain mid-ocean ridges for certain periods in the (post-mid-Jurassic) geological past. Successive polarity changes in the geomagnetic field are recorded simultaneously in two contrasting geological records: in the linear oceanfloor patterns and in sedimentary and volcanic sections. If SFS is uniform, then the width of each block of normal or reversed polarity in the oceanfloor profile is proportional to the duration of its generation. Other problems (assumptions) are that each polarity boundary is real (i.e. not superimposed later) and that a sequence of magnetozones can be

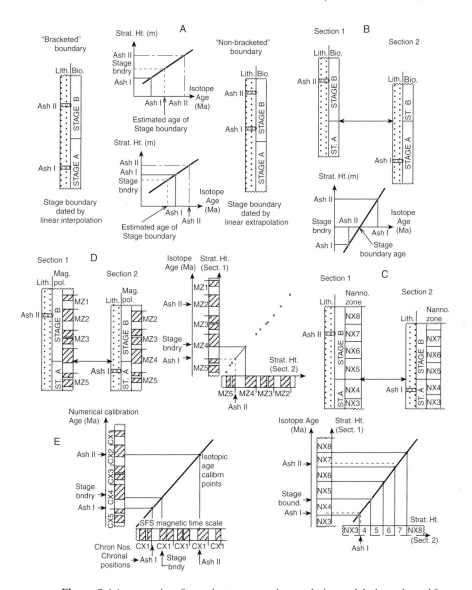

Figure 3.1 A composite of cases in geomagnetic correlation and dating, adapted from Hailwood (1989) with permission. A, single-section cases: ash horizons yielding dates, in one case bracketing the stage boundary in question and permitting graphic interpolation, the other not bracketing but permitting extrapolation. B, two sections, stage boundary extended by correlation, thereby exploiting the two ashes in the separate sections to give a numerical age. C, same, with nannofossil zones NX3–8, integrating chronostratigraphy (stages), biozones, and radiometric ages. D, also two sections, this time with magnetostratigraphic determinations (magnetozones MZ1–5). E, from magnetozones to magnetochrons: this time the rock section with its controls is graphed against the seafloor spreading scale (SFS).

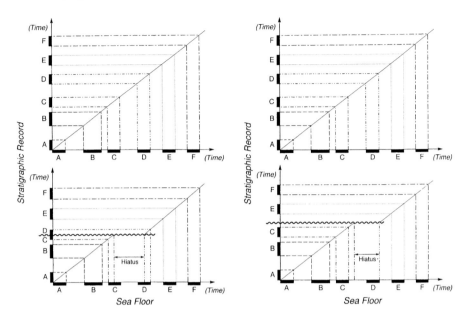

Figure 3.2 Relationship between magnetostratigraphy and magnetochronology, from Aubry (1995, Fig. 1, with permission). Upper two diagrams: A–F (black) are normal polarity chrons in time and normally polarized intervals in the stratigraphic record; ideally there is congruence between the stratal and the oceanic-crustal records. Lower two diagrams: the effects of unconformities. Left, two normal intervals are sutured or conflated into one. Right, two reversed intervals are sutured into one.

properly correlated with the GPTS. But the question of constant SFS rates is the important one here, being central to the generation of a *composite anomaly profile*.

Geomagnetic patterns in their SFS and stratigraphic manifestations can be compared graphically (Fig. 3.2), showing that polarity intervals can be conflated in the stratigraphic record due to hiatus. Graphic comparison of ridge segments shows that if sets of points fall on two or more linear segments, then the SFS rate must have changed on one or the other profile with the break in slope signalling the time of change (but not which profile has it) (Fig. 3.3). Repeating this exercise among profiles from the various spreading ridges in the various ocean basins identifies which profiles have the breaks, and it also yields particular segments where SFS rates must have remained effectively constant (Fig. 3.4). Thus by *iteration* can be assembled a composite overall profile which (by width ∝ duration) gives a relative SFS polarity timescale. To derive time from block width and a timescale requires an apparent spreading rate, which in turn requires radio- or astro-calibration. For the first polarity timescale Heirtzler *et al.* (1968) used essentially one profile from the South Atlantic and extrapolated from a Pliocene

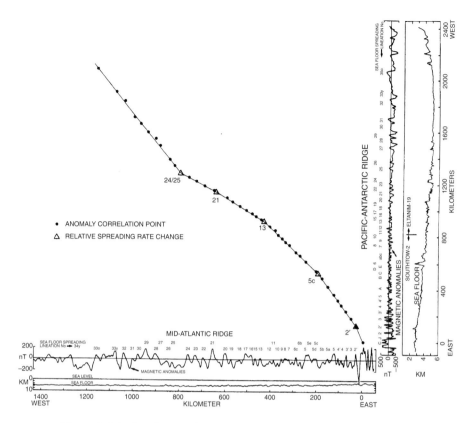

Figure 3.3 A plot of anomalies, Mid-Atlantic Ridge (Atlantis II-93 cruise) and Pacific-Antarctic Ridge (Eltanin-19 and Southtow-2 cruises), after Klitgord and Schouten (1986, Fig. 8) and Aubry *et al.* (1988, Fig. 2, with permission). The graph plots distances along tracks of the same magnetic lineations on the two ridges, so that the slope of the resultant line displays relative spreading rates between the two spreading centres.

ridge-axis-rate back to 80 Ma (late Cretaceous). Subsequent scales had more constraints (see especially Berggren *et al.*, 1985a–c). However, Hailwood emphasized that if the iteration is done correctly, then the relative (not quantified) durations of polarity intervals will be essentially correct regardless of uncertainties or errors in calibration points. By the same token, one cannot simply adjust a scale when new or improved calibration points become available; one must go back to the original profiles. Thus the GPTS is constructed on standardization to a model of (South Atlantic) spreading history (Cande and Kent, 1992, 1995). The next step is the integration of biostratigraphic events with the GPTS, and an ongoing reassessment of radioisotopic dates in their biostratigraphic context, in constructing the integrated magnetobiochronological Cenozoic timescale (IMBS or IMBTS). Four independent sets of data each has its own limits in

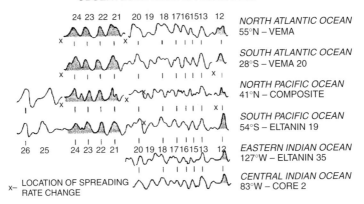

OBSERVED MAGNETIC ANOMALIES

Figure 3.4 Correlation of magnetic anomaly profiles from six oceanic regions, demonstrating spreading–rate changes and constant spreading rates between, after Aubry *et al.* (1988, Fig. 1, with permission). Segments between crosses display constant spreading rates and have been expanded by a constant amount here to match the anomaly spacing on the South Pacific profile.

precision and resolution, but by proper merging they are able to form a strong and mutually reinforcing scale, precisely because they are independent (Aubry *et al.*, 1988). Likewise, and contra Odin and Curry (1985), Aubry *et al.* showed that significant lengths of SFS record can be tested and *demonstrated* – not assumed a priori – to have formed at relatively constant rates, hence become candidates for standard reference sections.

The four sets of data are (i) biostratigraphic datums and zones, (ii) seafloor spreading magnetic lineation patterns, (iii) magnetostratigraphy from sedimentary and volcanogenic sections, and (iv) radiometric (isotopic) ages. (Cyclostratigraphy, the fifth and newest, is discussed below.) Biostratigraphy is correlated to magnetostratigraphy by *first-order* correlation where well-expressed datums occur together with well-defined and unambiguously identified magnetic anomalies (Fig. 3.5) and also where fossil dates are found in sediments immediately overlying oceanic basement, also with anomalies of that quality. The first-order correlations have been listed especially by Berggren *et al.* (1995a) and their importance to Cenozoic geochronology has been great. Notwithstanding further advances in the first-order correlation of datums with magnetochronology, Aubry (in Berggren *et al.*, 1995a) still felt obliged to warn that the number of sections with a reliable magnetostratigraphy is extremely small, with intervals such as the middle Eocene series where calcareous nannofossil bioevents are still poorly tied to magnetochronology. The main problem identified in making these first-order correlations concerns the

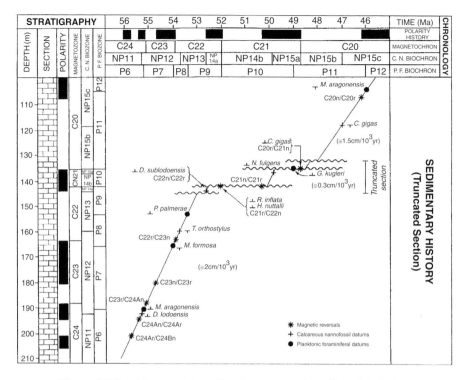

Figure 3.5 Graphic comparison of an Eocene oceanic section with the integrated geochronology, exposing a swarm of hiatuses that otherwise are largely cryptic (Aubry,1995, Fig. 5, with permission). Magnetostratigraphy and two biostratigraphic records in a calcareous section are plotted against magnetochronology and two biochronologies, giving a disrupted line of sedimentation rates.

temporal reliability of the bioevents: what do discrepancies in magnetobiostratigraphic correlations between sections actually reflect? The options are (i) diachrony through basin, province, or hemisphere, or (ii) problems in taxonomy and identification of 'species', or (iii) the presence of hiatuses in the sections truncating ranges. We return to that problem, below.

More contentious has been the calibration of the composite geomagnetic polarity succession to time and the matching of this chronology to the isotopic timescale (Aubry *et al.*, 1988; Hailwood, 1989). The field of Cenozoic geochronology was dramatically transformed by the virtual replacement of conventional ^{40}K–^{40}Ar dating with ^{40}Ar–^{39}Ar dating, in which technological advances can yield multiple ages, rapid and highly reproducible, with a very low standard error (*inter alia*, Berggren *et al.*, 1995a). For Aubry (1995, p. 214) the advent of the IMBS was more fundamental than a mere increase in rigour, resolution or accuracy: 'Geological time is thus embodied in two independent stratigraphic records. In the CTS, chronology is derived solely from the stratigraphic record.

For the CTS, it can be correctly argued that geochronology and chronostratigraphy are no more than "different aspects of a single procedure" (Harland *et al.*, 1990, p. 3). But the IMBS is radically different from the CTS in that it uses the sea floor lineation pattern as an independent chronometer, in which geochronology and chronostratigraphy appear as two fundamentally distinct disciplines. It is for this very reason that the IMBS constituted an unprecedented tool for stratigraphic analysis, although I would contend that its power has not yet been appropriately perceived.'

Homotaxy and diachrony

Shortly after d'Orbigny's and Oppel's epochal work in biostratigraphy, Huxley (1862) challenged the significance of faunal succession in establishing chronocorrelation. For it is one thing to demonstrate that a succession of faunas, floras or selected taxa can be confirmed from section to section, province to province, continent to continent – *homotaxy*; it is something else to show that such a match necessarily entails *synchrony*. In his characteristically colourful style Huxley suggested that 'for anything that geology or palaeontology are able to show to the contrary, a Devonian fauna or flora in the British Isles may have been contemporaneous with Silurian Life in North America, and with a Carboniferous fauna and flora in Africa'. Now, it is one of the most spectacular results of all palaeontological research that there is an eery repetition of forms and assemblages when conditions recur. Colonial sedentary skeletonized organisms forming build-ups on the seafloor, a set of teeth invented repeatedly for carnivory, photosymbiosis invented repeatedly in low-nutrient seas, the similarity of the Clarendonian (Miocene) chronofauna to the modern African mammal fauna (Webb, 1984), are but a few examples. But they are due to evolutionary convergence and they beg the question of homotaxy and synchrony. There remains the question of index fossils: how do we get from event *ordering* to event *dating*? 'The impasse in biostratigraphical theory is that the intrinsic palaeobiological and stratigraphical data for an individual event do not enable its isochroneity to be established' (Scott, 1985).

The standard response is twofold. First, that fossils signify irreversible events – itself worthy of some consideration. Kitts (1977) quoted the following from Grünbaum (1963): 'There is both a weak sense and a strong sense in which a process might be claimed to be "irreversible". The weak sense is that the temporal inverse of the process in fact never (or hardly ever) occurs with increasing time for the following reason: certain particular de facto conditions ("initial" or "boundary" conditions) obtaining in the universe independently of any law (or laws) combine with a relevant law (or laws) to render the temporal

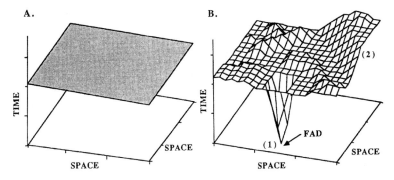

Figure 3.6 Hypothetical first-appearance surfaces in space–time (Dowsett,1988, with permission). A, a synchronous first-appearance surface. B, a diachronous first-appearance surface, in this case with two components. (1) The geographic centre of origin gives the first appearance datum. (2) Geographic dispersal occurred distinctly later, but even then there was a very patchy spread towards the area comparable to A – the undulations represent hindrances to dispersal and migration.

inverse de facto nonexistent, although no law or combination of laws itself disallows the inverse process. The strong sense of "irreversible" is that the temporal inverse is impossible in virtue of being ruled out by a law alone or by a combination of laws'. Geological conditions are irreversible in the weak sense because the conditions for reversal are never realized (Kitts, 1977) – the configuration is never quite repeated. But there are no laws to forbid that happening. Evolution is even less reversible statistically for similar but more powerful reasons: the chances of rerunning all the contingencies comprising the history of a species are zero and species are individuals in a sense that a sandstone or an anticline are not (see Chapter 8).

In the second response, biostratigraphic correlation depends on the evolutionary emergence of a species from its ancestor, its survival and its geographic expansion – one index of 'success'. The bottom surface of a species' envelope of spatiotemporal distribution might resemble the lumpy rather than the smooth sheet in Figure 3.6. The critical matter is the time elapsed from origin to geographic spread and the timescale of the exercise in correlation. Teichert (1958) cited the example of late Devonian goniatites: wherever conditions were suitable for goniatites, *Cheiloceras* is found in separate layers of strata above layers with *Manticoceras* – a century and a half's research in Europe and around the world had failed to falsify that succession which can, therefore, be used to determine sequence in layered rocks. And, wrote Teichert: 'As to time relationships, it would be nonsensical to assert that the zonal boundary between rocks containing *Manticoceras* and rocks containing *Cheiloceras* "transgress time". Such a hypothesis would require assumption of a highly unlikely pattern of faunal migrations, where swarms of species of *Manticoceras* are followed, every-where at the same distance and the same time interval, by swarms of species of *Cheiloceras*, the two waves preserving their separate identities on a staggered mass migration around the world, possibly throughout millions of years, without evolutionary changes and without ever becoming mixed. This picture is unreal. The only realistic conclusion is to assume that the boundary between the *Manticoceras* and the *Cheiloceras* zones is a true time plane.'

But, wrote Kitts, Teichert's theory of correlation requires that biological communities in different places be 'synchronized', or that biological signals be transmitted instantaneously, or both. Both assumptions are forbidden, as Teichert demonstrably was perfectly well aware; he and his colleagues were 'not doing a bad job of correlating' but they had gone seriously wrong in justifying theoretically what they were doing. The matter of the diachronous bottom surface of an index species' envelope was dealt with by Teichert and the standard texts as a timescaling problem. Thus Donovan (1966) wrote: 'Now, the mean duration of a Jurassic ammonite zone was about a million years, and most other zones were longer. It is clear that the time needed for wide dispersal is negligible compared with the duration of a zone, and the first appearance of a new species over a wide area can be regarded as simultaneous provided that the effect of facies can be discounted.'

Scott (1985) identified the central problem of diachrony and correlation – that there is no intrinsic way of assessing datums that 'behave themselves', i.e. datums that do not cross over in time when traced through space, other than to increase the density of biohorizons (Fig. 3.7). Consider Figure 3.8, displaying a late Neogene succession of planktonic foraminiferal events as found in two oceans. Compiled without palaeomagnetic or chemostratigraphic support, this sequence is a cogent example of detailed chronocorrelation. It surely fulfils Teichert's criterion for time significance in that the possibility of stately, consistent, diachronous shift from one ocean to the other is quite absurd – and yet, we can say nothing about shifts within the times *between* the datums. The main attack on this problem of homotaxy vis-à-vis correlation has exploited the rich oceanic record of the late Neogene, where translatitudinal diachrony has already been demonstrated for *Globorotalia truncatulinoides* and *G. inflata* (Kennett, 1970) and others (see Saito, 1984). The strategy has been twofold: (i) to employ extrinsic palaeomagnetic, strontium-isotopic, radiometric data and cyclostratigraphic data; and (ii) to clarify the ordination of events by graphic correlation.

In studying an equatorial array of sections, from the eastern Indian to the eastern Pacific Oceans, Johnson and Nigrini (1985) could sort radiolarian datums into synchronous and diachronous categories. By using available palaeomagnetics as control, they could so categorize 32 of 50 datum levels – enough to

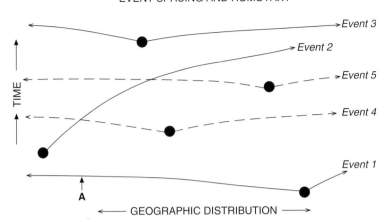

EVENT SPACING AND HOMOTAXY

Figure 3.7 Scott's (1985) illustration of how consistent homotaxial succession can be sustained within diachrony. Bioevents 1, 2 and 3 were consistently homotaxial until the discovery of 4 and 5 and their spread, which demonstrated in the vicinity of A an inversion of 4 and 2, then support from 5 that species 2 actually spread slowly and left a diachronous pattern (which is unscaled here).

make some meaningful generalizations. There was an asymmetry there, for fifteen of the nineteen synchronous horizons recognized are last appearances, whereas ten of the thirteen diachronous horizons are first appearances. The diachrony is on the order of 10^6 years, well above the nominal mixing time of the icehouse ocean of 10^3 years (and well above the believed dispersal time for ammonites, as above), and some of it indicates that some events occur in the Indian Ocean prior to their appearance in the Pacific – opposite to the strong westward flow. Whilst it has been intuitively likely and sparingly demonstrated for some time that diachrony is important across latitudes, Johnson and Nigrini seemed to demonstrate its strong pattern across longitudes within the tropics.

The further pursuit of diachrony in homotaxial biostratigraphic successions requires a sharpening of the calibration – the 'age model' in modern palaeoceanographic parlance – and the technique of graphic correlation (Shaw, 1964; Edwards, 1982a, 1984, 1989) is demonstrating its worth (Hodell and Kennett, 1986; Dowsett, 1988; Hazel, 1989, 1993; McLeod, 1991; F.X. Miller, 1977; Srinivasan and Sinha, 1991; Mann and Lane, 1995; Neal *et al.*, 1998). Graphic correlation (Fig. 3.9) yields a tightening of the spacing of events as well as the sorting-out of crossovers and the revealing of changes in relative rates of accumulation. A composite standard reference section is built iteratively by testing and adding-in events from logged stratigraphic sections. It took a quarter of a century, but graphic correlation is now established in the regime of

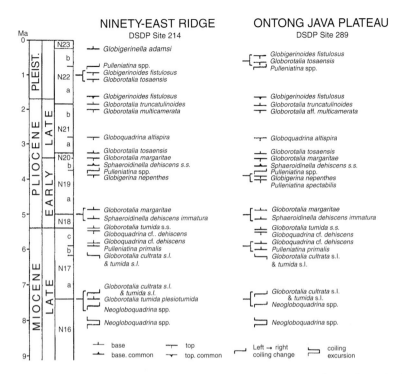

Figure 3.8 Homotaxial events in two oceans: Neogene, east Indian and west Pacific Oceans (Heath and McGowran,1984). (No independent controls, such as geomagnetics or cyclo- or chemostratigraphy.) At this timescale homotaxy between oceans is sustained in reversible events (coiling changes and gross abundance changes) as well as in irreversible events. There are apparent crossovers in the Pleistocene but the homotaxial succession overall is impressive *at this scale.*

oceanic micropalaeontology (Chapter 5). Srinivasan and Sinha (1991) estimated cross-latitudinal diachrony in the southwest Pacific in planktonic foraminifera, comparable in extent to the longitudinal shifts in radiolarians. Dowsett (1988) used the same set of DSDP holes for a mix of calcareous fossil datums, mostly last occurrences, to demonstrate marked diachrony in most, with an attempt to assess reliability. Hills and Thierstein (1989) extended the scrutiny in dataset and technique of investigation to twenty datums from thirty DSDP holes in all oceans, finding only eleven that seemed to be more or less isochronous and morphologically clear enough to be recognized consistently by most or all specialists. Their method for assessing those horizons was Hay's (1972), which tested the consistency of the ordinal relationship displayed by a pair of datums. By an iterative process of elimination, a subset is found of surviving datums that fall above a predetermined probability that the order is non-random. Ordinal reliability is bolstered by estimates of mean numerical age

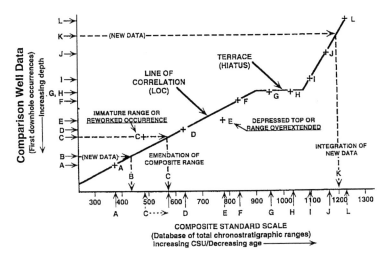

Figure 3.9 Graphic correlation (Neal *et al.*, 1988, Fig. 2, with permission). Here is their caption: 'Diagrammatic graphic correlation plot of datapoints that relate first downhole occurrences of particular fossils to their youngest CSU records in the composite standard. The overall chronostratigraphic relationship of the well interval is defined by the lines of correlation (LOC). Datapoints plotting off the LOC represent occurrences in the well that are younger (left) and older (right) than predicted by the database. The horizontal alignment of datapoints defines a terrace offsetting the LOC and indicating a chronostratigraphic break (hiatus) quantifiable in terms of composite standard units, which can be calibrated to any absolute chronobiostratigraphic timescale.' (Multisyllabic ebullience notwithstanding, there is no such 'absolute' scale.)

using the palaeomagnetic record in nineteen of the thirty holes. In assessing the reasons for the poor estimated reliability of most datums, Hills and Thierstein listed in decreasing order of importance three factors: diachronous occurrence; less-than-clear taxonomy plus gradual ancestor-descendant transformation; and postmortem reworking or destruction of material. The composition of the surviving subset of eleven – out of thirty starters – datums is interesting (Fig. 3.10). Seven of the nine calcareous nannofossil horizons survived the cull but only four of eleven planktonic foraminifera; and ten of the eleven are last appearance datums. The saga of *Globorotalia truncatulinoides* continued with an estimate of its speciation – the 'real' LAD – at 2.6–2.7 Ma at 20–35° S in the Indo-Pacific, marked diachrony in the Indo-Pacific at 15–40° S, and a consistent datum at about 1.9 Ma through the equatorial Pacific and the tropical and temperate Atlantic.

Using age models comprising other bioevents (diatom, calcareous nannofossil, and planktonic foraminiferal) and magnetostratigraphy, Spencer-Cervato

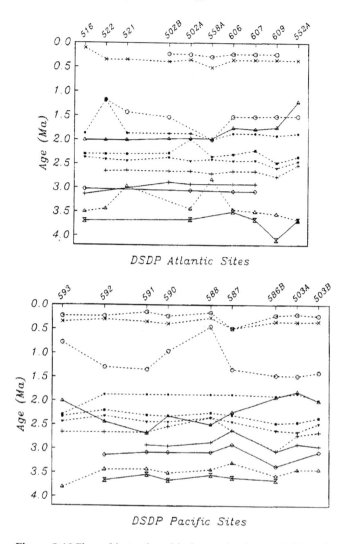

Figure 3.10 Eleven biostratigraphic datums in nineteen DSDP holes with magnetostratigraphy in the late Neogene in the Atlantic and Pacific Oceans (Hills and Thierstein, 1989, with permission). These datums were believed to have 'acceptably high ordinal reliability'.

et al. (1993) compiled new calibrations of radiolarian events in the North Pacific Neogene. They found that a large number of synchronous bioevents (both first and last occurrences) could be identified in the North Pacific, but that in many cases they preceded the corresponding events in equatorial waters by 0.5 to more than 3 myr, whereas those that are younger in the north are only slightly so (Fig. 3.11). They concluded that the evidence seemed to indicate migrations from the north to the equator, and extinctions at higher latitudes earlier than

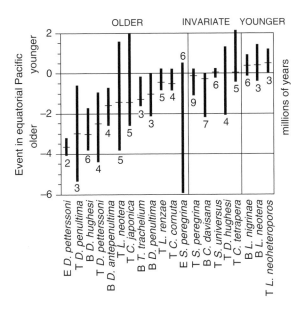

Figure 3.11 Calibration of Neogene radiolarian events in the north Pacific (Spencer-Cervato et al., 1993, with permission). Equatorial events (tops T and bases B) were set at 0. Black bars are north Pacific age ranges of named species and each number is the number of events included in age range (horizontal bar is suggested calibration of each event). More species are older, by 0.5 to >3 myr, in the north Pacific than are younger or about the same age.

at lower latitudes. In a broader survey (Spencer-Cervato et al., 1994) 124 biostratigraphic datums, culled by eliminating possible sources of error such as hiatuses and reworking, were given a global mean age and standard deviation by graphic correlation. Some of their analysis is summarized in Figures 3.12 and 3.13. They concluded that diachrony was frequent (only 53 of 124 events were demonstrably synchronous), more so among cosmopolitan than endemic taxa – thus giving a trade-off between the obtainable precision in age and the geographic extent of a bioevent – and that precision of age calibrations decreases with increasing age. Their most general conclusion was that the reality of common diachrony should be accepted more overtly by biostratigraphers, and that age calibrations should recognize this by being regional rather than global. Culling notwithstanding, the null hypothesis in this work was that bioevents are diachronous until proved otherwise – in contrast to the embedded biostratigraphic precept that they are synchronous until proved otherwise (as indeed they often are). Spencer-Cervato et al. considered four hypotheses for the inferred frequency of diachrony (71–124 culled bioevents). The first was based on the increase in latitudinal gradients during polar cooling (Kennett, 1982), one expected biotic response

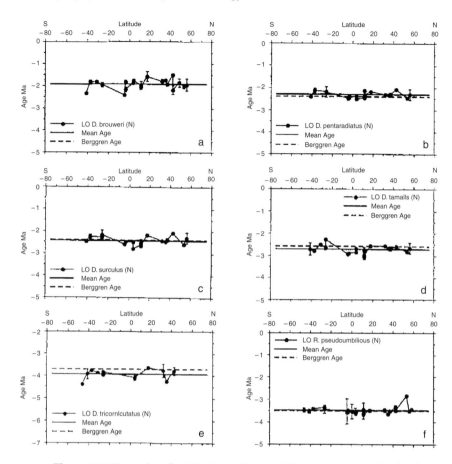

Figure 3.12 Examples of relatively synchronous bioevents among microfossils in oceanic facies (Spencer-Cervato *et al.*, 1994, with permission). FO and LO, first and last occurrence. Each event is plotted against latitude and the mean age is compared with the 'Berggren age' from Berggren *et al.* (1985a).

being a progressive replacement of cosmopolitan by endemic taxa. Supportive, but weakly, was the tendency for less diachrony among the latter, hence for an overall decrease through time. In their second hypothesis, evolutionary adaptation, they expected FAD diachrony as a species gradually expanded its geographic range through 'adaptive immigration'; likewise, LAD diachrony may result when 'certain adaptive capabilities of populations may be eliminated through geologic time'. Although Spencer-Cervato *et al.* cited diachrony for both, they made no case at all for the acquisition or loss of adaptations (which is hardly surprising, since non-tautological adaptation probably is untestable in the fossil record). Their third hypothesis is migration due to watermass changes which also is a response to polar cooling or warming. To the extent that patterns of diachrony are not

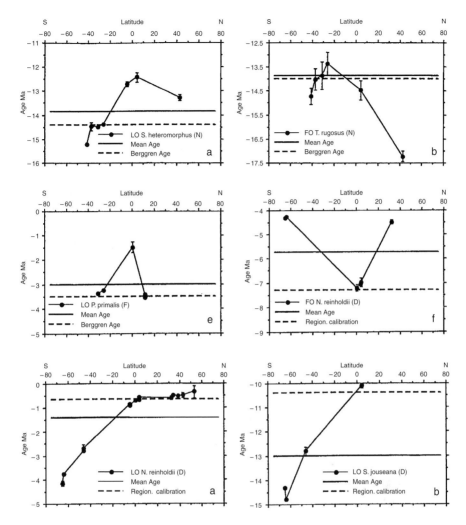

Figure 3.13 Examples of relatively diachronous bioevents among microfossils in oceanic facies (Spencer-Cervato *et al.*, 1994, with permission). FO and LO, first and last occurrence. Each event is plotted against latitude and the mean age is compared with the 'Berggren age' from Berggren *et al.* (1985a). Some events are hemispherical only, some are symmetrical about the equator suggesting bipolar cooling; both speciations and extinctions are included.

artefacts of hiatus or taxonomic bias or confusion (the latter is their fourth hypothesis), shifts in watermass boundaries as an intrinsic part of global climatic change would seem to be by far the most important factor. Figure 6.4 suggests third-order shifts of tens of degrees latitude. Diachrony in the Oligocene ocean was assessed by using the strontium-isotopic record as the extrinsic support

(Hess *et al.*, 1989). The results indicated mostly a sustaining of order through the geographic range sampled, with two obvious exceptions, but a fairly consistent longitudinal diachrony from the west to the east Pacific and thence into the Atlantic. An estimate of latitudinal shift was less securely based in correlations (as emphasized by Hess *et al.*); even so, there are greater shifts through time here than in any of the other studies.

What is the significance of this work on homotaxy and diachrony? One way of stating the objective is that it is to quantify the difference between the local biozonal boundary and the biochronozonal boundary (Fig. 2.11). Scott (1985) called for more datums – 'raising the density of events' – to make tests of homotaxis more rigorous. Instead, the density has remained constant and extrinsic evidence in the form of the GPTS has broken the impasse. Not a great number of events which occur in consistent order seem to be isochronous through their ranges. Last appearances are more reliable than first appearances for chronocorrelation. Where Blow (1970) reasoned – on somewhat limited cited evidence – that first appearances are intrinsically superior to extinctions as reliable events for correlation, and likewise for defining the P- and the N-zones, more recent research rather indicates the opposite. In the late Neogene, calcareous nannofossils are superior in chronocorrelation; the planktonic foraminifera are characterized by 'greater provincialism and more fluid ecophenotypic and evolutionary changes ... implying they are more interesting subjects for evolutionary studies but less suitable tools for biostratigraphy' (Hills and Thierstein, 1989). Perhaps it is also counterintuitive that longitudinal diachrony seems to be strong – as strong as the latitudinal effect which reflects the provincialism based in watermasses under temperature control. Last, there is the comparison of the late Paleogene with the late Neogene. Datums are in shorter supply in the Oligocene and diachrony might be somewhat stronger. But this comparison is betwen two dissimilar datasets; we need more from the earlier horizons. Some patterns exemplifying possible explanations of diachrony are discussed in Chapter 5.

Although care is taken to identify oceanic hiatuses as a distorter of apparent species' ranges (e.g. Spencer-Cervato *et al.*, 1994), hiatuses are not always apparent, even in well-studied well-sampled sections and especially when processing large amounts of DSDP and ODP data reported by numerous micropalaeontologists. Aubry (1995; in Berggren *et al.*, 1995a; Aubry *et al.*, 2000) in particular was sceptical of much such evidence of diachrony, warning that unconformities and hiatuses can produce similar configurations (Fig. 3.14).

It may well be the next step in studying diachrony to consider disjunct distributions and the concept of *allochrony*, or offsets, rather than the predeterminedly gradualistic *diachrony* (Chapter 5).

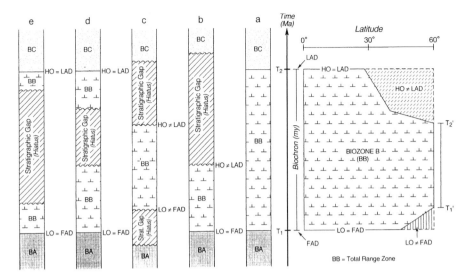

Figure 3.14 Alternative interpretations of the stratigraphic record (Aubry, 1995, Fig. 2 and 3, with permission). HO and LO, highest and lowest occurrences in each section; FAD and LAD, first and last appearance datums. Right, environmentally induced diachrony is usually regarded as the main problem in correlation. Duration T_1'–T_2' of the biozone at high latitudes is substantially less than the full duration of the biochron T_1–T_2. However, unconformities and hiatuses (left) can also reduce the apparent range of a taxon and mimic diachrony. Unconformities would produce an obvious effect in sections d and e but might cause problems by being overlooked in sections b and c.

Biostratigraphy and cyclostratigraphy

'Time's cycle' has been perceived at scales from the year to the giga-year in rocks and phenomena from the Archaean to the present (Fig. 3.15). From our biostratigraphy-centred position the rhythms and cycles in the Milankovitch band are of the most interest, and that interest has grown in recent years out of three questions (all of which were asked in the nineteenth century):

i.   Can the theory that solar-system dynamics modulate Earth's climate be tested in the geological record? Imbrie and Imbrie (1979) recounted how the answer turned out to be 'yes!'

ii.   Are there analogues of icehouse cycles in greenhouse worlds? 'Cyclical variations in the composition of pelagic and hemipelagic sediments ripple in an almost unbroken wavetrain from the Pleistocene Ice Age world into the warm Cretaceous Period' (Herbert *et al.*, 1995).

iii. Can such records not only be dated, but themselves be used for dating events and computing rates of sedimentation? Cyclostratigraphy

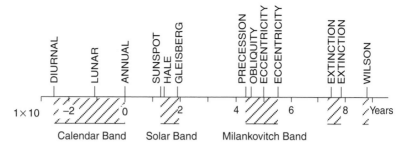

Figure 3.15 The spectrum of rhythms or cycles in geological time, plotted on a logarithmic timescale (Fischer and Herbert,1986, Fig. 1). Calendar frequency band, diurnal, lunar and annual cycles (e.g. varves, tree rings, stromatolites, coral colony growth) in the Earth–Moon system. Solar frequency band of cycles, sunspot, Hale (solar magnetic polarity reversal) and Gleisberg (intensity of aurora borealis). Of prime interest in cyclostratigraphy is the *Milankovitch* band: Earth's axial *precession*, peaks at ~19 000 and ~23 000 years; the axial *obliquity* cycle, varying through 3° at ~41 000 years period; short orbital *eccentricity* cycle at ~100 000 years period; long orbital *eccentricity* cycle at ~4 000 000 years. At longer timescales are the periodic major extinction events perceived by Fischer and Arthur (1997) and Raup and Sepkoski (1984) and the Wilson cycle, making and breaking supercontinents with concomitant death and birth of ocean basins.

plus astrochronology have added a new dimension, literally and metaphorically, to integrated geochronology. This is the 'breakthrough made in dating of the geological record' (Hilgen *et al.*, 1997).

Croll–Milankovitch astrophysics has determined 'orbital solutions' to the two-part system – Earth–Moon interactions determining the angle and orientation or the Earth's rotational axis (influenced by tidal friction and the shapes of the spinning bodies) and the orbit of that system around its Sun. Three perturbations were deconstructed from the two-part system – *eccentricity* of Earth's orbit from almost zero (almost circular) to 0.06 (slightly elliptical) with two main periods of ~100 000 and ~413 000 years; *obliquity* (tilt) of Earth's rotational axis to the orbital plane between 25° and 22° with a main period of 41 000 years; and *precession* of the rotational axis (precession of the equinoxes) with a period of 26 000 years in opposition to the eccentric orbit, giving a main period of ~21 000 (actually two peaks at ~19 000 and ~23 000 years).

These perturbations affect the reception of solar insolation, which reception varies as to global and latitudinal and seasonal distribution. Eccentricity is not a major factor in itself but it is highly influential in strongly modulating precession. The most obvious effect perhaps is the cyclicity of the Pleistocene ice ages via the variations at high latitudes, but the geological history of low-latitude monsoonal systems is also being scrutinized. As for cycles in warmer times than the late Neogene, one might see subtle variations in climate amplified in the sedimentary record, especially through bio-productivity cycles (carbonaceous, siliceous, and calcareous sediments) and carbonate dissolution cycles, but also in pulses in the delivery of siliciclastics and ferruginous materials and episodes in their winnowing (Rocc Group, 1986; Fischer and Herbert, 1986). Again from our biostratigraphy-centred position we can discern three broad developments in cyclostratigraphy: *interval* dating, astronomical *tuning*, and highly refined *integration* with and *calibration* of geochronological methods and scales.

Interval dating

(Weedon, 1993; Herbert, 1999). G. K. Gilbert in 1895 realized that strata contained astronomically forced cycles and he estimated the duration of the late Cretaceous by counting cycles between tiepoints (Fischer, 1980). House (1985) suggested that counting sedimentary cycles could yield the durations of bio-zones. *Durations* is the key word – one might be able to achieve great resolution and even accuracy in the time lapsed between two horizons without any reference to either *correlation* or *age determination*. This then is a floating timescale. Required in the first instance is a correct identification of the orbital frequency controlling the stratigraphic frequency. Weedon *et al.* (1997) spelled this out with an example: an average cycle wavelength of 1 m in oceanic sediments would imply a pelagic accumulation rate of 48 m/myr (precession as control), 25 m/myr (obliquity), or 10 m/myr (short eccentricity). The second requirement for a floating timescale is to recognize any gaps in a succession of cycles, best done in some cases by close comparison of sections in the same district (in oceanic drilling, by drilling A and B or C holes at the site: Ruddiman *et al.*, 1986), in others by scrutinizing ages of tie points, in still others by comparing a drilled section to an outcrop in the district. The best example of the latter was the demonstration by Fischer and Herbert (1986) of an excellent match of the densitometer and carbonate profiles in the Piobbico core (Late Albian, Appennines) with a surface section measured at Erma, 800 m to the east.

Recognizing ancient cyclicities

A. G. Fischer and colleagues (Fig. 3.16) deconstructed cycles in the Cretaceous of the Apennines, Italy – ~30 myr of Barremian–Cenomanian time lacking geomagnetic reversals and with few good radiometric dates. Although there was no astronomically derived external target curve as in the late Neogene – orbital solutions degrade in accuracy as one extrapolates them back in geological time – internal inferences could be derived from the beautifully preserved stratal patterns, which are based in limestone-marl-sapropel

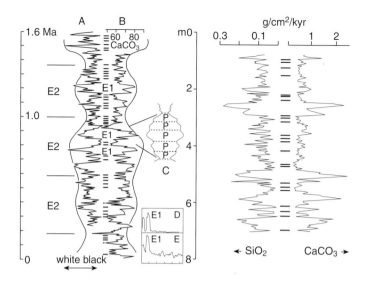

Figure 3.16 An 8 m section, estimated to span ~1.6 myr, of Albian age in the Scisti a Fucoidi, Piobbico, central Italy (Fischer and Herbert, 1986; Herbert and Fischer, 1986; Fischer, 1986; Fischer *et al.*, 1991). *Left*, A, densitometer scans of darkness. B, %CaCO₃. C, an expanded section of the carbonate curve, showing limestone-marl couplets with the marls periodically becoming laminated, less calcareous, black sapropels. D, E, adaptive multitaper spectra of carbonate curve and darkness curve, respectively. Darkness is essentially proportional to carbonate content, as seen by visual comparison of A with B and D with E. The couplets P are redox cycles inferred to be the precessional ~20 000 yr signal. They are grouped into bundles E1 (three of the ~17–18 E1 cycles are labelled) inferred to be the short, ~100 000 yr eccentricity signal. The bundles E1 are grouped into superbundles E2 by a modulation in the sapropels' thickness and incidence and corresponding (reciprocal) variation in limestone purity, inferred to be the long, ~4 000 000 yr eccentricity signal. *Right*, fluxes of carbonate and silica through the first 5 m, calculated by normalizing biogenic components to Al and using orbital cyclicity to quantify the fluxes, as described by Herbert *et al.* (1986, Fig. 5). Black bands, black shales, at the lows in productivity; biogenic SiO₂ (increases left) and biogenic CaCO₃ (increases right) covary, suggesting that the carbonate cycles are productivity, not dissolutional or diagenetic cycles. As in the Pleistocene icehouse cycles, sedimentary variance in these greenhouse cycles is dominated by the short ~100 000 eccentricity signal. With permission.

repetitions. The scale to the right is measurable; the scale to the left is inferred with some accuracy, *given* that the cycles are correctly identified as to their astronomical controls. Thus the 'central' patterns E1 are generated by densi-tometry (black–white) and carbonate percentages. Upscale, the E1 cycles can be bundled into E2 cycles. Downscale, the E1 cycles contain nested couplets which are redox cycles similar to some extent to the precession cycles of the late

Neogene. The biostratigraphic control was based on planktonic foraminifera (e.g. Herbert *et al.*, 1995) and a rough numerical calibration which, even so, could exclude the obliquity cycle from a nested hierarchy comprising precession P, short eccentricity E1, and long eccentricity E2. Herbert *et al.* (1995) also found that three planktonic foraminiferal datums held constant between two sections to within one bundle each (~100 kyr) – evidence of low or absent diachrony. On the identification of the long-eccentricity bundles the timespan of 1.6 myr becomes plausible and this cyclostratigraphy becomes available for interval dating. The sections were found to be quite complete and interval dating gave estimates for the durations of the Aptian and Albian Stages. Fluctuations in productivity, not dissolution or diagenesis, seem to underlie the carbonate cyclicity, and silica covaries with carbonate.

Herbert *et al.* (1999) summarized the evidence that these carbonate–marl cycles flavoured with silica and sapropel had several strong similarities to late Neogene cycles; thus, we are wrong to blur two notions. One is that the Cretaceous was more equable with lower equator-to-pole gradients compared to the Neogene; the other notion is of sluggish oceanic and atmospheric circulation and temporal invariance.

Cyclostratigraphic studies reveal temporal invariance in the Milankovitch frequency band in the Cretaceous similar to the late Neogene. The famous deep-sea anoxia was often driven by orbital variations (Schwarzacher and Fischer, 1982). Herbert and d'Hondt (1990), Herbert *et al.* (1999) and others showed that these Tethyan studies could be repeated in oceanic carbonates where precessional cycles modulated by short and long eccentricities were the drive. Note (Fig. 3.17) the correlation potential of bundled precessional cycles under one geomagnetic and two biostratigraphic controls.

Astronomical tuning

Emiliani (1955) demonstrated a cyclicity in deep-sea, oxygen-isotopic records for which he inferred astronomical forcing at the frequency of obliquity. Improved chronology showed that late Pleistocene $\delta^{18}O$ patterns were dominated by the short-eccentricity cycle with the obliquity and precessional cycles also showing through (Shackleton and Opdyke, 1973; Hays *et al.*, 1976). Hays *et al.* showed that small shifts in their timescale would make a very close match between their obliquity cycles and the calculated frequency of the astronomical obliquity (~41 000 years), and that that adjustment also brought shorter cycles into line with orbital precessional frequencies (19 000 and 23 000 years). This matching of earthly patterns with celestial patterns, independently of other chronologies of their events, is tuning. One must tune *to* something, of course, and the 'something' is the orbital solution.

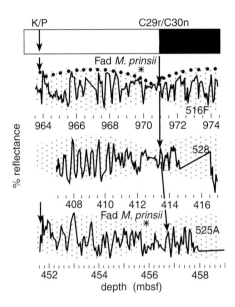

Figure 3.17 Close temporal matching of three late Cretaceous cyclical patterns in three DSDP sections on opposite sides of the South Atlantic Ocean (Herbert *et al.*, 1999, Fig. 9, with permission). The controls are biostratigraphic (FAD of the nannofossil *Micula prinsii* at two sites, Cretaceous–Palaeogene boundary at two sites) and magnetostratigraphic (Chron C29R–C30N boundary at three sites). Carbonate cycles based on reflectance are 20 kyr cycles which have their maxima (dots in Site 516F) modulated by ~400 kyr cycles (shaded).

Integration and calibration

The power of the Cenozoic IMBS lies in the incessant triangulating between bio-, magneto- and radio-chronologies against the linearizer, seafloor spreading. Astrochronology now is an integral part of this procedure; Berggren *et al.* (1995b) provided an excellent example of how a major synthesis could resolve revealed inconsistencies (Fig. 3.18). The integration of disparate fields of enquiry produces chronological structures stronger than the sum of the parts. Most of these components though are points or events, necessitating interpolation, in turn relying on assumptions of continuity, such as sedimentation rates or seafloor spreading rates. Cyclostratigraphy fills a special niche in an integrated astro-bio-magneto-radio-chronology in that one has, once the iterative matching has achieved a lock-in, a continuous scale. This acquisition adds markedly to the rigour and precision in locating bio-datums (see below) and pinning down major chronostratigraphic boundaries (Chapter 7). Cyclostratigraphy adds powerfully to interocean chronological correlation and resolution in such

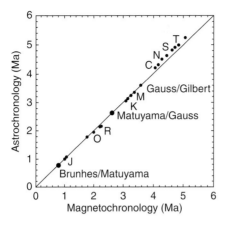

Figure 3.18 Magnetochronological–astrochronological comparison of geomagnetic reversal ages through the past five myr, using two scales from the early 1990s (Berggren *et al.*, 1995b, Fig. 3, with permission). There is good agreement back to base of the Gauss; astrochronological calibration points were already used (double circle). Divergence began in the Gilbert Chron, beyond which astro-ages were systematically older than geomagneto-ages by 150–180 kyr. Improved dating of the problematical timeslice between base Gauss and top Cochiti resolved the matter (Wilson, 1993). Chron boundaries indicated; Subchron boundaries are J, Jaramillo; O, Olduvai; R, Reunion; M, Mammoth; K, Kaena; C, Cochiti; N, Nunivak; S, Sidufjall; T, Thvera.

questions as palaeoceanographic change leading up to Northern Hemisphere glaciation – Tian *et al.* (2002) achieved these correlations and biostratigraphic refinements from an excellent section in the South China Sea.

Two astrochronological strategies: floating chronology or Pleisto–Pliocene umbilical cord

The first strategy begins with a cyclical pattern somewhere back in the stratigraphic record, disconnected from the umbilical cord of the Pleistocene–Pliocene celestial record; one extracts a signal in a floating chronology and spots the dominating orbital parameter. Uncertainties in astrophysical calculations make pattern matching uncertain, misleading or meaningless to a degree proportional to the temporal distance below the Pleisto–Pliocene. The second strategy entails working back from the Holocene into the Pleistocene and deeper into the Neogene, matching patterns with the targeted orbital chronologies (themselves improving further into geological time with better understanding of solar system dynamics: e.g., Laskar, 1999; Pälicke and Shackleton, 2000). This strategy began with theories of solar system dynamics predicting orbitals pacemaking global climatic oscillations especially the Pleistocene ice ages, it proceeded to the recognition of earthly cycles in various sedimentary and geochemical forms including oceanic oxygen isotopes, meanwhile astrophysically constructing successively better orbital curves back into Pleistocene then Neogene time, so that a match could be made between the celestial (orbital target) and earthly time series. In the Pleisto-Pliocene, we can assume that we have an adequate visual pattern of orbital insolation forcing. If we know the approximate age of a sequence, we can place it unambiguously by pattern matching using the recent as a fixed calibration point.

Late Neogene cyclostratigraphy of the Mediterranean region

Cyclostratigraphy with tuning was established within the most recent one million years of the Neogene, the Brunhes–Matuyama geomagnetic boundary being a crucial anchorpoint. Extension of investigations down the column, into the Pliocene then the Miocene, exposed discrepancies between astro- and magneto-chronologies in the region of ~4 Ma (Fig. 3.19) as well as within the construction of orbital curves (reviewed by Hilgen, 1994, 1999; Berggren *et al.*, 1995b). These problems were resolved in part by improved Ar–Ar dating, in part by revising a stretch of geomagnetic anomalies in the Pacific seafloor. Then: two numerical calibrations could be made to fit (Fig. 3.19). In one direction, strata with strongly cyclical characteristics could be mutually correlated and dated geomagnetically, employing a magnetic polarity scale itself revised as to both sequence of events and their dates. In the other direction the same strata could be mutually correlated and dated cyclostratigraphically, individual sapropels being correlated with the precessional pattern strongly and clearly modulated by eccentricity. The central claim of the paper was that the magneto-, chemo- and astro-chronologies were now concordant back into the latest Miocene.

Astrochronology is mediating time series and lines of enquiry that might not be directly comparable – meaning the first-order *comparison* where *correlation* is not necessary. Hodell *et al.* (2001) achieved a bed-by-bed correlation between the Messinian cyclostratigraphy in the Mediterranean (including the saline facies) and open-ocean time series in the North Atlantic on the Rockall Bank (ODP Site 982). They could extract two signals – a benthic oxygen-isotopic pattern, which they could tune by filtering to the obliquity pattern, and a bulk-density pattern tuned to summer insolation. This precessional pattern bundles according to eccentricity modulation ($\times 10^2$) (Fig. 3.20).

As in all interrelationships between biochronology and its partner chronologies there is extensive feedback between fossils and cycles. Fossils will supply a broad (10^{5-6} years) age range of, say, an eccentricity-modulated bundle of precessional cycles which, having been calibrated and dated, will return the bioevent with interest in the form of much improved precision and accuracy. Berggren *et al.* (1995b) presented 'a new global high-resolution biostratigraphy'

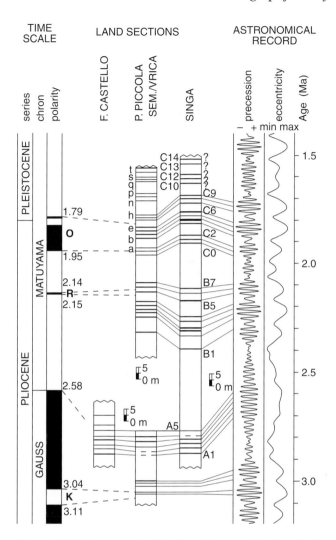

Figure 3.19 Astronomical calibration of sapropels in three land sections in the Narbone Formation in southern Italy (Hilgen, 1991; Berggren *et al.*, 1995b, Fig. 2, with permission). The age of the sections could be determined magnetostratigraphically at the resolution afforded by that method. Sapropels could be tied to the orbital time series in two ways by using late Pleistocene relationships – individual sapropels correlated with minima in the precession index and small- and large-scale sapropel clusters correlated with eccentricity maxima at 1000 ka and 400 ka, respectively. The target for correlation was the astronomical time series after the solution by Berger and Loutre (1991). Given the accuracy of this solution, bioevents and polarity reversals identified in the sections could be dated with accuracies of ∼5–10 kyr. Sapropels are coded in the Calabrian sections. Magnetic, O, Olduvi, R, Reunion, K, Kaena. Note the visual match between broad eccentricity cycles and clumping of the precession signal.

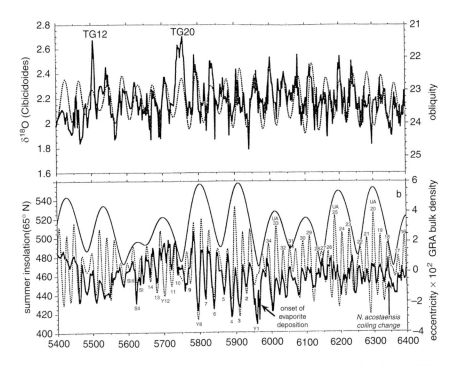

Figure 3.20 One-myr part of the stratigraphic section spanning time from 4.6 Ma (4600 ka, early Miocene) to 7.4 Ma (7600 ka, late Miocene), from Hodell *et al.* (2001, Fig. 4, with permission). The section was drilled at ODP Site 982 on the Rockall Plateau, North Atlantic Ocean (1134 m water depth). Depths were transformed to ages by interpolating age-depth pairs based on selected oxygen-isotopic, biostratigraphic and previously available astronomical events. There are two orbital tunings here. Upper panel, the benthic $\delta^{18}O$ (*Cibicidoides*) record was tuned by filtering at 41 kyr and matching the signal to the astronomical solution for obliquity – obliquity having been determined previously to dominate this part of the orbital record everywhere. Prominent oxygen-isotopic stages TG12 and TG 20 are identified. Lower panel, the gamma ray attenuation (GRA) bulk density record was tuned by filtering at 21 kyr and matching to summer insolation at 65° N (Laskar *et al.*, 1993). The filtering and pattern matching were done iteratively to achieve tuning. Tuning permits bed-by-bed correlation between the oceanic facies at Site 982 and the Messinian saline facies in Mediterranean basins. Events UA16 to UA34, Y1 to Y14, and SI to SIII are cyclic events in the Sorbas Basin in southern Spain. Bundling matches the eccentricity ($\times 10^2$). Glaciation intensified at 6.26 Ma, before onset of evaporite deposition at 5.96 Ma, and went through 18 oscillations controlled by the 41 000-yr obliquity cycle, terminating at stages TG12–11 at ~5.50 Ma.

anchored by the concordant magneto- and astro-chronologies for the Pliocene and Pleistocene, the bioevents (calcareous nannofossils and planktonic foraminifera) being placed against the magnetostatigraphy with a precision down to 10^4 years. The bioevents included evolutionary first appearances (FADs) and

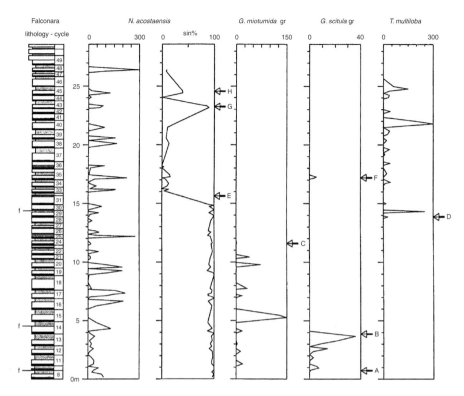

Figure 3.21 Quantified eco-biostratigraphic study of the Falconara section (Late Miocene, south coast of Sicily) of the sapropel-marl-diatomite cycles of the Tripoli Formation (Hilgen and Krijgsman, 1999, Fig. 2, with permission). Bioevents in stratigraphic order are A, *G. nicolae* FO, B, *G. nicolae* LO, C, *G. miotumida* group LO, D, *T. multiloba* first common occurrence (FCO), E, *N. acostaensis* S–D coiling change, F, *G. scitula* influence, G, *N. acostaensis* sinistral influx (up to 90%), H, *N. acostaensis* sinistral influx (up to 40%).

last appearances (LADs), fluctuations in presence or abundance of taxa (increase, dominance, absence, acme) and migrational events (FOs and LOs) in different biogeographic provinces. Several of these events are found in one ocean only, or in the Mediterranean Sea. Several nannofossil events could be demonstrated to be non-isochronous (allochronous, not diachronous) – *Ceratolithus rugosus* FAD (oceanic, 5.0–5.23, Mediterranean, 4.55 Ma); *Helicosphaera sellii* LAD (equatorial 1.47, midlatitude 1.22 Ma); reversal of dominance between *Gephyrocapsa caribbeanica* and *Emiliania huxleyi* (tropical-subtropical 0.09, transition 0.075 Ma).

Hilgen and Krijgsman (1999) took accuracy and precision a step further in the ongoing debate about the timing and forcing factors in the Messinian salinity crisis in the Mediterranean (Figs. 3.21, 3.22). Twelve planktonic foraminiferal events span less than one million years of Late Miocene time. Step 1,

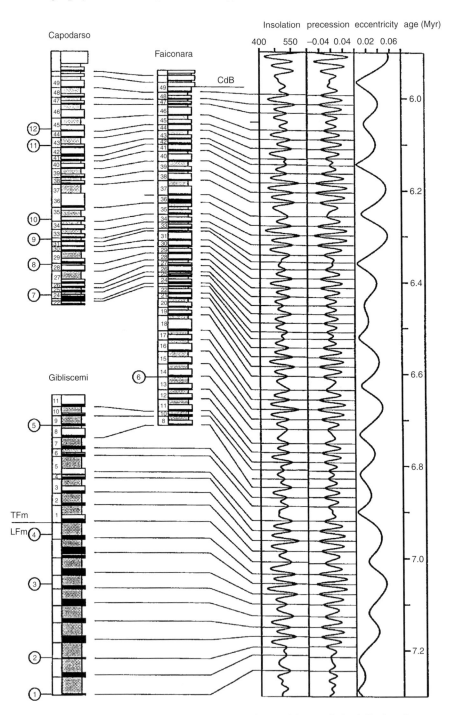

Figure 3.22 Hilgen and Krijgsman (1999, Fig. 3, with permission) displayed cyclostratigraphy and astronomical calibration in three Late Miocene sections in

biostratigraphic profiling of a section, is demonstrated in Fig. 3.21 which includes first and last events plus abundance events. Figure 3.22 demonstrates step 2 – correlation between three sections in Sicily – and step 3 – tuning to eccentricity-bundled precession and inflation. The starting point for this exercise was the two-part calibration of the first and last occurrences of *Globorotalia nicolae* – their consistent occurrences respectively in the Tripoli cycles T9 and T14 and their astronomic ages respectively of 6.829 and 6.72 Ma from sections in northern Italy and Crete. These ages for a six-cycle timeslice were used to tune all diatomite cycles to precession and summer insolation. Three of the events are abundance changes in sinistral-dextral cohorts of *Neogloboquadrina acostaensis* and one such switch (E in Fig. 3.21 = (9) in Fig. 3.22) is also seen isochronously in open-oceanic facies in Fig. 3.20. These bioevents are impressively rapid, and do not support an a-priori notion of pervasive diachrony in biostratigraphy at cyclostratigraphic timescales – let alone at standard-biostratigraphic timescales.

   Oceanic microfossils offer unsurpassed sampling opportunities and exemplify the effects of sampling density comparable to cyclostratigraphy – with the further similarity that there is no substitute for time-consuming counting of specimens in taxa (Fig. 3.23). This approach is quite different from the 'classical' approach outlined in Chapter 2, where progress was marked by finding the evolutionary pattern (speciation and extinction) responsible for producing 'total' ranges bounded by 'irreversible' first and last appearances. This is qualitative presence-or-absence data. The resolution constrained by cyclostratigraphy liberates a lot more detail in such 'reversible patterns' as abundance

Figure 3.22 *(cont.)*
Sicily, whose strong cycle patterns permit between-section cyclostratigraphic and biostratigraphic correlation and calibration to the astronomical record. Licata Formation, bipartite sapropel-marl cycles; Tripoli Formation, tripartite sapropel-marl-diatomite cycles, Calcare di Base, carbonate cycles. Planktonic foraminiferal events giving control are numbered in order: (1) *G. miotumida* group first regular occurrence (FRO), (2) *G. miotumida* influx conical types, (3) *G. scitula* last occurrence (LO) dominantly sinistral, (4) *G. scitula* group temporary disappearance, (5) *G. nicolae* FO, (6) *G. nicolae* LO, (7) *G. miotumida* group LO, (8) *T. multiloba* first common occurrence (FCO), (9) *N. acostaensis* S–D coiling change, (10) *G. scitula* influence, (11) *N. acostaensis* sinistral influx (up to 90%), (12) *N. acostaensis* sinistral influx (up to 40%). Astronomical curves are based on solution of Laskar *et al.* (1993). Beginning with (i) assuming that the sapropelic cycles are precession-controlled dry–wet cycles as in the Pleisto-Pliocene and (ii) precise dates from elsewhere in the Mediterranean of *G. nicolae* FO (6.829 Ma) and LO (6.72 Ma), diatomite cycles could be tuned to precession and summer-insolation time series.

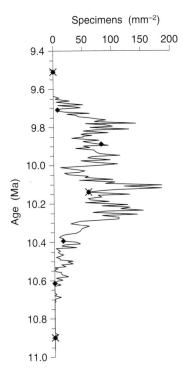

Figure 3.23 Backman and Raffi (1997, Fig. 1, with permission) demonstrated the effects of sampling resolution of an oceanic section (ODP Site 926) using the presence and relative abundance of the coccolith *Catinaster calyculus*, plotted per unit area (from fields of view of the smear slides) against time. Crosses, 1 sample–9.5 m ≈ 650 kyr; diamonds, 1 sample–3.0 m ≈ 650 kyr; small solid circles, 1 sample–0.1 m ≈ 8 kyr. The accumulation of data at a fine temporal scale is time-consuming but indispensable for high-resolution biostratigraphy.

changes and repeated comings and goings of a given taxon – patterns that have been little more than biostratigraphic noise until quite recently. Strongly calibrated bioevents now include abundance changes, coiling changes, and vertically disjunct distributions, as shown above. For Backman and Raffi (1997) and Raffi (1999) the presence-or-absence character of biostratigraphic ranges of taxa is not so objective as it might seem, for identifications entail 'an amalgamation of judgments involving taxonomic perception and abundance that, in turn, depends on such factors as productivity rate and preservation state'. They go on to suggest that the rigorous gathering of quantitative data 'would considerably open up the biostratigraphic black box'. Figure 3.24 samples numerous calibrated bioevents, all plotted against a curve for magnetic susceptibility, which cycles reflect precession cycles with increased susceptibility implying reduced carbonate values.

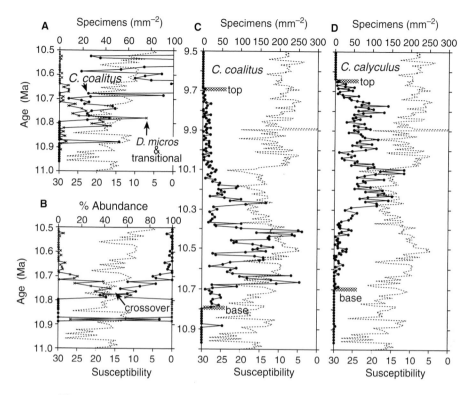

Figure 3.24 Backman and Raffi (1997, Fig. 10, with permission) compiled coccolith taxa abundances (specimens per field of view converted to per mm²) against time in ODP Site 926. Dotted lines in each panel represent magnetic susceptibility values (increased values imply decreased carbonate and vice versa) and cycles in susceptibility reflect precession cycles. C, D, abundances of *Catinaster coalitus* and *C. calyculus*. A, basal range abundances of *C. coalitus* and transitions from its ancestor *Discoaster micros*. B, same, this time locating the horizon of crossover in relative abundances, a useful class of bioevents (Thierstein *et al.*, 1977).

The importance of quantification is emphasized by the use of common occurrence, regular occurrence, acme, and coiling change in addition to first and last occurrence (Krijgsman *et al.*, 2004).

Backman and Raffi emphasized that this kind of study benefits biostratigraphy and biochronology in three ways. First, there are 'huge' improvements in calibration. Cenozoic bioevents have calibration uncertainties of the order of one to a few hundred thousand years due to relying on sedimentation rates and durations of the geomagnetic polarity zones – interpolation problems. Orbitally tuned scales remove the interpolation problem. In this study, 34 Miocene bioevents were calibrated with uncertainties from ±2 to ±28 kyr with an average of ±7 kyr. Second, Backman and Raffi distinguished between

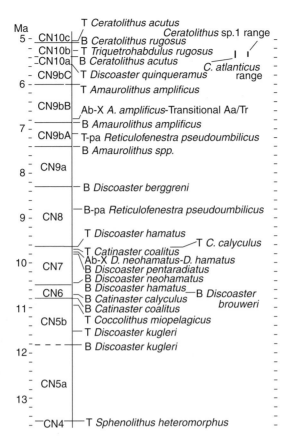

Figure 3.25 Summary of coccolith biostratigraphy in ODP Site 926 (Backman and Raffi, 1997, Fig. 15, with permission) using techniques illustrated in Figures 23 and 24 – dense sampling, standardized counting, and reference to susceptibility cycles inferred to reflect precessional cycles. Bioevents are shown against the zonal succession of Okada and Bukry (1980). T, top; B, base; Ab-X, abundance crossover; T-pa, top paracme; B–pa, base paracme; Transitional Aa/Tr, transitional between *A. amplificus* and *T. rugosus*.

depth uncertainty, mostly solved by closer sampling, and age uncertainty, which depends both on the accuracy of the age estimated for the bioevent and its isochrony through space. Cyclostratigraphy offers more rigorous tests of diachrony, as noted above. Thus, Backman and Raffi summarize: (i) *quantitative* methods permit deeper understanding of bioevents and their environmental context; (ii) closely spaced *sampling* will capture finer and perhaps crucial changes; and (iii) *calibration* is to an independent chronology, i.e. astrochronology.

Taking cyclostratigraphy back into the Miocene, Hilgen *et al.* (2000) integrated biostratigraphy and cyclostratigraphy across the Serravallian–Tortonian boundary. Of particular interest here is a test of three bioevents listed for the Ceara Rise in the western North Atlantic (Fig. 3.25). The first and last common occurrences of *Discoaster kugleri* and the last regular occurrence of *Coccolithus miopelagicus* were calibrated astrochronologically independently of the calibrations of the Ceara Rise horizons. The agreement between Mediterranean and tropical Atlantic is excellent.

Testing the Oligocene timescale

For the late Neogene there is a 'known' visual pattern of orbital insolation forcing. If one knows the approximate age of a stratigraphic section with cycles, then one may expect to place the cyclical succession against the insolation target by pattern matching and iterative tuning, thereby sharpening the dating of the bioevents or magnetic events in the age model that was the point of departure. Things are different in the late Oligocene (Weedon *et al.*, 1997; Shackleton *et al.*, 1999). Although parameters affecting insolation – ellipticity, tidal dissipation – seem to have remained close to present-day values over the past 25 myr and therefore present-day values can be used in Oligocene tuning (Pälicke and Shackleton, 2000), there is not a reliable target curve. These authors accordingly built an Oligocene cyclostratigraphic scale from oceanic sections on the Ceara Rise (tropical western Atlantic). Weedon *et al.* (1997) sought to oppose 'net sedimentation rates' (based on biostratigraphy from the IMBS) to 'pelagic sedimentation rates' (as inferred from cyclostratigraphy). Ignoring for the moment the estimated numerical ages of stratigraphic boundaries and bioevents and other events, this comparison is an interesting test of different kinds of timeslice dating. Weedon *et al.* extracted two highly cyclical patterns in two physical properties, magnetic susceptibility and percent reflectance, both manifesting variations in the degree of terrigenous dilution of pelagic carbonates (Fig. 3.26). There was a consistently strong single signal in power spectra from several cores from four ODP sites which varied but not drastically in wavelength (thickness). This main sedimentary cyclicity was inferred to control the entire stratigraphic pattern and to reflect the orbital obliquity signal – neither a shorter (precession) nor longer (eccentricity) signal could be made at all plausible using biostratigraphically anchored rates of sedimentary accumulation. A depth-to-time conversion could use the sedimentation rates derived from the average cycle wavelengths from spectral analyses (two examples in Fig. 3.26) and accepting that the stratigraphic cycles were obliquity cycles (taken as a rounded figure of 40 kyr, a little less than the modern 41 kyr, obliquity increasing slowly through geological time). Thus, 'cumulative time' could be calculated core top by core top, hung from zero, the reference datum (and proxy for the base of the Miocene), chosen as top occurrence of the calcareous

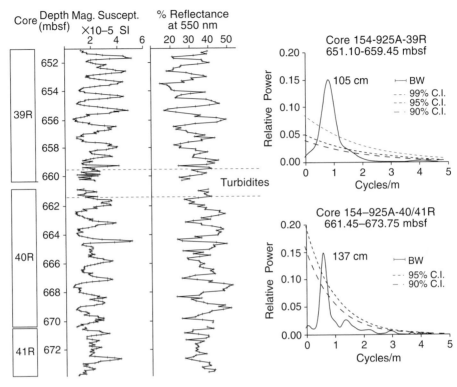

Figure 3.26 Cycles in oceanic facies of Oligocene age, ODP Site 925A, Ceara Rise (Weedon, Shackleton and Pearson, 1997, Fig. 2, with permission). Magnetic susceptibility and light reflectance both record the degree to which terrigenous material diluted pelagic carbonates. Power spectra display a strong single peak, the difference in wavelength probably resulting from accumulation rates (the range among four sites was 66–137 cm). Weedon *et al.* inferred that a single sedimentary cyclicity controlled sedimentation across the Rise and inferred further that it reflected the orbital obliquity signal. BW, bandwidth; CI, one–sided confidence interval.

nannofossil *Sphenolithus delphix*. Cumulative time was calculated independently for core top and bioevent at each site with no mutual biostratigraphic reference (except top *S. delphix*). Comparison of the sites against cumulative time showed some mutual similarities and some discrepancies due in part to faulting, and individual cumulative times were translated into a 'composite' timescale by visually matching data sets between sites using tie points (not shown here, but listed by Weedon *et al.*, 1997). The outcome (Fig. 3.27) shows reflectance data and biozones against composite age, with good agreement between sites for the Late Oligocene; base CP19A is anomalous. This construction could be compared with the Oligocene timescale of Berggren *et al.* (1995a) (Fig. 3.28). Weedon *et al.*

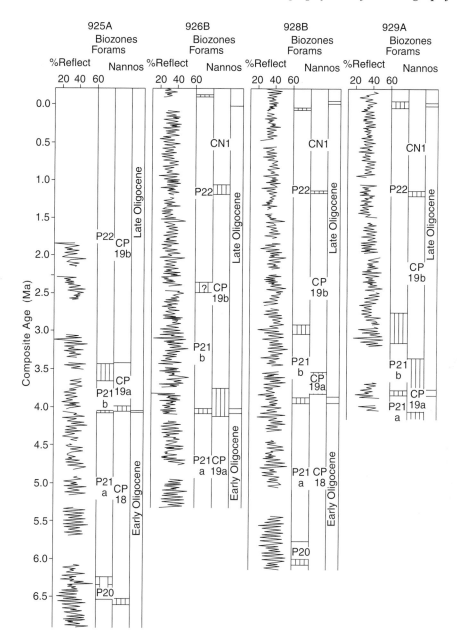

Figure 3.27 Following analyses as in Fig. 3.26, Weedon *et al.* (1997, Fig. 6, with permission) correlated core datasets between four ODP sites, zoned biostratigraphically using planktonic foraminifera and calcareous nannofossils (biostratigraphic boundary uncertainties shown by vertical hatching). They made a depth-to-time conversion by matching sedimentary cycles to obliquity cycles. No target (astronomical) curve was used for tuning and the data were not filtered. The reference datum (and proxy for base of the Miocene) was the top occurrence of *Sphenolithus delphix* – thus time zero in the composite timescale.

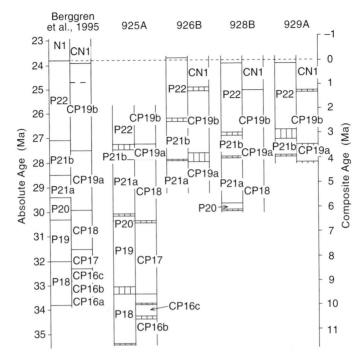

Figure 3.28 Weedon *et al.* (1997, Fig. 8, with permission) compared their mutually correlated and calibrated four sections (their Fig. 11) with the numerically calibrated Oligocene timescale derived from another direction – seafloor spreading rates in the IMBS (Berggren *et al.*, 1995a).

estimated cyclostratigraphic durations of the Late Oligocene of 4.06 to 3.88 myr from three sites, compared to 4.7 myr for the IMBS scale. There is also some mismatch between the two biozone sets: Weedon *et al.* (1997) pointed to base CP19a (base *S. ciperoensis*) falling below base P21a (base *G. angulisuturalis*) in the IMBS (where the two sets of bioevents were correlated to the geomagnetic succession), but close to base P21b (top *C. cubensis*) here.

Shackleton *et al.* (1999) extended this work on the Ceara Rise sections, reemphasizing that patterns of sedimentary accumulation rates are very different according to whether biostratigraphy or sedimentary cyclicity are the basis. A tuning target was now available although there were significant uncertainties in the orbital calculations compared to the Late Neogene. The main outcome in the present context is a list of biostratigraphic datums from Oligocene to Middle Miocene planktonic foraminifera and calcareous nannofossils. The tuned ages are close to the 'literature' ages (mostly from Berggren *et al.*, 1995a) in the Late Eocene but depart significantly for numerous events higher in the column.

4

Biostratigraphy and biohistorical theory I: evolution and correlation

Summary

This chapter is centred on the planktonic foraminifera as 'palaeobiological' entities in biostratigraphy – their taxonomy and classification, the nature of the species-level taxa, and macroevolution as revealed in lineage studies. Larger foraminifera and mammals are also discussed in this matter of evolution and biochronology.

Two approaches to evolutionary theory

A fertile research programme in evolution and the fossil record of biodiversity is based in the fluctuating distribution of taxa through time. The philosopher Gayon (1990) identified four common solutions to the problem of the origin of biotic diversity, of which the first three were: (i) the extrapolationist models of competition, coevolution, etc.; (ii) rehabilitation of the pre-Darwinian transcendental morphology (developmental constraints, embryology); and (iii) species-level processes analogous to but not extrapolated from microevolutionary processes (species selection, species drift). Solutions (i)–(iii) are biological – not quarantined from the physical world, but not crucially driven by it, either. In contrast, Gayon cited a fourth solution: (iv) 'much of the formation of new biota is due to major environmental changes caused by a dynamic earth resulting in major environmental change *beyond biological control or limitation* ... This last solution might well be the most "Darwinian" of all: It comes down to saying that microevolution is a utilitarian tinkering in a changing and hazardous world.' (Emphasis added.) (It is ironic that the appellation 'Darwinian' be ascribed to this last solution, because it has been said that Darwin moved away from interactions between organism and physical environment in his later thinking on evolutionary dynamics and moved closer to organism/organism interactions (Hallam, 1983).)

Thus, one research programme in evolutionary palaeobiology searches for chronological correlations between discernible changes in the fossil record and discernible changes in the environment. For marine microfossils this enquiry has been possible largely through the rise of the discipline of Late Phanerozoic palaeoceanography, itself a product of deep ocean drilling, of refined integrated chronology, and the reconstruction of time series in stable isotopes (see especially Zachos *et al.*, 2001b). But it has also extended Simpson's (1944) pioneering work on following diversity changes through time, now a well-known way of raiding the systematics monographs in what amounts to the taxic approach to evolution (e.g. Smith,1994). Permeating much of the discussion during the past approximately two decades has been a renewed consciousness of a dualism in evolution. As expressed by Eldredge (1979), the dualism is between:

i.   *transformational* evolution, concentrating on morphologic change (*sensu lato*), rejecting or neglecting speciation, descendant from Darwin's *descent with modification* via Dobzhansky's *change in the genotype of the population*; and

ii.   *taxic* evolution, in which the central problem is the other side of the Darwinian cleavage, the *origin of a species*, its reality and its significance as a historical singularity.

It has been said repeatedly that evolutionary palaeontology since 1859 has been dedicated virtually in its entirety to transformational evolution. The rise of the taxic view in recent years has been linked with increasing acceptance that morphological stasis is a widespread evolutionary phenomenon, at least among some clades. Allmon and Bottjer (2001) observed that stasis in a clade demands that long-term morphological patterns in that clade must be sought in patterns of species' originations and extinction, what with little happening in the interim. This is the taxic approach (to bioentities), in contrast to the transformational (to biocharacters), in which morphological trends are produced by gradual changes within species lineages.

We take the taxic/transformational dichotomy as one useful starting point in discussing fossils and biocorrelation.

Micropalaeontology and biostratigraphy: preconceptions and practice

The use of foraminifera in biostratigraphy was retarded for decades by the opinions of the 'English school' of the mid nineteenth century (Lipps, 1981;

Cifelli,1990). It happened through their attacks on d'Orbigny and their influence on Darwin, as described here by Lipps:

> Although d'Orbigny is now recognized as a generally good descriptive worker, he was ridiculed by English micropaleontologists for proliferating species and his theory of creations was resoundingly beaten by evolutionists. Thus, the founder of micropaleontology made no contribution to the use of microfossils in evolutionary studies. In fact, it seems that the English foraminiferologists were so incensed by d'Orbigny's attitudes that they too destroyed the potential of foraminifera in this type of work. Had d'Orbigny not clung so tenaciously to his version of successive creations and had the English micropaleontologists not later damned him so thoroughly, foraminifera (and micropaleontology) might have made sound contributions to stratigraphy and evolutionary paleontology very early on ...

Carpenter (Carpenter *et al.* 1862, pp. x–xi) arrived at a set of conclusions that dominated the study of foraminifera until the rise of industrial micropaleontology. Chief among his conclusions were: I. There was a great range of variation in foraminifera. II. 'The ordinary notion of species, as assemblages of individuals marked out from each other by definite characters have been genetically transmitted from original prototypes similarly distinguished, is quite inapplicable to this group'. III. 'The only natural classification of the vast aggregate of diversified forms which this group contains, will be one that ranges them according to their direction and degree of divergence from a small number of principal family-types'. IV. 'The evidence in regard to the genetic continuity between the Foraminifera of successive geological periods, and between those of the later of these periods and the existing inhabitants of our seas, is as complete as the nature of the case admits'. V. 'There is no evidence of any fundamental modification or advance in the Foraminiferous type from the Palaeozoic period to the present time'. In short, '... there is no indication of any tendency to elevation towards a higher type'.

Carpenter's views, especially regarding the longevity of species, were passed to Darwin who, in the *Origin*, was forced to consider Carpenter's assertions that there had been no change among the foraminifera since the earliest geologic period. Darwin reasoned that organisms should show continued advancement through time and, although he knew it was not a necessary outcome of evolution, it seemed at that time generally to be true. Foraminifera were one exception, and Darwin, accepting Carpenter's opinion, strained to reconcile such a record in the fourth edition of the *Origin*: 'It is not an insuperable difficulty that Foraminifera have not, as insisted on by Dr. Carpenter, progressed in organization since even the Laurentian epoch; for some organisms would have to remain fitted for simple conditions of life, and what could be better fitted for this end than these lowly organized Protozoa?' Thus, Darwin expressed the two most common and destructive views regarding foraminifera – that they ranged long without evolving, and that they were 'simple' organisms that we perhaps should expect not to evolve at all ...

Thus micropalaeontology, as a technology that was able to assist geology, was essentially killed by Carpenter's and Williamson's attacks upon d'Orbigny, and the acceptance of Carpenter's views, as with Darwin, eliminated foraminifera from serious consideration in evolutionary studies.

Small wonder that it took so long for the foraminifera to become established in biostratigraphy. There were more a-priori statements in the 1920s by Vaughan and Diener that foraminifera were of little use, although testing and experience were beginning by then to demonstrate otherwise, as mentioned in Chapter 2. All of these statements were evolutionary and theory-laden. Even Carpenter's opinion, that the foraminifera had not evolved since the Precambrian, which seems to be simply incorrect, was not so much observational in the 'objective' sense as consonant with the tradition of the times that fossil specimens be assigned to previously described extant taxa (Cifelli, 1990). That tradition is opposite to, but not greatly different in other respects from, the later tendency to distinguish nomenclaturally every departure from the type expected to have stratigraphic value (Cain, 1954; see below). In both cases the taxonomic programme is driven by the researcher's perceptions, not by any urge to work 'empirically' or 'objectively'. Finally, the stratigraphic importance of identified morphotypes was pointed out clearly in 1865 by A. E. Reuss but not generally appreciated for a long time afterwards (Glaessner, 1945).

The thoughts of Martin Glaessner

Glaessner (1945) clarified the theoretical development of the use of microfossils in biostratigraphy as follows: 'At present the *deductive* method of establishing biostratigraphic subdivisions on morphogenetic evidence is still in a preliminary stage of development and is yet far from replacing the usual *empirical* method of establishing ranges of species by compiling records of their occurrences' (emphasis added). Glaessner observed with approval the tendency for 'academic' as well as 'applied' micropalaeontology to turn from empirical to analytical methods, just as much later (Glaessner, 1966) he saw a healing of the split between biostratigraphy and palaeobiology by the application of bio-logical viewpoints to biostratigraphy. The 'deductive' strategy refers to the use of morphogenetic series in the arcane and difficult analysis of the larger foraminifera, which we defer pending a consideration of the plankton, which were treated more empirically and in the tradition of studies of the benthics in the biostratigraphy of petroleum geology, at least until the reconstruction of the *Orbulina* and *Globorotalia fohsi* bioseries (Miocene) in postwar petroleum exploration. However, it is useful to quote *in extenso* from this most thoughtful of micropalaeontologists, invited by the Co-editor, Angelina Messina, to write the leading article for the new journal *Micropaleontology* (Glaessner, 1955). Here is an extract from the section headed *biostratigraphy and taxonomy*:

> In most of Cushman's valuable monographs of foraminiferal families, the species are dealt with in stratigraphic order. This presupposes both reliability of age determinations and objective independence of taxonomic data from biostratigraphic successions. Unfortunately, both assumptions are often unwarranted. Even if the original age-determinations were correct, it would not be very helpful to describe, and to distinguish by differential diagnoses, all recorded Eocene representatives of, for examples, *Elphidium*, then all Oligocene species, and so on to Recent. As we are dealing with evolving populations, we can only develop a scientific taxonomy if we follow representatives of a species, a genus, or a family, through stratigraphic sequences in certain areas. This has become common practice in the study of corals, ammonites, belemnites (particularly of the Upper Cretaceous), lamellibranchs (Carboniferous, Permian and Jurassic), many other invertebrates, amphibians, reptiles (of the Karroo system) and mammals (horses and elephants). In the study of foraminifera, this approach is still exceptional, being confined to larger foraminifera and such Cretaceous forms as *Globotruncana*, *Bolivinoides*, and *Neoflabellina*. We have to face the fact that in the early stages of the approach to evolving populations, the *nomenclature* tends to become confused rather than clarified, but our primary concern is with taxonomy, the results of which are then expressed in the agreed terms of zoological nomenclature. Any attempt to put *nomenclature* first can only lead to a spurious taxonomy in which *scientific names* and their bearers, the *type specimens*, are classified instead of the natural phenomena which we wish to arrange in a system.
>
> The stratigraphic position of a taxon is not a valid element of its *definition*. This must be purely morphological; in other words, two foraminiferal populations which cannot be distinguished morphologically, but which are of different ages, must be considered identical. The stratigraphic position, however, enables us to *evaluate* morphological characters, and in this sense it must be taken into

consideration. The differences between *Bolivinita* and *Bolivinitella*, and between *Lepidorbitoides* and *Nephrolepidina*, illustrate this point.

The main problems are, on the infraspecific and specific levels, the distinction between nontaxonomic individual variants and modifications, geographic subspecies, and transients in time or chrono-subspecies; on the generic and higher levels, the proper grouping on the basis of significant *character combinations* and the balance between horizontal and vertical classification (Simpson, 1944).

The rise of micropaleontology was connected with the discrimination of minor morphological differences which made it possible to distinguish strata of different ages on the basis of foraminifera alone. Now that there can be no more doubt about the changes of foraminiferal faunas in time, there is no need to continue blindly in the direction of minute 'splitting' between any two individuals which may be of different ages and therefore, as names in faunal lists, helpful in biostratigraphic zoning. Enough material has been accumulated to turn to the analysis of populations, which is the only legitimate scientific practice in taxonomy. When that is done, we can follow the example of Tan Sin Hok's morphogenetic studies, or test the 'plexus' concept (George, 1956, and references), or the proposals of Sylvester-Bradley (1951) and others, relating to the use of subspecies. In no other branch of paleontology is it so easy to obtain large numbers of individuals (or samples of successive populations) from unbroken sequences of strata. The lag in the application of modern concepts to foraminiferal biostratigraphy can only be explained by historical reasons, but it cannot be justified [emphasis in original].

Later, after another blast at biohistorical vandalism committed in the name of nomenclatural purity:

Genera are particularly important in stratigraphy, firstly, because they have wider geographic ranges than species and are therefore valuable for long-range correlations, and secondly, because their names are part of the binomina of species. The practical stratigrapher has to know his genera, though he may temporarily record his species by numbers or letter designations. Any change in generic concepts and nomenclature is therefore a step to be undertaken with a sense of responsibility, not for formal reasons, but with the purpose of clarifying the evolutiom, and the distribution in time and space, of groups of species which can be reasonably claimed, on the basis of morphological and stratigraphic evidence to be related.

> Genetic relationships are seen in paleontology as morphological relations in time. In this sense, taxonomy and stratigraphy are interdependent.

Although this statement lacked a definite opinion on the nature and recognition of fossil species, Glaessner was more than clear on the relationship between taxonomy in its broadest and most meaningful sense and biostratigraphy in the narrowest sense of its own taxonomic needs. There need be no conflict and there was no justification for a narrow pre-evolutionary typology which ignores the potential of protistan palaeobiology.

Planktonic foraminifera in correlation: bioentities or biocharacters?

Can we detect any resonance between these views of Glaessner and the precepts and practices of the most active biostratigraphers of the time? Here is Subbotina (1953) on the taxonomy of her systematic monograph on Palaeogene planktonic foraminifera:

> As a starting point for taxonomic differentiation, the author has used the concept of a species as a group of individuals possessing definite morphological peculiarities (qualitative individuality according to Lysenko) and which share a more or less homogeneous bionomic situation. The author has tried to relate the changes in morphological characteristics with changes in the environment, although in our present state of knowledge this is not always possible because of difficulties relating to paleontological material generally.
>
> Intraspecific variability in many pelagic foraminifera of the Paleogene leads to the formation of a series of distinct forms in many species. In other words a pelagic foraminiferan may sometimes be represented by a series of varieties which can be described as forms of a single species. One of these varieties may predominate in terms of numbers of individuals and this is then regarded as the typical form for a given locality at a particular time; other varieties are less abundant and less typical. However the designation of a 'typical form' is inappropriate unless account is taken of the whole range of diversity which occurs among members of a particular species. Typical examples at one paleogeographic location may be quite atypical of the species elsewhere. The erection of 'typical forms' of a species for the whole range of paleozoological situations in which it occurs is a task for the future and cannot reasonably be undertaken at the present time. It is still difficult to explain, for

example, why among the fossil foraminifera we have examined from Paleogene deposits of southern USSR we have found an abundance of shells belonging to certain varieties but far fewer individuals belonging to others. Forms which are typical of the species at the time of its first appearance, i.e. the primary or precursor forms, are not always typical of the species as a whole, because they are represented by very few examples and differ essentially from the more abundant, subsequently encountered, 'usual' forms of the species …

In determining species, we have not only taken into account morphological differences and bionomic characteristics, but also the time factor with reference to the stratigraphical position of the forms we have described. In this respect, the author is in complete agreement with D. L. Stepanov who considers that individuals which, although very similar morphologically, are found to occur regularly in different stratigraphical horizons are unlikely to belong to a single species.

The same consideration must obviously apply to taxonomic units above the species level. Thus it often happens that, with the accumulation of new facts, the original concept of the genotype as being the most typical species of the genus has to be changed. In the case of the genus *Globorotalia*, for example, Cushman (1927), who established the genus, designated the present-day form G. *menardii* as the type species. Among fossil forms, however, those possessing smooth shells like those of the genotype are very much rarer than others with more inflated shells furnished with well-developed sculpturing. Nevertheless, the concept of the genotype as the species initially established as being typical of the genus is still of considerable value and the retention of a particular species as the genotype, even when it is known to be no longer typical, is of value in enabling us to trace the historical development of the particular genus.

'In order to denote varieties we have used the trinomial system of nomenclature. Where we are describing species having several varieties we begin by giving the general characteristics of the species as a whole, taking account of all its constituent varieties, and then we go on to give a short account of each variety separately.'

US *National Museum Bulletin* 215 (Loeblich *et al.*, 1957) was one of the most important documents in all micropalaeontology; its planktonic foraminiferal biostratigraphy becoming the prime reference for the subsequent expansion of micropalaeontology and palaeoceanography. The 'philosophy' then is of more than desultory interest. How did they go about their taxonomy? What are their species? Loeblich wrote the Preface; he began with the observation that there are two camps – perhaps as always – the one complaining bitterly about increased taxonomic splitting, the other enthusiastically doing the splitting. But there have been 'many different geologically and ecologically restricted species and genera masquerading under a single name': that must be rectified, and stability and the conservation of taxa cannot be sustained at former levels for their own sake whilst techniques improve and new assemblages are studied (which is not to foreshadow 'the immediate and indiscriminate erection of a multitude of new names'). Loeblich continued (emphasis added):

> Part of the difficulty lies in the lack of sufficient experimental data on living populations to allow a determination of the truly important taxonomic characters. As a result, one specialist may place the greatest taxonomic emphasis on wall structures, another will consider the apertural position of prime importance, while others will use chamber arrangement, presence of particular internal characters, or even surface ornamentation as generic or family characters. Yet any of these proposed bases of classification might be considered useless by another equally sincere worker. Each individual is entitled to his own opinion, provided it is based on facts and logical assumptions from these facts; but it is obvious that all workers, given the same set of facts, will not always arrive at identical conclusions; therefore, there is no insistence that all the papers here included use the same terminology or bases of taxonomic classification. We do feel it necessary, however, to ask that reasons be given for placing a genus or species in synonymy, or for subdividing a previously known genus or species, and to ask that means be presented for distinguishing the new forms from other similar forms. In addition, it seems advisable that a *general taxonomic philosophy be accepted – that certain characters be considered of higher taxonomic value than others and be used similarly throughout the classification.* Where new taxonomic units are proposed in the included papers, this is done.

There followed a stern warning, necessary because most micropalaeontologists are only secondarily taxonomists or zoologists: '*Specimens placed in each species must be like the original type specimens*, and if this necessitates a new name for a form widely but erroneously known by an old and classic name, *sentiment cannot intervene*' (emphasis added). The first italicized passage is the only statement in the volume concerning the recognition and discrimination of species of foraminifera.

In revising the classification of the planktonic foraminifera, Bolli *et al.* (1957) had to order and arrange the materials accumulated during the first burst of modern biostratigraphy. Their classification was noteworthy for its thoroughness and for the rigorous application of the Rules of Nomenclature. But it was characterized too by a primitive and pre-evolutionary taxonomic philosophy: the characters were ranked as of family, subfamily, generic or specific value, and the taxa fell into place. As knowledge of the planktonic (and benthic) foraminifera advanced, successive classifications were produced (Banner and Blow, 1959; Loeblich and Tappan, 1964a, b; Lipps, 1966; Pessagno, 1967) but, as I once put it (McGowran, 1971): 'There is no significant variation discernible in the taxonomy of these authors. Classification has changed as continued scrutiny has revealed more "basic" characters which are accorded increased significance, and the importance of wall structure and its external manifestations undoubtedly will increase further. In the way in which the data are assessed, there is a consistency of approach in that characters of a designated hierarchical value are used to define taxa of successively lower rank. Although the fact of evolution is invoked repeatedly in stating the need to use 'basic' and 'less adaptive' characters – those less likely to have been affected by external, environmental influence – this taxonomy is nonevolutionary. It is *a priori* and deductive. Mayr (1969) has distinguished five theories of classification, not necessarily unmixed: essentialism, nominalism, empiricism, cladism and evolutionary classification. In these terms the taxonomy under discussion has much in common with pre-Darwinian essentialism or Aristotelianism (see also Cain 1960). Its most important flaw is that there is no way of determining in advance which characters are the most stable, and no way of proving that importance in physiology and function should be correlated with importance in classification. There is 'no logical or *a priori* primacy of any character. The assumption that a classification based on chemical or physiological characters is "better" or "more natural" than a classification based on morphological or behaviour characters has no more intrinsic validity than Aristotle's scale of weighing characters (Mayr 1964, p. 22).' Sporadic attempts at evolutionary classification (McGowran, 1968a; Steineck and Fleisher, 1978) did not stimulate foraminiferal taxonomy and the group, unmatched in the completeness of its fossil record, played little part in the *Sturm und Drang* of two decades of uproar in taxonomy recounted by Hull (1988). A noteworthy exception was the cladistic revision of the Neogene planktonic foraminifera (Fordham, 1986).

But what of the entities actually utilized in biostratigraphy? Cain (1954) described the palaeontological habit of naming, binomially, individual variants 'that could not possibly be called species in neontology' but which have been found to be more restricted in time than are the more abundant and long-ranging forms with which they intergrade. The habit has been widespread in foraminiferology. Consider the *Orbulina* bioseries, the phyletic succession

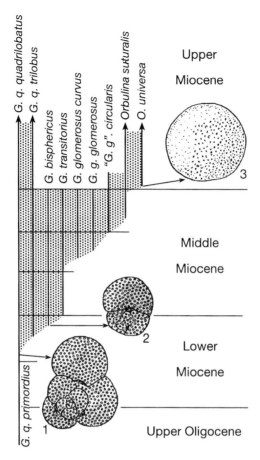

Figure 4.1 The *Orbulina* bioseries: evolution of *Orbulina* from *Globigerinoides*, from Glaessner (1966, with permission) after research and figures by Wade (1964, 1966). Figured specimens 1–3 are *G. quadrilobatus quadrilobatus*, *G. bisphericus*, and *O. universa*, respectively. Stippling indicates continuous variation between previously named morphotypes and their ranges. Horizontal lines delimit seven zones in southern Australia.

leading rapidly to the so-called *Orbulina* surface (Finlay, 1947; LeRoy, 1948), 'points of lowest stratigraphic occurrence of *Orbulina* in a continuously deposited, open sea, deep water globigerine facies sequence, fall on or in close proximity to an equivalent time horizon within Middle Tertiary sections in tropical and subtropical zones of the world' (LeRoy, 1948). The evolution of *Orbulina* was demonstrated by Blow (1956) as leading to two possibly end-members, *Biorbulina* and *Orbulina*, and Wade (1964, 1966) showed the virtually complete intergradation between the morphotypes, named as morphospecies in three genera (Fig. 4.1).

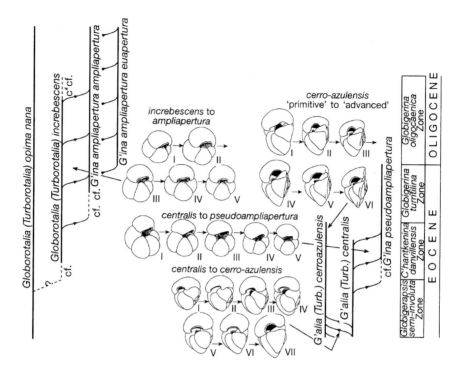

Figure 4.2 Bioseries from *Globorotalia centralis* to *Globorotalia cerroazulensis*, from *Globorotalia centralis* to *Globigerina pseudoampliapertura*, and from *Globigerina increbescens* to *Globigerina ampliapertura ampliapertura* and *Globigerina ampliapertura euapertura* (composited from illustrations by Blow and Banner, 1962, with permission). Vertical lines are ranges of named morphotypes and arrows indicate full intergrading over the biostratigraphic interval shown (see zones at right). The same kind of continuous variation is shown by arrows within three of the clusters of illustrated specimens; the fourth displays a shift from 'primitive' to 'advanced' within the taxon *Globorotalia cerroazulensis*. Each of the four clusters of illustrated specimens was selected from a single sample and the positions of the four samples are indicated approximately.

Another bioseries recognized early in the development of planktonic foraminiferal biostratigraphy was the evolution of the *Globorotalia fohsi* lineage (Bolli, 1950, 1957b, 1967) (see below). Blow and Banner (1962) described other lineages including the examples shown here (Fig. 4.2). (They were revised subsequently but that is unimportant here.) Not only is there the same extensive intergradation between named morphotypes as before, but Blow and Banner demonstrated that intergradation with series of illustrated specimens taken from the same samples. For the sake of biostratigraphic correlation, the formalizing of morphospecies is taken far beyond the limits that Cain (1954) and others regarded as even remotely tolerable in neontological systematics.

Blow (1969) explained in part this typological strategy: '... the writer notes the type reference of the taxa and refers to illustrations which indicate the writer's concept of the taxon. These concepts have been rigidly applied stratigraphically to the extent that the writer might be taken to task for having been too much of a "taxonomic splitter". Nevertheless, it would seem preferable to be a "splitter" in biostratigraphy rather than a "lumper" for it can be left to the individual biostratigrapher to form his or her own conclusions as to which, if any, of the taxa discussed below might be "lumped" together.' In a more expansive and ambitious explication (1979) Blow aligned himself with the taxonomist Blackwelder (e.g. Blackwelder and Boyden, 1952; Blackwelder, 1967), basing much of his 'basic philosophy' on reference to Blackwelder's *five acceptances* – statements which Blackwelder (1967) assembled as being virtually unchallenged in modern systematics but which taken together 'form an entirely false picture of what taxonomy is today'. Blackwelder's acceptances were: (1) that taxonomy in the past was exclusively 'morphological' and that the New Systematics differs in being 'biological'; (2) that a so-called biological species concept is superior in taxonomic work and is now in widespread use in zoology; (3) that species are different in nature from other taxa; (4) that there can be a direct basis of classification in phylogeny, or that the aim of classification is to reflect phylogeny, or that taxonomists must study the origin of the taxa which they distinguish and define; and (5) that only a phylogenetic classification is a natural one.

For the 1950s and the early 1960s that 'entirely false picture' was quite a reasonable taxonomic manifesto for an evolutionary taxonomist. The agenda behind Blackwelder's rejection of it was to be found in his defence of the essentialist or typological theory of classification (Mayr, 1969). Blow's approval of the greater part of Blackwelder's philosophy was due to *his* seeking some respectability for his strategy of distinguishing finely split taxa for biostratigraphic purposes. His criteria were precision and consistency and there was not, for Blow, any seeking out of species that exist in nature. Where Linnaeus's dictum stated that 'it is the genus that gives the characters, and not the characters that make the genus' (a dictum still held, even though evolution has replaced Aristotelian logic as the reason: e.g. Mayr and Ashlock, 1991), Blow's copious writings indicate his answer as the opposite. Blow is by no means alone among foraminiferologists in either his overall taxonomic essentialism or his typological species concept. A subsequent monograph (Brönnimann and Resig, 1971) recognized very finely split species. Two tests or criteria recur in foraminiferology. One is the pragmatic, expressed bluntly by Jenkins (1973): 'The true acceptance test is empirical: if the new species or subspecies has any value taxonomically, stratigraphically or paleoecologically then other operators will use them. Useless taxa will either be ignored or

placed in synonymy at a future date.' The other criterion encouraging acceptance of a taxonomy – highly typological and essentialist or not – is consistency: a large monographic study can be quarried for macroevolutionary
analysis because, says this reasoning, the species comprising its database are
mutually consistent (e.g. Jenkins, 1968; Berggren, 1969b). Indeed, Blow's great
monographic treatment of the Cenozoic planktonic foraminifera has attracted
several workers in the recent revival of macroevolutionary foraminiferology
for that very reason with no real qualms expressed over the nature of the
entities ('species') in the database.

If we are dealing with a primitive pre-evolutionary typology in the taxonomy of
the planktonic foraminifera, then two interesting and non-trivial questions arise:
(i) How come the development of biostratigraphy to the level of circumglobal
phylozones and chronozones has been so successful? And (ii) could the same result
have been achieved by showing more respect for the actual species of nature?

The first question can be answered with reference to the macroevolutionary
distinction between transformational and taxic evolution. With that distinction
in mind we can interpret the following extract from Blow (1979) as a very clear
example of the transformational worldview:

> What matters in taxonomy and classification is the constancy of the
> morphotypic expression of the character and to what extent such a
> character, or difference in character, pervades a presumed natural
> group. Again, and as a corollary, the importance of the difference of a
> character (or characters) in taxonomy and classification must also be
> assessed in terms of what happens to the morphological difference
> throughout space and time within the subjectively assessed taxon or
> taxa-group. From this, the concept of the persistence in time of a
> character, or the regular modification of the same character in time,
> arises as one of the more fundamental, less subjective, means of assessing
> the validity of a character within a taxon at any classificatory or
> systematic level. It matters not, in the writer's opinion, whether the
> character is large or small, or whether it needs an electron-microscope or
> a hand-lens to see it, and the only test to be applied is purely the one of its
> persistence in space and time allied to its demonstrability. Thus, the
> character must be demonstrable, capable of some sort of measurement
> and record, and must be capable of objective definition and scientific
> treatment. No constraints as to size are acceptable since the assessment of
> any demonstrable character is subjective in terms of its genetic validity.

So there we have it. Taxa have very little to do with biospecies; they are
entirely subjective, in Blow's opinion, so that taxonomists can only say

'I believe … ' or 'I think … ', not 'I demonstrate … ' or 'I show … ' It is the character, not the species, that counts as observable, measurable, and most of all as objective. This is a prime example of the transformational approach to macroevolution. It matters not whether you be a splitter or a lumper, whether you search for real species that exist independently of the systematist or you are an unashamed typologist; and communication is not hampered between two biostratigraphers with vastly different views on the study of organic diversity. Just so long as the first appearance of a character can be found in a few specimens in one of a series of samples, we can deal in phylozones without either sinking in the morass of an archaic taxonomy or losing any resolution in biostratigraphic division or correlation. Table 4.1 illustrates this equivalence of character datums and taxonomic datums.

In 1971 I commented on the second question as follows:

> It is generally agreed that the basically sexual animal species is made up of populations between which there is more or less gene flow; interfertility (inclusive) and reproductive isolation (exclusive) are the criteria for defining this biological unit. Difficulties and the need for various qualifications arise from the structure of populations as dynamic complexes that evolve in space and time. Even before we reach the problem of recognizing species in foraminifera there is the matter of how sexual many foraminifera are, including planktonics. Beyond this is the task of forcing the four-dimensional 'biological reality' into the framework of Linnean nomenclature. Evolutionary divergence subsequent to speciation becomes manifested in due course in anatomy, and discontinuities within fossil assemblages (other than proloculus size, coiling direction and the like) are the starting point for recognizing phena and therefore species-group taxa. Complaints about the impossibility of recognizing species without the breeding criterion, and doubts about the 'reality' of species, are due to confusion of concepts (Mayr, 1969). Problems inherent in fossil assemblages (mixing of populations, loss of facts on water-column stratification, differential solution, etc.) should not be underrated; nor, however, should they be an excuse for abandoning the effort to approximate reality. A less negative reason for a narrow typology in foraminiferal taxonomy has been the alleged value of index species defined in this way. This practice became very widespread but its necessity has never been demonstrated. The possibility of recognizing the 'unitary evolutionary role' of a lineage on the basis of preserved characters, as advocated by Simpson (1961), was demonstrated by Wade (1964) in planktonic foraminifera

Table 4.1 *A 1960s vintage chart of datums signalling major Cenozoic divisions, but this time including the morphological characters as datums (FADs and LADs) as alternatives to the taxa that they help to define. Thus, we can sidestep refractory systematic problems if necessary without losing out biostratigraphically. (G., Globorotalia; Sph'opsis, Sphaeroidinellopsis; FAD, LAD first and last appearance datums.)*

Geochronological division	'Transformational': datum based on morphocharacter	'Taxic': datum based on genus or species
Pleistocene		
	marginal keel	FAD *G. truncatulinoides*
Upper Pliocene		
	'teeth'; single aperture	LAD *Globoquadrina*; LAD *Sph'opsis*
Lower Pliocene		
	second aperture	FAD *Sphaeroidinella*
Upper Miocene		
	coiled–biserial test	LAD *Cassigerinella*
Middle Miocene		
	spherical test	FAD *Orbulina*
Lower Miocene		
	second aperture	FAD *Globigerinoides*
Upper Oligocene		
	planispiral test	LAD *Pseudohastigerina*
Lower Oligocene		
	tubulospines	LAD *Hantkenina*
Upper Eocene		
	surface texture	LAD *Morozovella*
Middle Eocene		
	tubulospines	FAD *Hantkenina*
Lower Eocene		
	planspiral test	FAD *Pseudohastigerina*
Upper Paleocene		
	angular chambers	FAD *Morozovella angulata*
Lower Paleocene		
	double keels; apertural plates	LAD *Globotruncana*;
Maastrichtian		LAD *Rugoglobigerina*

without forfeiting biostratigraphic precision. Ultra-typological 'species' may have the value claimed for them, but the important changes in population samples with time on which they are based can be expressed equally well (for the biostratigrapher) as successional taxa. Studies of evolutionary rates as possible environmental trends based on 'species' as currently defined are self-consistent when the taxonomy is uniform (Jenkins 1968, Berggren 1969b), but will be even more meaningful when intergrading sympatric associations are treated as such.

Consider Wade's analysis of the *Orbulina* bioseries (Fig. 4.1), interpreted as an anagenetically expanding lineage which then contracted cladogenetically into two branches. The incoming of successional morphotypes, named as species and subspecies, gives us high biostratigraphic resolution. But Figure 4.1 shows seven faunal units (zones) which in this example could be matched using taxa that take full account of the intergrading variation; that is, by recognizing successional species or subspecies in a lineage, the taxa being marked off by the incoming of new morphotypes. Admittedly, there would be nomenclatural confusion for a time whilst foraminiferology was joining the biohistorical main-stream. But Blow was quite sanguine about an exuberant efflorescence in nomenclature which would be sorted out and culled in due course by testing and experience. That tactic need not be cornered by typology.

Stainforth *et al.* (1975) came closer than did most to grappling with the problems of recognizing and dealing taxonomically with evolving lineages. They are cautious, sceptical about the actual configuration of some lineages as reconstructed (' ... discussed by some authors so dogmatically as to make one forget that their opinions are only educated guesses'). The central problem for Stainforth *et al.* was an interesting mixture of their perception of phyletic gradu-alism in the planktonic foraminifera and the need for a pragmatic treatment of nomenclature – meaning biostratigraphic utility: the monograph was conceived initially and prepared by the Exxon Production Research Company. The authors stated their precepts for lineage studies prior to the establishment of taxa. Thus:

1. Obviously essential is appearance of supposedly primitive forms before presumed descendant (stratigraphically higher) advanced ones ...
2. Individuals are subordinate to whole populations in discussing and assessing all postulated lineages. Populations found at successively higher stratigraphic levels may contain virtually the same range of morphologic variants but in somewhat different proportions. The mode of populations differs with time and follows discernible trends at measurable rates. Corollaries to this precept include the following:
    a. The establishment of lineages should be based on, or at least backed by, detailed statistical analyses ... seldom the case ...
    b. To an appreciable extent experience leading to a developed sense for recognizing faunal changes may substitute for formal statistical studies ... In contrast, a lineage synthesized solely from type and other figures in the literature generally has little value.

    c. As in all statistics, the smaller the sample the greater the probable error. The evolutionary status of species cannot be assessed reliably from sparse or single specimens.

3. Empirical studies suggest that saltatory modifications of planktonic foraminiferal populations are exceptional; incipience of *Globigerinatella* is one of the few suspected cases. Aside from such rare exceptions, *an essential of evolutionary change is that it be gradual*. Whatever trends of morphologic change may be postulated, specimens should be readily available which differ barely perceptibly, one by one, yet provide an unbroken lineage between extreme forms. When evolution is slow, a wide range of variants may occur in a single assemblage … [emphasis added].

4. Evolutionary processes are affected by ecologic and climatic factors, among which temperature is especially influential …

5. Homeomorphy is readily demonstrable among planktonic foraminifers, as is to be expected when unrelated groups evolve towards ideal adaptation to a free-floating existence …

6. Hypothetical but not ignorable in evaluating postulated foraminiferal lineages is the fact that these protistans are very simple organisms. Each test may be viewed as a simple geometric design susceptible of description by a few mathematical symbols suitable for expressing form of the spiral through midpoints of its chambers, their number in each whorl, and rate of their size increase … Small changes in these primary parameters, *whether rooted in natural variability or in progressive evolution of the species*, can lead to pronounced secondary differences between individuals of the same stock … [emphasis in original].

Their final point is the observation of evolutionary convergence as seen in skeletal morphology, and the shade of orthogenesis lurks in their startling conclusion: 'The empirical suggestion emerges that these trends represent response to an inherent, irreversible life force in foraminifera. Consequently, reasonable doubt is justifiable when an author postulates evolution in an opposite direction.'

    Figure 4.3 is Stainforth's notion of the species concept (Stainforth *et al.*, 1975). It is a splendid rendering of a plexus changing gradually (gradualistically?) in time, illustrating the difficulty of applying binomial nomenclature and representing the only advance on Glaessner's discussion of two decades before. So how did Stainforth *et al.* actually tackle the identified problem of planktonic foraminiferal species? The short answer is that they did not. They listed practices published previously:

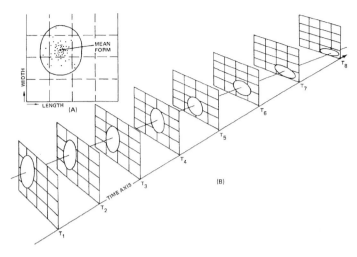

Figure 4.3 Stainforth's notion of the evolving species (Stainforth *et al.*, 1975, with permission) explicated in his caption: Diagrammatic illustration of species concept. Characteristics of species tend to change gradually and progressively through time so that a purely objective definition of a species may be extremely difficult. (A) Characteristics of a species at one time indicated here by points denoting the length/width ratio of somewhat differing adult individuals; the concentration of points in a central position indicates the mean form and the oval outline indicates the interpreted range or maximum divergence. (B) The mean form (connected by line) and maximum divergence of individuals (oval) for such a form is indicated for successive times T_1, T_2, T_3, ...; although gradual change (evolution) results in only slight differences between adjacent times, early forms at T_1 and T_2 differ completely from late forms at T_7 and T_8.

1. Include all variants under one species name and designate their status informally (e.g. primitive, simple, median, advanced, etc.).
2. Same, but define successive stages more precisely, designating them by a code (e.g., as forma Alpha, forma Beta, etc.).
3. Recognize successive species and divide them into sequential sub-species, preferably designating them by names which convey a sense of evolution ...
4. Essentially the same, but give all recognized variants the rank of species.
5. Indicate morphologic development by use of generic and subgeneric names

However, the monograph is biostratigraphic in its intent and the discussion of lineages and species concepts is much more a prescription for future research programmes than an introduction to their review of index species.

The systematics of planktonic foraminifera has not developed during the past four decades in the inductive way that one would intuitively have expected. In the '*ab initio* study of new assemblages', Blow (1979) characterized *taxometry*: 'the conscious act of identification and sorting of the members of fossil assemblages and populations usually involves, as a first step, the discovery of the extent of morphological similarity (or dissimilarity) without a programmed intention of first discovering phylogenetic affinity'. Zoning and correlating transformationally – using characters not taxa – as we have been doing, we have avoided taking the first steps toward a thoroughgoing taxic research programme. The first step of all is to sort each sample into groups of specimens, or phena. If the *phenon* is a 'sample of phenotypically similar specimens; a phenotypically reasonably uniform sample' (Mayr, 1969; Mayr and Ashlock, 1991), then 'reasonably uniform' does *not* turn on how far from the holotype we may venture in our 'species concept'. Instead, the prime question is: where are the morphological discontinuities in a sorted, large array of specimens from a single sample? – a very different question. The phena having been sorted on the basis of those discontinuities, the next step is to group them into species, still within the sample. In foraminifera, there are the more obvious groupings of phena separated by e.g. coiling direction and proloculus size and the less obvious ontogenetic changes; early stages of growth carry significant information. Perhaps the most important potential problem in discriminating phena and phenotypically recognized fossil species within single samples concerns the spatial separation of populations within the pelagic water column. With the exception of isotopically differentiated phena that information is lost.

Have we retrieved the foraminifera for palaeobiology? A major conclusion of this discussion has been that rapid biostratigraphic progress has been achieved, typological essentialism notwithstanding, and nothing has been lost in that respect because taxonomic datums actually are character datums. The healing of the rift between biostratigraphy and palaeobiology (Glaessner, 1955, 1966) has still to be completed, but has been recast in sequence-stratigraphic terms (Holland, 1999, 2000). Meanwhile, consider for a moment the supraspecific classification of the planktonic foraminifera. Luterbacher (1964) justified the classifications of Bolli *et al.* (1957) and their successors, in which characters are ranked a priori, as being advantageous from a practical point of view: since the biostratigraphic unit is the 'species', the genus and higher categories can be neglected and need not confuse the issue – the issue being exclusively zonation and correlation. There was some reaction to this brutal coupling of taxonomy with correlation. Bandy (1972) erected several phylogenetically defined subgenera of *Globorotalia*, a move that received limited support until a phylogenetic atlas of Neogene planktonics strongly supported the approach (Kennett and Srinivasan, 1983). There was a change in the conservative central European attitude: Toumarkine and Luterbacher (1985) placed within the same genus all species believed to belong to the same lineage. I find that a comment decades ago still holds (McGowran, 1971):

> ... a similarity in the results of various classifications, even when the introductory remarks reveal substantial differences in taxonomy, seems to be a measure of our advancing knowledge of evolution in the planktonics. Bandy *et al.* (1967) have put an extreme view: "the stratigraphic occurrence of many planktonic forms is now so well documented that occurrence in geological time is valid as a critical factor in classification". Their *only* criterion for distinguishing *Neogloboquadrina* from *Globoquadrina* is that "it developed from a different lineage in the later Miocene and is therefore genetically unrelated". This is far indeed from apriorism, but perhaps is too extreme for a group showing such strong parallelisms in relatively few characters ...

Microfossil and micro-organismal contributions to macroevolutionary topics have increased markedly in recent years and some are touched on below. Fortunately, systematics and monographing also continue with two particulary noteworthy bursts in Palaeogene planktonic foraminifera (Olsson *et al.*, 1999; Pearson *et al.*, 1999 and in press) and a third to follow.

Planktonic foraminifera: biogeography and stratification

The structure of the modern ocean reflects both Earth-crustal history and the realities of planetary dynamics. As to the former: the main differences from the Mesozoic ocean are valve closures at lower latitudes – the closure of low-latitude Tethys as Africa and India collided with Eurasia (Australia soon to follow) and the isthmus of Panama rose later, and valve openings at higher latitudes – the opening of a circum-Antarctic oceanic throughway and Arctic-Atlantic passages. As to the latter: rotation and the highly unequal latitudinal receipt of solar energy determine patterns of evaporation–precipitation and the major wind systems. The combination of these determinants produced some eighteen major surface water masses in the global ocean (as well as marginal seas) and the annual surface production of organic carbon follows the pattern of water masses quite well in several respects (references in Norris, 2000). Prominent features include the high production in the equatorial and marginal current systems and low production in the big subtropical gyres. McGowan and Walker (1993) and Norris (2000) suggested that the basic number of watermasses has probably remained fairly constant for the past 20–30 myr – broadly, the Neogene Period – if not longer, and that the major distribution

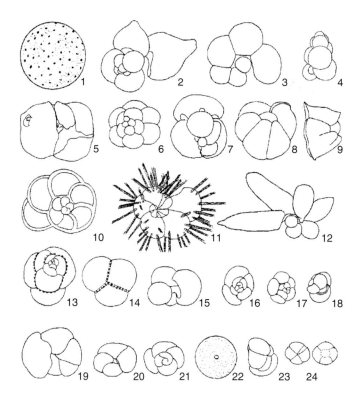

Figure 4.4 Pelagic foraminifera – three climatically zoned groups of species indicated a diversity gradient from tropical (high) to polar (low) (Murray, 1912). This work was a major outcome of the *Challenger* expedition in the 1870s. The groups (nomenclature updated) were distinguished as follows. Tropical forms: 1, *Orbulina universa*; 2, *Globigerinoides sacculifer*; 3, *Globigerinella siphonifera*; 4, *Globigerinoides ruber*; 5, *Sphaeroidinella dehiscens*; 6, *Turborotalita quinqueloba*; 7, *Globigerinoides conglobatus*; 8, *Pulleniatina obliqueloculata*; 9, *Globorotalia truncatulinoides*; 10, *Globorotalia menardii*; 11, *Hastigerina pelagica*; 13–14, *Candeina nitida*. Temperate forms: 12, *Globigerinella digitata*; 15, *Globigerina bulloides*; 16 & 23, *Globorotalia inflata*; 19–21, *Globorotalia scitula*; 22, *Orbulina universa*. Polar forms: 17–18, *Neogloboquadrina dutertrei*; 24, *Neogloboquadrina pachyderma*.

of biogeographic regions correspond reasonably well with the distribution of watermasses.

Considerations of the global diversity gradient and latitudinal provinciality go back to A. R. Wallace and other eighteenth-century biogeographers. Murray (1912?) referred to planktonic foraminiferal bipolarity, diversity gradient across latitudes, and provinces (tropical, temperate, polar: Fig. 4.4). The number of living species has remained at once stable and uncertain for several decades at ~40–50 species (Hemleben *et al.*, 1989) with from ~20 (10 indigenous) species in

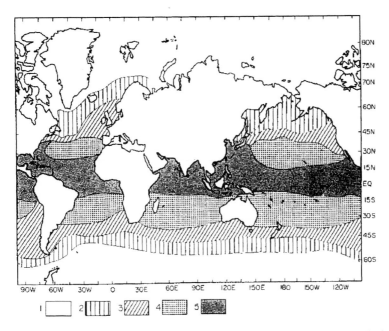

Figure 4.5 Biogeographic provinces in modern planktonic foraminifera, displaying the pronounced bipolar pattern (Arnold and Parker, 1999, after Bé, 1977, with permission). 1, Polar, 2, Subpolar, 3, Transitional, 4, Subtropical, 5, Tropical provinces.

the tropics to ~5 species (one indigenous) in the polar regions up to the ice. The biogeography summarized by Bé and Tolderlund (1971) and Bé (1977) still holds (Figs. 4.5, 4.6). There are nine provinces in a symmetrical bipolar pattern, shifted north into the warmer hemisphere. For Arnold and Parker (1999) the provinces do not map cleanly on to defined physico-chemical oceanographic properties; they found the provinces easier to describe than explain. However, provincial spreads seem to have some relationship to the watermasses. The southern polar and subpolar provinces are in the Antarctic circumpolar current system; the southern transitional province in the south lies between that and the subtropical gyres; the subtropical provinces are in the zones of the subtropical gyres and the tropical province is in the equatorial zone of high surface production of organic carbon. Where provincial boundaries cut most strongly across latitudes, major currents are the ready explanation. The provinces are more numerous than the 'large-scale functional ecosystems' (Arnold and Parker, 1999) whose overlaps largely define the transitional zones.

Low-latitude watermasses tend to be more strongly stratified near their surfaces than polar and subpolar waters and seasonal variations are strongest at the mid latitudes. Stratification is seen in the various clines in the upper several tens of meters of the water column – temperature, salinity, nutrients,

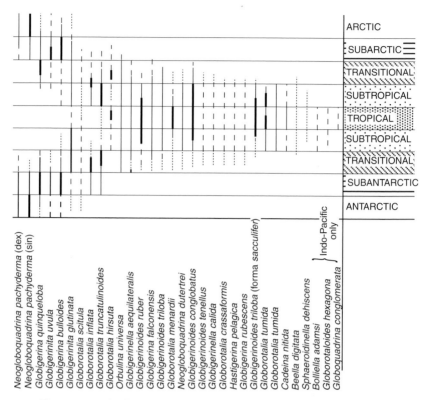

Figure 4.6 Latitudinal range of modern planktonic foraminifera, showing the same bipolarity as in Figure 4.5 (Arnold and Parker, 1999, with permission).

light. Changes in physical properties are sharp enough to form surfaces – hence the notion of aqueous stratification. The various planktonic foraminiferal species arrange themselves in terms of environmental variables: temperature (including latitude), feeding strategy (grazing and bacterial, carnivory, photosymbiosis, opportunistic) and depth and frequency of reproductive cycles (Figs. 4.7, 4.8).

Planktonic foraminifera and stable isotopes

Studies of the stable isotopes of carbon and oxygen in planktonic foraminiferal calcite have fallen into three broad categories in the past half-century and especially the past quarter-century – (i) isotopes helping us construct a broad environmental template changing through geological time; (ii) isotopes indicating depth distributions and habitats of the species; and (iii) isotopes signalling the ecological strategy of photosymbiosis and its iterative acquisition

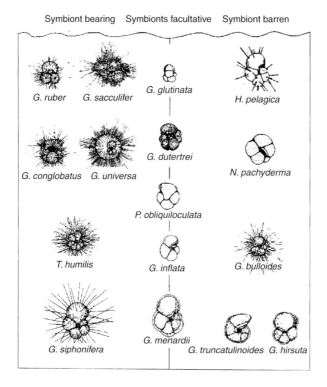

Symbiont bearing   Symbionts facultative   Symbiont barren

G. ruber   G. sacculifer

G. glutinata

H. pelagica

G. conglobatus   G. universa

G. dutertrei

N. pachyderma

P. obliquiloculata

T. humilis

G. inflata

G. bulloides

G. siphonifera

G. menardii

G. truncatulinoides   G. hirsuta

Figure 4.7 Depth distribution of modern planktonic foraminifera, also showing the relationship with symbiotic algae (Hemleben *et al.*, 1989, with permission).

in different phyletic lineages at different times. Rohling and Cooke (1999) summarized the chemistry, potential and problems in this expanding field. Pearson (1998a) discussed the use of stable isotopes in deconstructing the evolution of the planktonic foraminifera.

Oxygen ratios $^{18}O/^{16}O$, expressed as $\delta^{18}O$, reflect in biocalcification the reservoir ratio, especially as global ice accumulations wax (reservoir values become heavier) and wane (the reverse), plus the temperature at the time and place of crystal growth. The global hydrological cycle is the general control. There are various vital and diagenetic effects. Diagenesis can distort the primary oxygen signal in superficially well-preserved oceanic planktonics (Pearson *et al.*, 2001). The oxygen signal became broadly understood more rapidly than did the carbon signal. The ice-volume effect revolutionized our understanding of the late Neogene ice ages (Emiliani, 1955; Shackleton and Opdyke, 1973). Profiles of $\delta^{18}O$ through the Cenozoic Erathem have become prime references of global environmental change (Shackleton and Kennett, 1975; Savin *et al.*, 1975; Miller *et al.*, 1987; Wright and Miller, 1993; Zachos *et al.*, 2001b). They have been invaluable proxies for palaeotemperature even though the thermometer is rubbery given

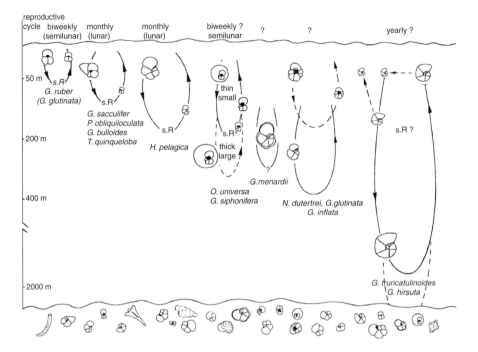

Figure 4.8 Depth distribution of reproductive cycles in planktonic foraminifera (Hemleben *et al.*, 1989, with permission).

that polar icecaps are of uncertain volume or even presence in the earlier Cenozoic. Fortunately the icecap effect trends in the same $\delta^{18}O$ direction as the temperature effect on skeletonization – we can see warmings and coolings even if not reliably the numerical values of temperature. Planktonic foraminiferal assemblages are affected by more subtle factors than merely water temperature – nutrient, light intensity, advection (references in Thomas, 1999).

Carbon ratios $^{13}C/^{12}C$, expressed as $\delta^{13}C$, reflect the workings of the global carbon cycle including the activities of the biosphere. There are between-reservoir effects, such as between the ocean and carbon pools outside the ocean, e.g. the growth of rain forests and the burial of organic carbon – both preferentially fix ^{12}C so that foraminiferal calcite is heavier in response. There are within-reservoir effects in the ocean – high fertility and exuberant productivity in surface waters is signalled by heavy values in planktonic foraminifera at the surface; the return of that carbon to the ocean at depth by respiration and degradation produces lighter calcitic values. Thus $\Delta\delta^{13}C$, the difference between surface and deep signal, is a productivity signal (e.g. Berger and Vincent, 1986). At the same time though different watermasses have different histories and different carbon ratios.

Technological advances permitted isotopic resolution and differentiation between deep-dwelling and shallow-dwelling plankton (Douglas and Savin, 1978). Figure 4.9 displays the $\delta^{18}O$–$\delta^{13}C$ fields in terms of planktonic foraminiferal ecology and palaeoecology. We see that there is a consistent distinction between the shallow-living plankton and deep-living plankton and the benthos, living and calcifying kilometers below the photic zone. We see too that those distinctions are sustained for 20 myr in all six signals in oxygen and carbon. This is a powerful indication that the ocean is interacting, both within itself (within the surface mixed layer and between it and the bottom of the psychrosphere) and as an entire reservoir vis-à-vis other reservoirs, such as the substantial incorporation of organic carbon into sediments. Hence the essence of the Monterey effect – the whole oceanic reservoir responds to carbon burial with the positive shift at MC_i (the Monterey carbon excursion) and the implied CO_2-drawdown is manifested by an inferred reverse-greenhouse effect at the positive oxygen shift, at least in the bottom waters (AA_i).

The third category of stable-isotope studies in planktonic foraminifera addresses the phenomenon photosymbiosis. Photosymbiosis, a splendid way of conserving and recycling resources in low-nutrient environments, has been emphasized as a major source of evolutionary innovation (Margulis and Fester, 1991), perhaps contributing to diversification and species longevity (Norris, 1996). Well known as a recurrent response to stable, nutrient-depleted waters, photosymbiosis has been found in less than 10% of ~150 extant families of foraminifera, but the strategy is well established in the two most active carbonate producers, the tropical larger foraminifera and the planktonics (Hallock, 1999). Hallock et al. (1991) observed a strong chronological correspondence during Palaeogene time among the diversities of the large benthics – alveolinids and nummulitids – on the one hand, and the planktonic acarininids and morozovellids on the other, the inferred common ground being photosymbiosis in response to oligotrophy coevally in the pelagial and the neritic. Photosymbiosis is known in about one-quarter of living planktonic species which host either but not both chrysophytes or dinoflagellates (Bé, 1982; Hemleben et al., 1989). Photosymbiotic species become most common in waters with stable mixed layers and a thermocline below the euphotic zone thoughout the year (Ravelo and Fairbanks, 1992; Andreasen and Ravelo, 1997).

Stable isotopic signatures offered the possibility of inferring photosymbiosis in the fossil record and identifying its onset in a lineage or clade (Spero and DeNiro, 1987; Spero, 1992; D'Hondt and Zachos, 1993). Norris (1996) listed five empirical patterns in Neogene species tending to distinguish them from coexisting asymbiotic species. (i) They have the most negative $\delta^{18}O$ of coexisting species (seeking the brightest habitat, gaining the warmest too). (ii) There is no

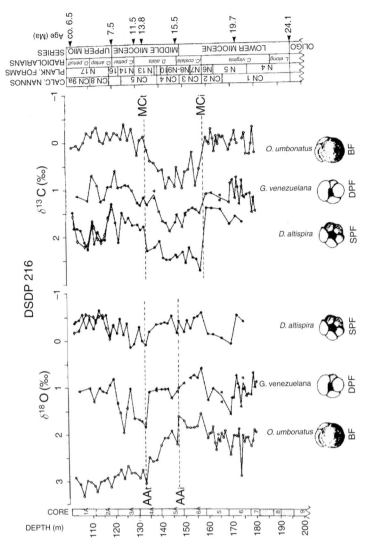

Figure 4.9 Stable isotopes and stratification in foraminifera, Miocene carbonate section, DSDP Site 216, Ninetyeast Ridge, Indian Ocean (Berger and Vincent, 1986, with permission). BF, benthic foraminifera, SPF and DPF, shallow and deep planktonic foraminifera. AA_i and AA_t, initiation and termination of spurt in growth of Antarctic icecap (chilling of tropical deep waters). MC_i and MC_t, initiation and termination of Monterey carbon excursion – the positive shift in the three profiles suggests that light carbon was accumulated outside the ocean as a whole (in oil source rocks and brown coals), not between surface and deep waters. Note the clear separation of all six profiles throughout the ~18 myr of the Miocene Epoch. The temporal pattern of the six single-species profiles suggested cause and effect – carbon shift in carbonate-carbon~burial of organic carbon~CO_2 drawdown~threshold in reversed greenhouse~chilling and icecap growth~global cooling, fall in sea level~return of light carbon to ocean.

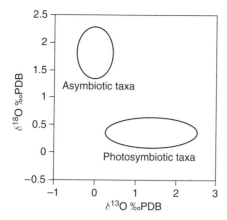

Figure 4.10 Recognizing photosymbiosis in fossil planktonic foraminiferal shells in the oxygen/carbon isotopic field (Norris, 1996; Pearson, 1998a). Higher carbon numbers in the calcifying shell imply withdrawal of light carbon by phytosymbionts in the cytoplasm. Meanwhile, lower oxygen numbers imply growth in the upper – warmer, better lit – oceanic mixed layer.

change in $\delta^{18}O$ with increasing size (remaining in surface waters, unlike the deeper-growing asymbiotic taxa). (iii) They have the most positive $\delta^{13}C$ of coexisting species (because their photosymbionts take up ^{12}C preferentially). (iv) They display a large range in $\delta^{13}C$ with increase in shell size (variations in vital effects?), and (v) a steeper slope in $\delta^{13}C$ against size than among asymbiotics (increase in symbiont density and activity). An empirical model of the $\delta^{13}C$–$\delta^{18}O$ field of photosymbiotic species as against asymbiotic is shown in Figure 4.10. These criteria indicated that the mid-Cretaceous radiation produced deeper-growing species and surface-water species, but no indications of photosymbiosis (Norris and Wilson, 1998). The same criteria showed in strong contrast that photosymbiosis was invented by a diverse array of species in the Late Cretaceous radiation, all of which clades went extinct at the end-Cretaceous catastrophe so that photosymbiosis had to be reinvented in the Palaeogene (d'Hondt and Zachos, 1998). Those authors used this as an example of *historical contingency* in planktonic foraminiferal evolution wherein the range of potential ecological roles was not affected whilst the pool of available clades available to fill those roles was.

MacLeod (2001), promoting an explicitly phylogenetic approach to planktonic foraminiferal systematics and ecology, was sceptical of these isotopic inferences. He made highly reductionist comparisons of three character-state pairs in a small selection of modern species (habitat shallow/intermediate/deep, spines absent/present, symbionts absent/present), and statistical analysis failed to reject random links between the pairs – no deterministic link could be shown between photosymbiosis and shallow dwelling. The available phylogenetic and ecological data suggested to MacLeod that the living species hosting photosymbionts do so as an ancestral condition.

Planktonic foraminifera and molecular systematics

The systematics of the foraminifera are sharing in the spectacular advances in molecular systematics based on the analysis of DNA sequences, with far-reaching implications (Kucera and Darling, 2002; Darling *et al.*, 2004). The foraminiferal advances are a 1990s achievement rooted in obtaining and authenticating pure foraminiferal DNA, as reviewed by Pawlowski (2000). There are four overlapping categories of new insights that are of interest here – the origins of major clades; the likelihood that stable morphospecies are species clusters; constraints on and insights into biogeography, ecology, and speciation; and comparisons and contrasts in rates of evolution, phenotypic and molecular respectively.

The planktonic foraminifera have been regarded by recent consensus as being monophyletic or minimally polyphyletic, beginning from small, low–diversity, 'generalized' (globigeriniform) founder populations in the Jurassic, whence sprang the two-pulsed radiation of the Cretaceous (praeglobotruncanid-rotaliporid, succeeded by globotruncanid), as in the ecological model advanced by Caron and Homewood (1983). Following the terminal Cretaceous extinctions, the Paleocene recovery and Palaeogene radiation was grounded in one, two or three small, 'generalized', opportunistic, survivor species (Berggren and Norris, 1997; Pearson, 1998c; Olsson *et al.*, 1999), contributory recolonization of the planktonic habitat from the benthos being a possibility but not advocated enthusiastically. The Palaeogene radiation was not terminated at all cleanly, as was the Cretaceous radiation. However, the Neogene radiation (Hemleben *et al.*, 1989; Stanley *et al.*, 1988; MacLeod, 2001) is twofold – the spinose globigerinid clade (Globigerinidae) and the non-spinose globorotaliids (Globorotaliidae) both rooted in the Palaeogene – plus the microperforates (Candeinidae), an outgroup and much smaller hangover from the Palaeogene.

The planktonics are polyphyletic

Several phyletic uncertainties notwithstanding, there is a sense of unity and closure about the Globigerinida, configured as this succession of radiations punctuated by Danian and Oligocene interregna. They look like a coherent group, low in diversity and strong in allochronous convergence due to iterative evolution. Perturbing this neat scenario, Pawlowski (2000) summarized studies of molecular systematics as suggesting the following hypotheses. The Globigerinida are polyphyletic. Molecular comparisons confirm the morphological and palaeontological separateness of the three clades Globigerinidae, Globorotaliidae and Candeinidae. But the three groups do not cluster together in the SSU rDNA tree. *Globigerinita glutinata*, the single studied representative of the Candeinidae, is 'unambiguously' placed far from the other planktonics and within the benthic Rotaliida. The globorotaliids branch *either* as a sister group of the globigerinids *or* within the Rotaliida, but the latter hypothesis is favoured by several homologous regions in the segments shared by the globorotaliids and rotaliids but not by the globigerinids. Moreover, Darling *et al.* (1999) found that the only two globorotaliids in their samples, *Neogloboquadrina dutertrei* and *Globorotalia menardii*, do not cluster together (as do the globigerinids) but branch off deeply within the benthic group and separately from each other – providing 'conclusive support' for the polyphyletic origins of the planktonics.

Planktonic foraminiferal species become species clusters, deeply cleaved in geological time

A second outcome of molecular systematics is the likelihood that marine diversity may have been underestimated by an order of magnitude (Knowlton, 1993, 2000, with references). The 'new systematics' (Huxley, 1940) as an integral component of the synthetic theory of evolution emphasized the polytypic species and excessive splitting was condemned as ultra-typological. The present situation and advancing perception are quite the reverse – that excessive lumping characterizes the present systematics situation in many groups of marine organisms. In the case of the planktonic foraminifera it may be that most of the 40–50 extant species are actually clusters of cryptic or sibling species. This is one pungent response to a persistent question – how come such a seemingly successful pelagic group, tracing its ancestry(ies) back into the Mesozoic, never attain the global diversities of other plankton?

On combined molecular, ecological, and morphological evidence *Globorotalia truncatulinoides* is found to be a complex of four genetic species adapted to particular oceanic conditions, two in the subtropics, one in the Subantarctic Convergence and one in Antarctic waters (de Vargas *et al.*, 2001). *Orbulina universa* comprises three cryptic species, also distributed according to oceanic provinces and particularly to chlorophyll *a* concentration at the sea surface (de Vargas *et al.*, 1999), or even four species (Darling *et al.*, 1999). *Globigerinella siphonifera* comprises five types in at least two sibling species, also distinguished by isotopic signature, shell porosity, and photosymbiotic species (Darling *et al.*, 1997, 1999; Huber *et al.*, 1997). *Globigerinoides ruber* split into two lineages as long ago as 22 Ma, with subsplits of each 11–6 Ma and two extant crown groups, *G. ruber* in three genotypes in one and two *G. conglobatus* plus a fourth *G. ruber* in the other (Darling *et al.*, 1999). Not only is there a profusion of cryptic groups, probably species, but they have remained that way for millions of years, contradicting the notion that sibling species imply recent separation. Indeed, the four lineages crowned respectively by the spinose globigerinids *Globigerinella siphonifera*, *Orbulina* plus *Globigerinoides sacculifer*, *Globigerinoides ruber-conglobatus*, and *Globigerina bulloides*, mutually diverged in the relatively narrow window in the late Oligocene, ∼30–27 Ma (Darling *et al.*, 1999, Fig. 2).

Darling *et al.* (2000) analysed three high-latitude species – *Neogloboquadrina pachyderma, Turborotalita quinqueloba, Globigerina bulloides* – finding that each consists of at least 3–5 genetically distinct variants that might be cryptic species. The three species groups are bipolar and disjunct, not ranging latitudinally through the tropics (see Figure 4.6) and some of the genetic entities are known only in the Arctic or the Antarctic. However, a few of each cluster reside in both hemispheres and Norris and de Vargas (2000) opined that there has been genetic exchange between the poles at some time in the past 200 000 years.

> *Biogeography, ecology and speciation – 'toward a new view of the planktonic foraminiferal past and present diversity'* (de Vargas *et al.*, 2002).

The species of planktonic foraminifera have huge populations with huge geographic ranges and experiencing, as it now appears, huge gene flow. Two implications of this work are (i) that the speciations are adaptive, not some stochastic happening as promoted by the anti-natural–selection strand of evolutionary theory, and (ii) that some apparent examples of anagenetic gradualism will turn out to be species successions (see below). De Vargas *et al.* (2002) studied *Globigerinella siphonifera* by broad geographic sampling. Not only did they find four strictly homogeneous and different genotypes – the four (types I–IV) have four overlapping but largely distinct ecologies: I, oligotrophic, preferring shallow waters; II, cosmopolitan and may be adapted to the deep chlorophyll maximum; III, preferring highly productive waters (upwelling and cold); and IV, mesotrophic. Thus we have experienced a nice progression in our knowledge of *Globigerinella siphonifera* – from seeing two morphogroups in the 1960s, to recognizing two different chrysophycophyte endosymbiotic algae (and attributing them to two separate hosts) in the 1980s, to perceiving two groups on physiological, morphological, and genetic criteria in the 1990s, to increasing that number to four groups with new ecological dimensions.

There are now some seven living planktonics genetically analysed and displaying three or four different genotypes. As to three of them (*Orbulina universa, Globorotalia truncatulinoides, Globigerinella siphonifera*), De Vargas *et al.* (2002) argued fourfold reasons for identifying these genotypes as species:

i. The groups are tight, highly homogeneous genotypes arising through a process of concerted evolution.

ii. There is a considerable genetic distance between the four genotypes of *Globigerinella siphonifera* comparable to distances among organisms clearly divergent for millions of years.

iii. The genetic analysis is consistent with differences in life span, growth, photosynthetic pigments, symbionts, shell chemistry, porosity, coiling ratios, and test form detected morphometrically.

iv. They are adapted to different niches – genetic isolation seems to have been accompanied or followed by adapting to specific hydrological conditions ('ecogenotypes').

De Vargas *et al.* (2002) observed that the *Globigerinella* and *Orbulina* lineages produced similar patterns of adaptive radiation, implying similar responses to oceanic productivity and stability, independently and at different times. The next step is to address the most urgent problem – how to distinguish ecogenotypes in the fossil record long after the direct evidence has vanished. This will require identifying morphological and chemical characters through three ways: more combined genetic and morphological study; extensive geographic sampling; and a tight relationship of sampling to ocean structure and chemistry.

Speciation in biostratigraphically important lineages

Speciation?

The planktonic foraminifera have enjoyed a resurgence in palaeobiology in recent years. Some of the preceding discussion implied that the delay in that happening was due to extreme splitting and ultra-typological or essentialist thinking in 'applied' palaeontology, but there were other factors. Evolutionary biology has been dominated by terrestrial metaphytes and metazoans, more comprehensible as to their functional including reproductive biology than marine zoo- and phytoprotists which are capable of producing complex and beautiful shells but masking their adaptive significance (if any). At any rate, the population genetics and reproductive mechanisms are very different from most metazoans. The terrestrial environment offers spectacular variations in the physical environment and geography, also in barriers separating populations. Marine populations and more uniform marine environments are very different from the terrestrial situations utilized to develop modern notions of species and speciation (Mayr, 1942). Mayr noted that the study of geographic variation in marine animals was made difficult by their pronounced phenotypical reaction to water conditions, and their apparently less variation in space may have been in part a function of inadequate knowledge. However, the so-called cosmopolitan species were an exception to the prevalence of geographic variation; another distinctive characteristic is the existence of so many bipolar species. Speciation in marine animals moves 'at a snail's pace' compared to terrestrial.

Mayr and Ashlock (1991) listed several plausible mechanisms of population polytypy and breakup leading to speciation. (i) In allopatric or geographic speciation, spatially isolated populations become reproductively isolated either by splitting the range (dichopatric; popularly known as vicariance) or by budding and rapid differentiation of very small peripheral isolates (peripatric). (ii) In (contentious) sympatric speciation, contrasts develop between local populations adapting to contrasting habitats, perhaps by becoming host-specific plant-feeders and parasites. (iii) In (doubtful) parapatric speciation, isolating mechanisms might build up across a cline that has followed an environmental gradient. (iv) In speciation through time, a species (which is also a phyletic lineage) might change genetically so that allochronic populations belong to different species. Mayr and Ashlock clearly preferred the allopatric situations for most animal speciations.

Equally clearly, biogeography is a crucial parameter. How will models developed in the terrestrial realm export to the marine realm? The marine areas are huge and monotonous in contrast to the terrestrial and the populations of organisms such as skeletonized microplankton are characterized by very large populations – estimates of protozoans and animals produce numbers like $\sim 10^{13-16}$ individuals (Lazarus et al., 1995, gave a rather conservative estimate of 10^{13} individuals) – and very large areas of distribution, for which it is difficult to imagine isolates remaining distinct for long enough to differentiate and speciate peripatrically (McGowan, 1986; Lazarus, 1983). Planktonic foraminiferal biogeography corresponds reasonably well with the properties and confugurations of watermasses but is not constrained by them – for there are nine provinces covering eighteen major watermasses. Gene flow is very strong, may follow the global circulation pattern at the ocean surface (Darling et al., 1999), keeps disjunct populations at higher bipolar latitudes in frequent contact with each other (Darling et al., 2000), perhaps by using upwelling cells as stepping stones (Norris and De Vargas, 2000), a marine analogy of island-hopping – and yet genetic differentiation and most probably speciation are not less exuberant but more exuberant than all this might imply.

Norris (2000) assessed recent studies on planktonic foraminiferal speciation. Recent genetic evidence of very high gene flow at global scales makes its shut-off during isolation unlikely in allopatry or vicariance employing such isolating (dispersal-limiting) mechanisms as strengthened oceanographic fronts, sea level changes, or tectonic barriers. Scenarios such as the Tasman Front (SW Pacific Ocean) becoming a genetic barrier between Pliocene populations of *Globoconella* (Wei and Kennett, 1988) are plausible but not conclusive. The major tectono-physical changes, such as the opening and closing of oceanic gateways, were not simple isolating mechanisms or barriers to dispersal and promoters of endemism, as is believed, reasonably enough, but influence plankton evolution by changing the structure of the pelagic environment. Parapatry, such as depth parapatry, and sympatry, such as seasonal sympatry where reproduction might be cued by different environmental signals, are the more likely possibilities: Norris favoured such mechanisms for achieving speciation in the face of sustained, strong gene flow. However, there is a shortlived (10^{4-5} years), peculiar fauna of acarininids and morozovellids associated very precisely with the latest Paleocene thermal maximum. Kelly *et al.* (1998) presented evidence for two scenarios each invoking steepening of clinal gradients during extreme oligotrophy – peripheral isolates and peripatry; and extreme ecophenotypic variants arrayed along an intensified ecological gradient.

Cladistics or stratophenetics?

Cladistics aims to identify holophyletic groups (clades) comprising all the descendants of a common ancestor. In a dichotomy, two holophyletic groups descended from a common ancestor are sister groups. These objectives of cladistics apply to nested taxa up and down the taxonomic hierarchy, and correctly reconstructing the succession of cladogenetic branching points is crucial to the entire exercise. *Defined* as holophyletic, the clade is *diagnosed* by new evolutionary features or characters, synapomorphies. Synapomorphies are homologous shared characters, inferred to be present in, and only in, the nearest common ancestor.

There are two points particularly pertinent to biostratigraphy (e.g. Norell and Novacek, 1992; Norell, 1992; Padian *et al.*, 1994; Smith, 1994; Wagner, 1995, 2000; Sereno, 1997). First, there is a succession of cladogenetic branching points, 'out there' in the real world, awaiting discovery cladistically in the taxonomic data. Likewise, there is a succession of first appearances to be discovered biostratigraphically in the fossil-stratal record. In an excellent fossil record and with an accurate phylogeny, the clade rank (branching order along the spine of a pectinate cladogram) and the age rank (order of first appearances) should correspond closely. In a perfect world, they would be completely congruent. Thus arose the notion of 'missing' ranges and lineages. Since sister taxa must share a common temporal origin, the temporal range of the younger taxon must extend backwards in time to match the earliest record of the older sister taxon. Again, an inferred ancestor–descendant relationship will show up gaps in the fossil record. In each case the missing range is a ghost lineage (Norell, 1992) or range extension (Smith, 1994) and the missing taxon is a ghost taxon (Fig. 4.11). The sum of the durations of missing ranges is the 'stratigraphic debt'. The second pertinent point is that it is vital that only monophyletic taxa be considered – for paraphyletic groups will 'confound biostratigraphic refinement' (Padian *et al.*, 1994).

Comparing age rank with clade rank should expose gaps in the recovered fossil record. Applying these principles to the dinosaurs seems to expose the absence of tangible – i.e. fossil – evidence for numerous lineages inferred as ancestral to higher taxa. Gaps in the fossil record of lineages can amount to millions or tens of millions of years (Sereno, 1997, 1999). The balance of Norell's (1992) discussion was that cladistic analysis could 'correct' the observed stratigraphic ranges of individual taxa to conform with predictions of phylogeny, more than vice versa. Norell asserted, further, that phylogenetic corrections are required of even the best fossil records: even the best stratigraphic records underestimate the ages of clades. Others (Smith, 1994; Wagner, 1995) are more inclined to see the biostratigraphic/cladistic challenge as flowing in both directions, and 'stratocladistics' (Fisher, 1994) actively assesses cladistic inconsistencies vis-à-vis stratigraphic inconsistencies.

The hope that inferred branching order take preference over observed appearance in the fossil record (Schaeffer *et al.*, 1972; Norell, 1992) is not the first such strange inversion. Gould (2002) devoted considerable attention to the late-nineteenth-century ideas of evolution such as phyletic life cycles (and the 'biogenetic law' that ontogeny recapitulated phylogeny) and innate tendencies to progressive development – such ideas were known as orthogenesis and associated most strongly with the American palaeontologist Alpheus Hyatt. Hyatt elevated stage of development in a lineage above stratigraphic succession, even in relatively dense samplings such as of the fossil molluscs. He was not the only palaeontologist of the times to infer stratigraphic order from presumed phyletic 'stage' taken from ontogenetic recapitulation of ancestral stages. 'Consider the immense confidence that a scientist must be willing to invest in the validity of a chosen surrogate to substitute any other criterion for the eminently available (and obviously meaningful) stratigraphic order of time as the measuring rod for vertical position in phyletic charts' (Gould, 2002, p. 376) – a confidence that Gould himself observed in modern cladists.

There is a large literature on phylogeny and the fossil record (e.g. Wagner, 2000) including the estimation of taxic ranges from the fossil record (e.g. Marshall, 1997). It is probably fair to say that much of the discussion of cladistic analysis in the fossil record is based on taxa with fair-to-mediocre fossil

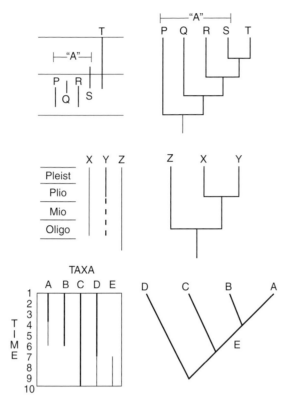

Figure 4.11 Fossil ranges and biostratigraphy in a cladistic light. *Upper* pair of diagrams as explained by their authors (Padian *et al.*, 1994): How paraphyletic groups confound biostratigraphic refinement. (*Left*) The stratigraphic ranges of the individual taxa in A are not congruent: P, Q and R span various ranges in time period 1, while S and T are found in both periods 1 and 2. Thus, the range of taxon 'A' (P + Q + R + S) is hardly distinguishable from that of T, because both 'groups' are found in both time periods. (*Right*) However, when a cladistic approach is taken, S is the sister taxon of T, which suggests that T has been traditionally included in 'A' on the basis of primitive features, not synapomorphies (shared derived features). In fact, no features characterize group 'A' that are not shared by T. 'A' therefore has an artificially extended range. *Second* pair of digrams (Padian *et al.*, 1994): X and Y are sister taxa and Z is their outgroup. The fossil record of X begins well before Y's, implying that the latter should be extended down. *Third* pair of diagrams (Norell, 1992): A, fossil record of four monophyletic taxa A, B, C, D (arbitrary time units and solid bars); B, phylogeny of same. B, if well founded, 'corrects' the ranges in A by claiming two ghost lineages (thin extensions to ranges A and D) and a ghost taxon (E). With permission.

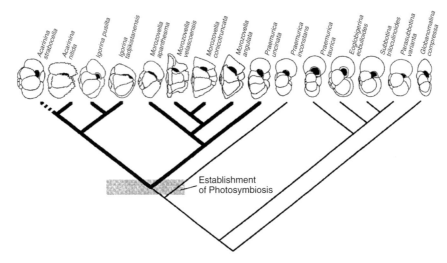

Figure 4.12 Paleocene trochospiral planktonic foraminifera: cladogram illustrating the suggested phylogenetic relationships (Norris, 1996, Fig. 3; Berggren and Norris, 1997). The shaded area indicates the position, inferred isotopically according to the criteria in Figure 4.10, at which photosymbiosis becomes an important ecological strategy in the early Palaeogene radiation. Heavy lines: species identified as possessing that strategy (no data for *A. strabocella*). Berggren and Norris noted that no isotopic or stratigraphic data were used in constructing this cladogram. With permission.

records – generous chunks of lineages missing and taxon ranges unreliable. Marine protists have a comparatively excellent record which has received little cladistic attention (d'Hondt, 1991; MacLeod, 1993, 2001; Norris, 1996; Berggren and Norris, 1997 (Fig. 4.12)). Whilst not suggesting that correlation and age determination would be improved cladistically but addressing phylogeny and relationship, Padian *et al.* (1994) admonished students of the protists in these stern terms:

> In protistan paleontology, stratigraphy and morphology provide the framework on which most systematic and evolutionary studies are done. This is not surprising given the relatively fine resolution that is attainable within samples. However, a strictly stratigraphic approach can lead to circular arguments regarding relationships. For example, the appearance of distinct morphological sequences through time, combined with the presence of possible morphological intergrades, is often interpreted as an evolutionary series connecting ancestral and derived taxa. How characters change is also determined from these sequences, and independent, non-stratigraphic evidence is wanting. It is neither sufficient nor robust to: (a) recognize a relationship between

two morphologically similar taxa based on their sequence in the geological column, (b) describe the character transformations based on this sequence, and (c) then construct an evolutionary scenario to explain the change. While the protist fossil record is the best available, its resolution does not abrogate the need for independent assessment of the relationships among protist taxa.

The relatively high quality of the marine-protist record hardly encourages that kind of analysis, but it does encourage the detailed tracing of similar morphologies through successive stratigraphic levels, which is stratophenetics (Gingerich, 1979, 1990; Wei, 1994; Pearson, 1998b; see Figure 4.13). Stratophenetics has two core precepts: (i) morphological similarity implies phylogenetic propinquity, and (ii) stratigraphic records are sufficiently complete and continuous to permit the reconstruction of phyletic transition and cladogenesis. There are three steps in the stratophenetic reconstruction of phylogenetic trees:

i.   Within-sample organization and recognition of phena, groups of specimens morphologically similar, within a more or less continuous range of variation, each group distinct from the others in the sample;

ii.   stratigraphic organization which is the usual stratigraphic procedure of ordering the local succession by superposition and making correlations independently of the lineages under study;

iii. stratophenetic linking into time series. These procedures can be based both in qualitative but testable perception and quantitative assessment of similarity.

To illustrate variation in degree of preservation of lineages in different groups of well-studied fossils, compare and contrast Figures 4.13 and 4.14. There are gaps, disagreements and problems in this stratophenetic reconstruction of the phylogeny of early Palaeogene planktonic foraminifera, but there is no invoking of any equivalent of the cladists' ghost lineages. In contrast, the well-collected (in vertebrate-palaeontological terms) and much mulled-over record of horse evolution seems to have numerous and significant gaps.

Stratophenetics in micropalaeontology

For all of the six to seven decades of planktonic foraminiferal studies, morphospecies have been linked phenotypically and biochronologically into theories of pattern, i.e. evolutionary-genealogical trees or bushes (Glaessner, 1937; Bolli, 1957a; Berggren, 1968, 1969b; McGowran, 1968a; Blow, 1979; Kennett and Srinivasan, 1983; Pearson, 1998c; *inter alia*). These endeavours were not quantified (Hoffmann and Reif, 1988) but were stratophenetic, even though

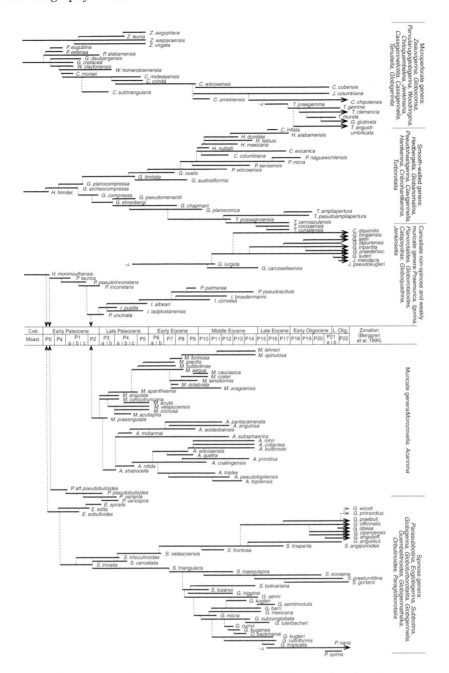

Figure 4.13 Phylogeny of Palaeogene planktonic foraminifera (from Pearson, 1998c, with permission), compiled stratophenetically after six decades of intensive, biostratigraphically and palaeoceanographically driven search, collection and analysis. There are three uncertainties (question marks) but no ghost lineages. (Which is not to imply that there are no interesting and important problems remaining; the Late Paleocene shape of the smooth-walled clade is not congruent with Figure 4.15, for example.)

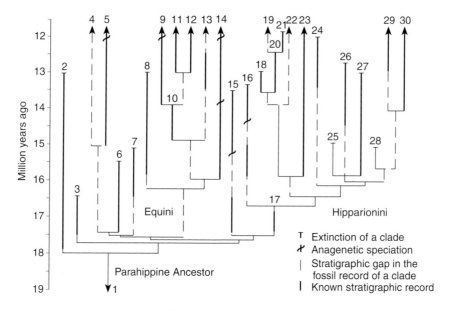

Figure 4.14 Phylogeny of horses: intra-Miocene branching giving rise to the clades of the Equini and Hipparionini (MacFadden and Hulburt, 1988, Fig. 2; Hulburt, 1993; Gould, 1996, with permission). A classical example of evolution and the fossil record since the early 1870s (Huxley, Kovalevsky, Marsh), that record, cladistically treated, seems to be very incomplete with gaps of the order of millions of years. Contrast with the stratophenetic reconstruction of a much more complete fossil record in Figure 4.13.

numerical rigour was intrinsic to Gingerich's (1979) original use of the term. In the past two decades there have been several quantified enquiries informed by three broad kinds of objective: (i) the shape of evolutionary radiations, patterns of extinction, etc.; (ii) lineage studies in such dialectics as punctuated versus gradual transformation and cladogenetic versus anagenetic speciation; (iii) the influence of discoverable environmental change, itself emerging as strongly steplike global palaeoceanographic and environmental change.

Paleocene phena in the Globanomalina *clade*

In Figures 4.15 and 4.16, *Globanomalina* samples are shown as boxed sketches, the contents of each box comprising one phenon – a phenotypically continuously varying sample – and the phylogeny is suggested stratophenetically. In the phenon labelled *ehrenbergi*, we see specimens variously like the older *compressa*, clearly foreshadowing *pseudomenardii* (but *not* including that morphospecies), and identical with *chapmani*. Thus this cluster of co-occurring specimens '*ehrenbergi*' has a broader morpho-range – more morphotypes – than do any of the other three. The succeeding *chapmani* and *pseudomenardii* not only

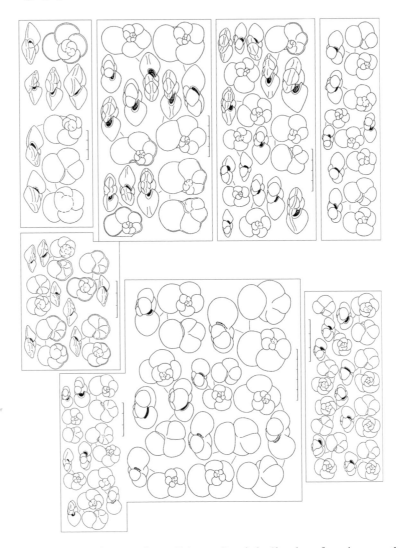

Figure 4.15 Paleocene phena: *Globanomalina* clade. Sketches of specimens gathered to show variation within samples (McGowran, unpublished). For ranges and reconstruction of relationships, see Figure 4.16. Clockwise from middle left, boxes contain *Globanomalina pseudoscitula*, *G. pseudomenardii*, *G. chapmani*, *G. ehrenbergi*, *G. compressa* (upper right), *G. imitata*, and *G. ovalis*. Scale divisions, 0.1 mm.

show less variation but there is no doubt about the assignment of every specimen to one or the other. There was cladogenesis there, with a phenotypically cleancut speciation to *pseudomenardii* and a gradual pseudospeciation to *chapmani* by the survival of that morphotype only.

The other branch of the Paleocene *Globanomalina* clade includes the homogeneous *imitata* succeeded by the more variable *simplex*, the smaller specimens

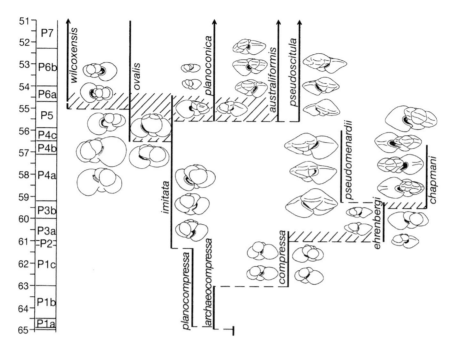

Figure 4.16 Paleocene radiation and stratophenetic tree: *Globanomalina* clade in the planktonic foraminifera. In terms of morphological convergence *G. australiformis* and *pseudoscitula* replace (not displace) *ehrenbergi* and *pseudomenardii* respectively. Shading shows transitions. It is noteworthy that the turnover in this clade is coeval with the evolutionary events associated with the Late Paleocene hyperthermal – in tropical planktonic foraminifera (Kelly *et al.*, 1998) and in neritic Tethyan, larger benthics (Orue-Etxebarria *et al.*, 2001). (See Fig. 7.9.) (McGowran, unpublished.)

at the margins of which population anticipate *wilcoxensis* and *planoconica* respectively. From latest Paleocene to earliest Eocene we have an array of morphospecies. Timing of the expansion of intergradation (shaded) vis-à-vis the extinction of *chapmani* and *pseudomenardii* does not prove but surely does suggest cause and response.

Middle Miocene Orbulina *and* Fohsella *bioseries*

By Neogene times strengthening climatic gradients in a cooling global ocean were forcing stronger faunal differentiations and planktonic foraminiferal biostratigraphy reflected this in the formalization of three parallel zonations for the Miocene (Fig. 4.17) – tropical–subtropical, temperate (transitional), subantarctic–antarctic (Berggren *et al.*, 1995a). A strong warming reversal in the early-to-middle Miocene meant, though, that the famous *Orbulina* bioseries could be used in the temperate as well as the tropical successions, the three key first appearances of *Praeorbulina sicana*, *Praeorbulina glomerosa* and

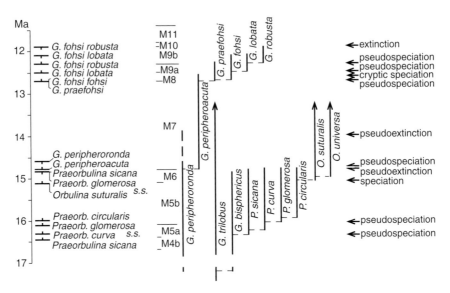

Figure 4.17 Middle Miocene bioseries: *Orbulina* (Pearson *et al.*, 1997) and *Fohsella* (Hodell and Vayavananda, 1993; Norris *et al.*, 1996). Numerical timescale, datums (first and last appearances) and M-zones: after Berggren *et al.* (1995a). For the different kinds of speciation and extinction, see Figure 4.38. Cryptic speciation inferred from a sudden shift in ecological strategy, itself inferred isotopically. It is cryptic because it is not seen or marked in the changes through time in the fossil phenotype (i.e., the shell).

Orbulina suturalis being logged *exactly* at the same times (that is, at the supra-Milankovich dates 16.4, 16.1, 15.1 Ma, respectively) in the different provinces together spanning tens of degrees' latitude in each hemisphere. As well as that, these bioevents occur in homotaxial succession in central Paratethys in the north and in the shallow to very shallow seas in southern Australia in the south, up to 600 km over the continent from the shelf edge. There is no whiff of diachrony, no hint of peripheral isolates as a source for peripatric speciation, in this successional pattern throughout a huge sector of the global ocean. How come? Haynes (1981) suggested that different populations might segregate into different depth zones, a notion which Lazarus (1983) christened 'depth-parapatry'.

It is clear that the *Globigerinoides trilobus* lineage broadened its morpho-range by adding the morphotype *bisphericus*, but the next broadening was much more pronounced with the rapid addition of four *Praeorbulina* morphotypes to the morphological continuum. The last additions were the orbulinids *suturalis* and *universa*. This broadening, accommodating six then eight named morphotypes in a continuum, existed throughout the biogeographic range for ∼1.6 myr (16.4–14.8 Ma), after which the continuum was gutted, leaving each end of the morpho-range and henceforth two lineages. From this, we infer cladogenesis terminating anagenesis. The two lineages have similar dinoflagellate photosymbionts if not the same species of *Gymnodium*, but *trilobus* has a slightly higher limit than *universa* to the lower end of its range of temperature tolerance. By comparison with the spatial dimension, the vertical dimension of the planktonic habitat displays steeper gradients which can be tracked by stable-isotopic signals, offering a powerful outgroup to the distribution and change in phenotype (viz., modern and fossil skeletons). Pearson *et al.* (1997) tested this possibility of depth parapatry for the *Orbulina* bioseries. The controls were $\delta^{18}0$ and $\delta^{13}C$ plots for shallow-dwelling *Globigerinoides ruber* and deep-dwelling *Globoquadrina venezuelana*. On the $\delta^{18}0$ evidence there was no discernible change in depth for any morpho-type in this *Globigerinoides–Orbulina* clade: all evolution occurred in shallow mixed-layer habitats. Subtle ecological shifts remain just possible but did not show up in any consistent offsets. Likewise, $\delta^{13}C$ patterns showed no change and no offsets, which tends to confirm the modern sharing of the same or closely related symbionts by *Globigerinoides trilobus* and *Orbulina universa*. Thus, none of the isotopic evidence confirms any discernible change in depth of symbiotic association, therefore of depth parapatry, just as no biogeographic or morphological evidence supports peripheral isolation and recolonization – which leaves us with some kind of sympatry, perhaps a seasonal staggering of reproductive rhythms, itself mysterious: 'It is impossible not to be impressed by the profound morphological changes which have occured in the evolution of *Orbulina*, for which there is still no satisfactory explanation' (Pearson *et al.*, 1997).

Meanwhile, the *Globorotalia (Fohsella)* lineage made few and small phenotypic changes during its first ~9 myr before a burst (< 1 myr) in the tropics/subtropics preceding its extinction (Fig. 4.17). Again, morphotypes were plucked out of a morphological continuum and chronocline to define zones. Using the $\delta^{18}0$ record of the shallow-water-dwelling *Globoquadrina altipira* as control, Hodell and Vayavananda (1993) could demonstrate a pronounced and abrupt isotopic shift in the *Fohsella* lineage, from consistently lighter than coeval *D. altispira* to consistently heavier. The crossover occurred just before the emergence of *G. robusta* and was interpreted as an increase in the depth habitat of *Fohsella* from near-surface mixed waters to waters at intermediate depths near the thermocline. Hodell and Vayavananda found no consistent isotopic differences between coexisting morphotypes either before or after the abrupt shift. However, the most rapid changes in morphology occurred within a ~0.3 myr interval coinciding with the isotopic shift. Norris *et al.* (1996) scrutinized this transformation further, exposing two seemingly independent changes. On the one hand, morphometric analysis confirmed that shell shape, the silhouette in eigenshape analysis together with the development of a marginal keel, records an almost unbroken anagenetic trend, confirming qualitative observations of an additive chronocline from *peripheroacuta* to *robusta*. On the other hand, the lineage underwent a sharp change in ecology, from both growth and reproduction (and gametogenic calcification) in the mixed surface layer to reproduction (and gametogenic calcification) near the thermocline. Norris *et al.* argued that this innovation in reproductive ecology is strong evidence for a biospeciation, a cladogenetic budding followed by the extinction of the ancestor. Since there was no statistical relationship between reproductive ecology and either shell outline or the evolution of the marginal keel, the speciation was cryptic: we do not see it in the fossilized phenotype (i.e. the skeleton).

Even so, the succession of named morphotypes remains empirically robust in biozonation!

Late Neogene Globorotalia *bioseries*

To Lazarus *et al.* (1995) recent studies of speciation using marine microfossils have not conclusively distinguished between sympatric and parapatric modes of speciation. They made a case for sympatric cladogenesis giving rise to the late Neogene index species *Globorotalia truncatulinoides* from *Globorotalia crassaformis*. They found a pattern in two DSDP holes in the southwest Pacific (Tasman Sea, more than 1000 km apart) that indicated simultaneous speciation during the continuous geographic co-occurrence of ancestor and descendant forms in a population estimated conservatively at 10^{13} individuals. The patterns were based on morphometric, principle component and discriminant analysis and display 'very gradual' cladogenesis through $\sim 0.5 \pm 0.2$ myr. This event was followed by morphological, apparently anagenetic change, very little in *crassaformis* and significantly more in *truncatulinoides* which, however, could be either true phyletic change or climatically induced shifts in a morphocline. Most of the evolution of *Globorotalia truncatulinoides* took place ~ 2.8–2.5 Ma, implying forcing by global cooling towards late Neogene glaciation. A third biostratigraphically important morphotype, *Globorotalia tosaensis*, seems at no time to be a separate lineage, at most representing an early stage in the emergence of the *G. truncatulinoides* lineage. Such arguments have to be reevaluated against the indications that *Globorotalia truncatulinoides* is a complex of four genetic species (de Vargas *et al.*, 2001). Meanwhile, biocharacters discriminated taxonomically for biostratigraphic purposes are visible enough but do not always diagnose valid species.

The *Globorotalia (Globoconella) puncticulata-inflata* clade of the southwest Pacific has received attention since Malmgren and Kennett (1981) used it to corroborate phyletic gradualism over punctuated equilibrium. Wei (1994) used quantitative stratophenetics and allometric heterochrony and Schneider and Kennett (1996) and Norris *et al.* (1994) comparative isotopic signals.

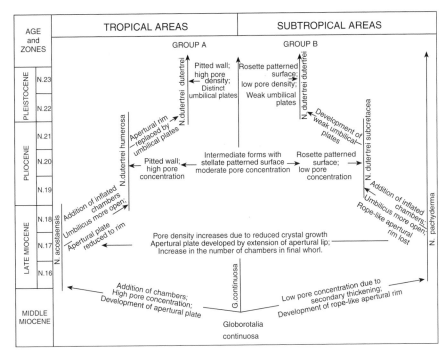

Figure 4.18 Reconstructed evolutionary and phenotypic relationships in the late Neogene plexus *Neogloboquadrina dutertrei* (Srinivasan and Kennett, 1976, Fig. 2, with permission). Two bioseries are recognized, culminating respectively in *N. dutertrei dutertrei* Groups A and B. Phenotypically intermediate forms existed at all times, implying an evolutionary–biogeographic line.

The *Neogloboquadrina dutertrei* plexus displays a skilful interpretation of changes across latitude through the late Neogene (Kennett and Srinivasan, 1976), the morphotypes *dutertrei* and *pachyderma* today inhabiting mutually exclusive provinces (Fig. 4.18). Allopatry also holds within *N. pachyderma* itself on coiling of the test, $> 90\%$ sinistral forms in colder waters ($< 9\,°$C), a narrow mixed zone within the (southern) subpolar province, and $> 50\%$ dextral forms in the subpolar-transitional provinces (~ 9–$18\,°$C) (Fig. 4.19). That these adaptations in turn hold through time was shown when coiling ratios proved to be a useful ecostratigraphic tool in the late Pleistocene, subsequently extended to Late Neogene palaeoceanographic cycles (Fig. 4.20; Bandy, 1972). It now appears that coiling variants are species – in retrospect, probably an easier explanation of the steady consistent responding by morphotypic ratios to environmental shifts than alternative conjectures, such as a strong genetic linking of shell coiling to (unknown) adaptations.

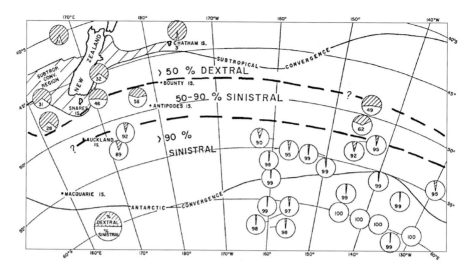

Figure 4.19 Coiling distributions, *Neogloboquadrina pachyderma*, South Pacific (Kennett, 1968).

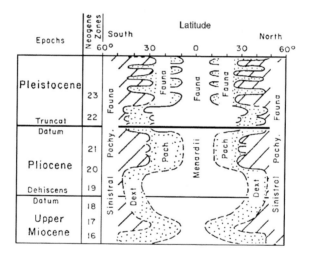

Figure 4.20 Palaeoceanographic cycles, equatorial to polar, generalized through the late Neogene (Bandy, 1972, with permission). Inferred water temperatures: dominantly sinistral *Neogloboquadrina pachyderma* fauna, <9 °C; dominantly dextral *N. pachyderma* assemblage, ~9–18 °C; *Globorotali menardii* group, tropical waters, >18 °C.

Cryptic species: anagenesis or cladogenesis or bifurcation? Gradualism or punctuation?

For two decades the fossil phenotypes of marine protists indicated gradual anagenesis whilst animals were suggesting more punctuated cladogenesis.

Gradual anagenetic trends were shown by Malmgren and Kennett (1981), Arnold (1983), Lohmann and Malmgren (1983), Malmgren *et al.* (1983), Malmgren and Berggren (1987), Hunter *et al.* (1988), and Kucera and Malmgren (1998) with a density of sampling through time rarely approached in metaphyte and metazoan fossil records. Benton and Pearson (2001) were inclined to repeat the generalization that marine invertebrates and vertebrates show punctuated patterns, with periods of rapid speciation alternating with periods of stasis, whereas marine microplankton tend to show more gradual speciation.

In reviewing and defending the Eldredge–Gould theory of punctuated equilibrium (relatively rapid speciation followed by stasis in the species), Gould (2002) was impressed by the robustness of some of these indications of gradualism. He developed two somewhat contrary arguments here. (i) Speciational gradualism is linked to bifurcation and anagenesis whereas punctuated equilibrium and stasis are linked to cladogenesis. The dominant Neogene planktonic foraminiferal clades Globorotaliidae and Globigerinidae (Stanley *et al.*, 1988) display cladogenesis as the overwhelming phylogenetic pattern, far ahead of anagenesis and bifurcation as shown by Wagner and Erwin (1995). This is one of the more spectacular examples of a major shift in our perceptions – from Simpson's (1944) famous estimate of some 10% of evolutionary change by speciation (cladogenesis) and 90% by anagenesis to approximately the reverse. It also argues for punctuated equilibrium as the dominating mode of evolution in Neogene planktonic foraminifera, the rigorously documented examples of gradualism (above) notwithstanding. (ii) Gould revived an earlier speculation (Gould and Eldredge, 1977) that a stronger tendency to gradualism in protists compared to metazoans might reside in the contrast between asexuality and sexuality. Thus foraminiferal 'species' are analogues of the metazoan lineage or clade, *not* the metazoan species. The supposed foraminiferal 'species' is not an entity but a collectivity, and the perceivedly gradualistic change in evolution is actually a series of short steps. However, the typical foraminiferal life cycle is characterized by an alternation of sexual and asexual generations but only sexual reproduction has been observed in the planktonics (e.g. Hemleben *et al.*, 1989; Goldstein, 1999). The speculation seems to be superfluous.

Two recent developments bear strongly upon this topic – inferences of behaviour shift in (the fossil record of) planktonic foraminifera (depth of reproductive cycle, acquisition of photosymbiosis, both from isotopic arguments); and genetic species clusters with subtle, or little, or no shell-phenotypic expression. In some cases of gradualist anagenesis the trend may be non-instantaneous replacement of one species by another, following relatively rapid splitting by reproductive isolation but little phenotypic divergence. Norris (2000) modelled the generation of an apparent gradual anagenetic trend by sampling different proportions of two end-member morphologies through the time series. He suggested that the following studies may be false anagenetic series: *Contusotruncana contusa* in the Maastrichtian (Kucera and Malmgren, 1998); the punctuated anagenetic shift from *Globorotalia plesiotumida* to *G. tumida* in the Early Pliocene (Malmgren *et al.*, 1983, 1984); and *Globorotalia truncatulinoides* in the Pliocene (Lohmann and Malmgren, 1983). Other series seem to be interrupted by rapid behavioural shifts implying cryptic speciation – the *Fohsella* case; and the acquisition of photosymbiosis in Paleocene *Praemurica* leading (in due course) to the *Acarinina–Morozovella* radiation (Kelly *et al.*, 1996; Norris, 1996; Quillévéré *et al.*, 2001) (Fig. 4.12).

Thus, gradualistic, morphotypically expressed anagenesis is being supplanted to some considerable degree in our perceptions by punctuated, cryptic cladogenesis.

Coiling in trochospirals

Although Bolli (1950, 1971) discovered that populations with proportionate coiling (~50% dextral and sinistral) tended to evolve into dextral or sinistral forms, the topic did not flourish outside the use of stable coiling ratios in *Neogloboquadrina pachyderma* as a palaeoceanographic tool, as discussed above, and some ecostratigraphic uses. Things have changed with the demonstration that differently coiled populations of species such as *N. pachyderma* and *Globorotalia truncatulinoides* have discernibly different ecological preferences and the likelihood that they are separate species. Norris and Nishi (2001) found that each of the major radiations from the Cretaceous to the Neogene was founded on clades with proportionate coiling and that biased coiling developed iteratively. They found too, contradicting previous interpretations of environmental control, that coiling patterns are heritable and that bias becomes fixed until the clade expires.

Stage of evolution: the larger foraminifera

Although not all large foraminifera possess photosymbionts and some of the smaller forms do, the symbiotic strategy seems to be the main force driving the repeated evolution of larger benthic groups (Hottinger, 1982, 1983; Hallock, 1985, 1987; Hallock *et al.*, 1991; Brasier, 1995). Some of the main biological characteristics of the larger foraminifera, in contrast to asymbiotic smaller benthics, are: (i) long life cycles, up to two years in *Marginopora*, implying an extensive ontogeny constructing complex shell structures and attaining exceptionally large body sizes with maximized surface area:volume ratios; (ii) refined reproduction cycles, often including protection of juveniles; relatively suppressed sexual phase; dimorphic form with larger megalospheric tests; (iii) symbiosis protected by shells; thinned upper walls, numerous spaces formed by structural complexity; (iv) protoplasmic differentiation in canaliferous foraminifera; (v) abundance and diversity in regions of low primary productivity where these organisms and other oligotrophic calcifiers generate carbonates in warm-neritic environments. Typically they live in warm and well-lit waters on continental shelves or volcanic pedestals at low latitudes, with a tendency toward very large and close-packed populations, often monospecific, which become geologically rock-forming. An additional very important fact is that the appellation 'large', meaning large, fusiform or discoidal, and internally complex, is accorded to a structural, functional and evolutionary grade of protozoan: they constitute a highly polyphyletic nontaxonomic group displaying between the late Palaeozoic and the Recent some of the most impressive examples of allochronous convergence that one could hope to see. Hottinger has interpreted their evolution as repeated responses to the pressures of K-selection (Fig. 4.21). With a generous array of morphocharacters recording change through time, the larger foraminifera have long been attractive candidates for phylogenetic and biostratigraphic studies; the down-side has been that their anatomical analysis and recognition in hard limestones is a difficult discipline in which few toilers have been expert at any one time. Fossilized assemblages of larger foraminifera frequently do not contain planktonics, and vice versa, making for rare cross-links (e.g., Adams, 1970, 1984). Luterbacher (1998) reminded us of another contrast: 'The two types of zonations are based on different approaches. The planktic biozones are mainly based on – possibly somewhat heterochronous … – first and last appearances of index taxa and have therefore an "event-stratigraphic" connotation. Larger foraminiferal zones are ideally based on successions of biometric populations within phylogenetic lineages; species are essentially morphometric units … ' Abrupt changes and telescoping are clues to reduced rates of accumulation or non-deposition (Fig. 4.22).

Tan Sin Hok (1932, 1939a, b, c) was a particularly eloquent early advocate of phylogenetic analysis of the larger foraminifera. Not only was that strategy much superior to 'typological' morphologic analysis, but 'it will be far more fascinating by its revealing some directing laws of organic life'. Thus: 'The author's opinion concerning the biostratigraphic value of phylomorphogenetic researches is that *they mean a methodical search for index fossils, and that they also enable to establish a series of time markers in which the past geologic time is recorded in a gapless manner*, as the consecutive terms of the same bioseries represent a rational and continuous sequence. Such a palaeontological chronometer is of utmost importance, as by means of it other forms can be dated independently of the stratigraphical observation' (1939b; emphasis in original).

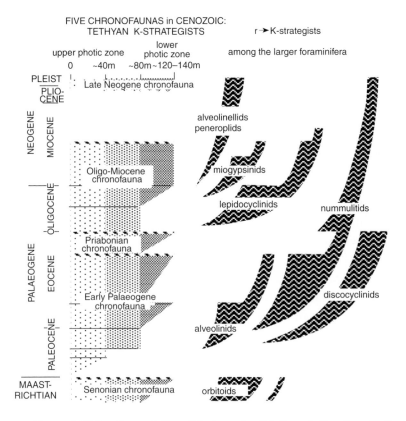

Figure 4.21 Iterative evolution of the larger benthic foraminifera, from McGowran and Li (2000) based heavily on the work of Hottinger (1982, 1997), especially the notion that larger foraminifera iteratively adopted ecological K-strategy in the neritic realm (diversity, size, internal complexity and number of spaces, other photosymbiotic adaptations). *Left*: blocks represent distribution of large foraminifera within the photic zone through time, using generic lists in Hottinger (1997, Table II) where time is divided into sub-epochs, space into three parts of the photic zone. Some lower-photic boxes are half-filled to indicate the presence of only a few genera. Heavy cable pattern: major extinction horizons of K-strategists, terminating the evolution and expansion of chronofaunas (Chapter 6). *Center and right*: trajectories of the major groups independently adopting neritic K-strategy, adapted from Hottinger (1982) with permission.

Tan introduced detailed morphogenetic analysis into micropalaeontology (Glaessner, 1943, 1945). The method was based on the statistical treatment of single characters in adequate samples from successional populations, and the evolutionary stages of each character made up a bioseries. The most famous of Tan's bioseries are the examples of *nepionic acceleration* (Fig. 4.23) – the reduction through time of the post-embryonic stage in ontogeny, and a form of tachygenesis

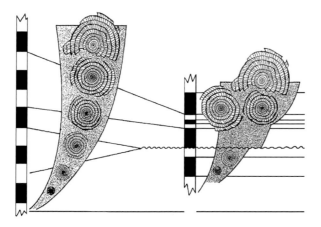

Figure 4.22 Telescoping of an evolutionary lineage in *Nummulites* (Luterbacher, 1998, Fig. 8, with permission). 'Full' lineage at left; telescoping at right with intervals of reduced or no sedimentary accumulation. Biostratigraphic boundaries 'very frequently correspond to these intervals'.

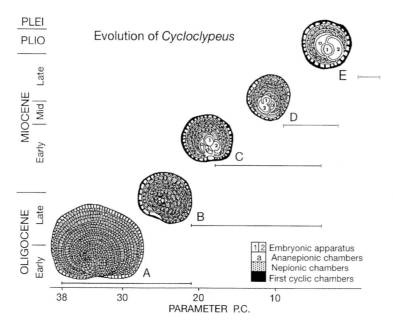

Figure 4.23 Nepionic acceleration in *Cycloclypeus* (Adams, 1983, with permission). The 'heterostegine' coil (stippled; *Heterostegina* was the ancestor) in the primitive Oligocene species is reduced through time and cyclical growth (black) is hastened (only the first annular chamber is shown and most of the shell is omitted from these sketches). Adams observed that the direction but not the rate of evolution was constant. Parameter P.C., number of nepionic chambers with ranges at stratigraphic levels A–E.

(Adams, 1983). Figure 4.23 illustrates nepionic acceleration in *Cycloclypeus*. Tan concluded repeatedly from the study of different and unrelated larger foraminifera that the direction of evolution was constant although the rate was not. He stressed too that classification had to be based on combinations of character-bioseries, and he foresaw that, as Glaessner (1945) put it, 'with an increase in material filling the gaps on which classification largely depends, it becomes increasingly difficult to devise an adequate taxonomy for the species and for their subdivisions and larger groupings. Taxonomy and nomenclature have yet to be adjusted to the new methods and new results of morphogenetic work. Until this is done the old, admittedly inadequate, classification "*per genus et speciem*" will still be, used. And Tan distinguished sharply between the 'typological point of view' in which nepionic bioseries are 'saltatory' as he himself thought in his early work, and the insight that comes from examining large samples.

> Supposing that we have the following three consecutive bioserial stages viz. *a*, *b*, *c*. In chronological sequence, we shall find e.g. populations of the following types, viz. (1) *a*; (2) (*a* + *b*); (3) *b*; (4) (*b* + *c*); (5) *c*. And never e.g. the sequence (1) *a*; (2) *c*; (3) (*a* + *b*). It means that the variation curve apparently shifts in a definite (i.e. rectigrade) sense; the primitive stages become gradually extinct, meanwhile the younger ones appear, and gradually attain greater profusion. (In essentials *Cycloclypeus* shows this type of orthogenesis too [Tan meant: as well as the groups based on *Lepidocyclina* and *Miogypsina*]).
>
> In population (b) stage *b* represents a mutation (in Waagen's [i.e., palaeontological] sense of 'mutation') of stage *a* of population (a). Accordingly a systematic differentiation is accounted for. But in population (*a* + *b*) stage *b* may also be a mere fluctuation, in which case it does not deserve differentiation. As it is impossible to discern *mutation b* from *fluctuation b* in the population (*a* + *b*), one is compelled to draw a boundary between *a* and *b* in population (*a* + *b*) as well; in other words to record the presence of every bioserial stage in every case. In this manner we determine the total time range of a morphologic feature. From these remarks the importance of the investigation of a great quantity of material is obvious.
>
> As a matter of fact coexistences are very tiresome for the practice of Biostratigraphy. It is e.g. not exact to conclude that a locality which has yielded stage *b* must be younger in all cases than another with preceding stage *a*. This is true only if it is sufficiently demonstrated that the respective strata contain these stages *exclusively*, but not if our evidence is based on a few specimens.

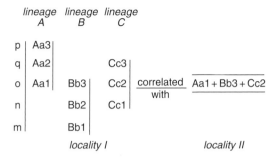

Figure 4.24 Correlation using bioseries (Glaessner, 1945, after Tan Sin Hok). The assemblage from locality II can be matched with a particular level (o) in the succession at locality I.

> Though these coexistences are complications in the biostratigraphical sense, from the phylogenetic point of view they are important, as they prove that two different forms belong exactly to one and the same lineage. Hitherto this has been one of the few means of demonstrating the phylogenetic purity of our material … '

The use of words such as 'orthogenesis' and 'rectigrade' suggests obvious comparisons with Osborn's distillations of the mammal fossil record, as considered below. Tan probably did not believe in some internal drive or predestined outcome even though he assembled abundant evidence for unidirectional trends; nor, however, did he overlook the possibility that the rate of change from one stage to the next in a bioseries could be due to local environmental factors. He tested the nepionic development in *Miogypsina* in southern Europe and the Indo-Pacific region and found that the same stage of evolution was reached at the time of extinction of *Eulepidina*, *Spiroclypeus* and *Miogypsinoides* in the two regions. 'Since it is improbable that evolutionary changes varying in speed under the influence of environment can produce, under different regions, the same combinations of bioserial stages in a number of co-existing species (populations), the actual occurrence of such combinations would strengthen the case for the stratigraphic application of the morphogenetic method' (Glaessner, 1945). Glaessner went on to sketch Tan's method of biostratigraphic correlation by stage of evolution (Fig. 4.24).

Tan's themes were gradual unidirectional evolution in bioseries, as revealed by the study and statistics of adequate samples, and the value of within-lineage changes for correlation and age determination. The challenges were to achieve accurate anatomical reconstructions and to integrate the phylomorphogenesis and biostratigraphy of the larger foraminifera with the rest of Late Phanerozoic stratigraphy. Those challenges are still with us, as outlined in Chapter 7. The

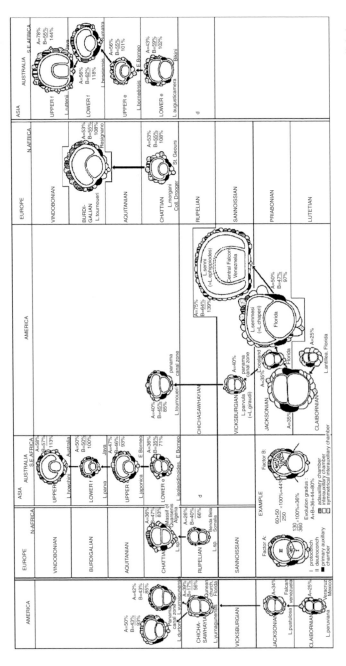

Figure 4.25 'Morphogenetic analysis of bioseries' (van der Vlerk, 1959, with permission from the Geological Society) – parallel changes in the early ontogenetic stages of the orbitoidal genus *Lepidocyclina*. There are two lineages each shown in three biogeographic provinces, late Middle Eocene to Middle Miocene: the X-lineage (three left columns) and Y-lineage (three right columns). There are two quantifiable factors, both measured on equatorial sections: A, the degree to which the second chamber embraces the initial; and B, what percentage of the circumference of the second chamber is taken up by adauxiliary chambers; A + B gives a percentage 'that may be regarded as a grade of evolution' because it increases through time in the different lineages and different provinces.

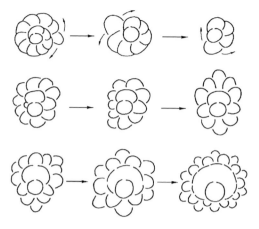

Figure 4.26 Nepionic acceleration in orbitoidal foraminifera (Drooger, 1993, with permission). This is accomplished by, respectively: upper row, shortening of the ancestral spiral; middle row, increasing symmetry; bottom row, increasing the number of accessory chambers.

theme of lineages and correlation is taken further here as examples of stage-of-evolution.

Van der Vlerk (1959) outlined a method of measuring evolutionary change in the embryonic-nepionic ontogenetic stages of *Lepidocyclina*, as revealed in equatorial sections of the test (Fig. 4.25). 'Factor A' shows to what degree the second chamber embraces the initial chamber. 'Factor B' shows what percentage of the circumference of the second chamber is taken up by adauxiliary chambers. Their sum gives a percentage 'that may be regarded as a *grade of evolution*' (emphasis added). Having expressed gloom over the use of the Lyellian percentage method, the use of index fossils, and the assemblages of genera, as in the letter classification (Chapter 7), Van der Vlerk was very cautious about the meaning of the parallel trends in the different provinces. He says of stage-of-evolution: 'The use of bio-series may be more successful. Provisional morphogenetic investigations of the foraminiferal genus *Lepidocyclina* look promising' – a restrained comment, indeed.

Drooger (1963) built on Tan Sin Hok's (1936, 1937) 'remarkable advance' in a major study of the orbitoidal genus *Miogypsina*. Subsequently (1993) Drooger drew together all the available evidence to construct a chart of evolution and correlation in the *Miogypsina* clade between the American, Mediterranean and Indo-Pacific regions (provinces) (Figs. 4.26, 4.27). Also shown here (Fig. 4.28) is Adams's (1983, Fig. 2) version which acknowledges Drooger as a major source. Here too we are in the realm of nepionic acceleration, as can be seen at a glance at the sketches, which show the disappearance of the simple coil of chambers

Miogypsina clade in three provinces

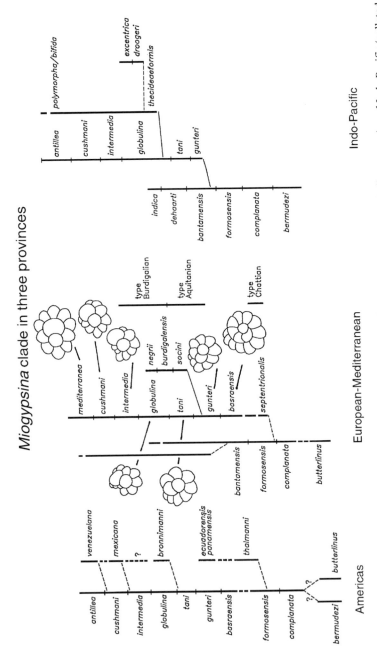

Figure 4.27 The *Miogypsina* clade in three provinces – Americas, western Tethys (European–Mediterranean), and Indo-Pacific (collated from Drooger, 1993, Figs. 56, 61, 64). Sketches were added (Drooger's Figure 59) of embryonic apparatus for seven successive Mediterranean species. There is a 'long-lasting central lineage' in which nepionic acceleration was rapid and the average duration of ten successive morphometrically discriminated species was of the order of one million years per species. Side-lineages developed separately in the three provinces, with permission.

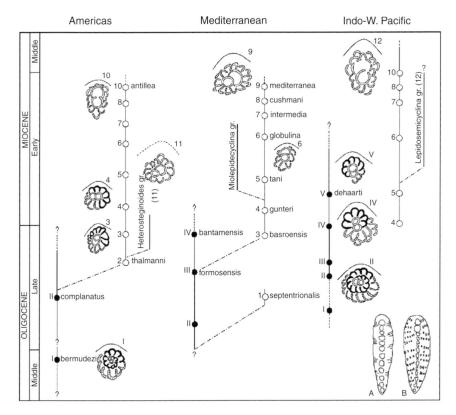

Figure 4.28 Evolution of *Miogypsina* in three provinces (Adams, 1983, with permission). Drawings of embryonic apparatus: protoconch and deuteroconch and deuteroconch, clear; spiral chambers with single aperture, black; primary and secondary spiral chambers with retrovert apertures, stippled; third and fourth spirals, clear. Closed circles: relative positions of *Miogypsinoides* species I to V (lacking lateral chamberlets: vertical section A). Open circles: relative positions of *Miogypsina* species 2 to 12 (possessing lateral chamberlets: vertical section B).

inherited from the ancestral *Pararotalia*. Of the other skeletal characters that must be considered, the most obvious is the development of lateral chamberlets, an adaptation to increase the plots of photosynthesizing symbionts and the means of distinguishing *Miogypsina* from *Miogypsinoides*. Note that *Miogypsina* evolves twice in the Mediterranean region and that *Miogypsinoides* attains successively more advanced and younger stages from west to east. And note too the remarkable parallelism in the successional stages – species? – in *Miogypsina*. Thus each province hosted the mainstream evolution – a 'long-lasting central lineage' – of *Miogypsina* whilst each province produced its own phyletic side branches. Those side-branches notwithstanding, the lineages of larger foraminifera are characterized much more by change within the lineage: anagenesis not cladogenesis.

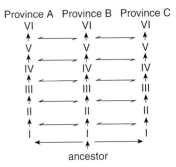

Figure 4.29 Parallel evolution in larger benthic foraminifera or repeated interchange? The polarized alternatives.

Hottinger (1981) distinguished between two kinds of generic character: qualitative or unchanging characters and quantitative or changing characters. The acquisition of lateral chamberlets in *Miogypsina* exemplifies the former and the measurable changes in early ontogeny the latter. Adams (1983) distinguished further between the classical method and the morphometric method of evolutionary and taxonomic discrimination. The classical method 'depends largely on the subjective appreciation of a large number of variable characters', holistic and clarifying similarities between members of different lineages, and more applicable at higher taxonomic levels. The modern morphometric method 'is based upon the mathematical expression of a small group of characters or even of a single character'; it demonstrates bioseries in Tan's sense – unidirectional changes in characters – but tends to obscure other similarities and is applied at lower taxonomic levels where it is the most convincing method of arguing phyletic speciation. Clearly, the two methods are complementary and hierarchical. Adams made the most important point that novelties support the notion of punctuated evolution whilst all the biometric work from Tan Sin Hok's in the 1930s onwards supports the notion of phyletic gradualism.

We come to the question of the mainstream lineages of *Lepidocyclina* and *Miogypsina* in the three provinces. The parallels established or at least suggested by Tan, Drooger, van der Vlerk and others focus on an ancient evolutionary question but with particularly interesting data: does the succession of morphogenetic stages in at least approximate synchroneity (plus the emergence of novelty allochronously) indicate repeated interchange across interprovincial barriers or a form of orthogenesis? Figure 4.29 puts the polarized alternatives schematically. Adams noted that the evolution of the *Cycloclypeus* test as a whole may be said to demonstrate Haeckel's principle of recapitulation, in that the first two growth stages represent the ancestral genera *Operculina* and *Heterostegina*; and that the reduction in the initial spire in *Miogypsina* and in *Cycloclypeus* and the progressive changes seen in *Lepidocyclina* can all be regarded as examples of orthogenesis, in the descriptive sense and perhaps better termed *unidirectional evolution*. Then, wrote Adams,

> if the reality of recapitulation and orthogenesis (unidirectional evolution) is acknowledged, it is easy to understand why some genera are represented by similar species in different provinces while others are not. Dispersal is difficult for all species of large foraminifera, but especially for those spending no part of their life cycle on weeds since they cannot be drifted on floating vegetation. However, it is unnecessary to postulate repeated trans-oceanic crossings if the passage of just one species can produce the same evolutionary result in each region. Unidirectional evolution could ensure the production of similar end members in lineages on either side of major barriers provided they shared a common ancestor. This seems to have occurred both with *Miogypsinoides* and *Lepidocyclina*, although Freudenthal (1972) postulated a separate origin for Tethyan and American species of *Lepidocyclina*. The occurrence of *Miogypsinoides complanatus* (Schlumberger) and *Miogypsina gunteri* Cole [Fig. 4.27] in all three provinces could have ensured the subsequent appearance of similar descendants in each region provided local extinction did not supervene. Regional side branches could, of course, be expected and indeed produced (the *Heterosteginoides* group in the Americas; the *Lepidosemicyclina excentrica* group in the Indo-Pacific; and the *Miolepidocyclina* group in the Mediterranean). Representatives of these subgroups do not seem to have gained access to other provinces, thus further indicating that dispersal was difficult.

The evolution of the alveolinids in the Palaeogene exemplifes the ascendancy of anagenesis over cladogenesis. Hottinger (1981) distinguished as *generic*, characters that are qualitative or unchanging: thus, the genus *Alveolina* has a test of but one kind of structure throughout its spatiotemporal range. Within *Alveolina*, Hottinger recognized two sets of quantitative characteristics: one, unchanging in function or through geological time (shape of shell; spire); the other, changing through time and including especially the size of the proloculus and the index of elongation (the ratio of axial diameter to equatorial diameter). To make sense taxonomically of extensive parallelism within *Alveolina*, one must reconstruct lineages – very closely related species coexisting in the same environment cannot be recognized otherwise. Figure 4.30 illustrates this startling notion. Hottinger has plotted three timeslices through three lineages evolving in parallel, but at

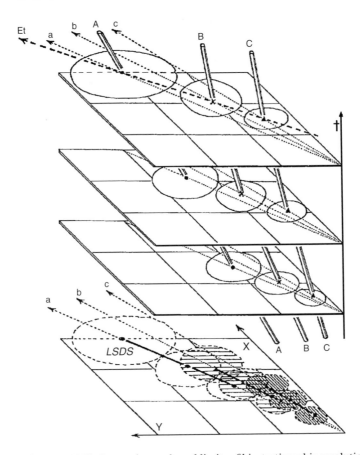

Figure 4.30 Phylogenetic trends and limits of biostratigraphic resolution in larger foraminifera, based especially but not exclusively in the alveolinids (Hottinger, 1981, with permission). Three horizontal planes are three timeslices (vertical axis t = time). Horizontal axes X and Y represent morphological characteristics changing quantitatively and measurably with time. Data from the three timeslices are projected onto the plane at bottom. A, B and C are mean values of successive species interpreted as phylogenetic *lineages*, and circles define specific variation at the give time. Phylogenetic *trends* extracted from the lineage reconstruction are denoted by a, b and c, different for each lineage. Et marks trends independent of time, usually reflecting ecological gradients. 'The similar directions of phylogenetic and ecological trends correspond in reality in many groups of larger foraminifera; they are not fortuitous and have a functional significance.' In the projection plane, LSDS is the least significance difference for successional species in a lineage. There is clear discrimination here between (i) neighbouring species within each timeslice (between-lineage), and (ii) successional species between timeslices (within-lineage). Two implications are (i) species can be separated and identified only when the timeslices can be separated, and (ii) the spacing of these timeslices represents the limit of biostratigraphic resolution.

different rates, as tracked by the same measurable characters, the proloculus size and elongation index.

By projecting the intra-lineage variation (circles) on to the common plane below, not only is the inter-lineage variation in evolutionary rate demonstrated, but allochronous overlap is shown to be so extensive that one cannot distinguish members of separate lineages without having the time planes available a priori. One *cannot* identify a single species from an alveolinid limestone of unknown provenance, but one *can* distinguish and identify two or three species in the same sample! The combination of measurable variation and measurable change through time within the lineage sets the limits on biostratigraphic resolution, and Hottinger introduces the *least significant difference* in mean values for successional species. Within that constraint, 'Concurrent lineage-zonations … based on parallel evolutionary lineages have the double advantage to be subjected to a mutual control of simultaneously changing morphologic characters in several lineages where the index fossils of each lineage supplant each other in successive time levels' (Hottinger, 1981).

Stage of evolution: the Cenozoic mammals

For Henry Fairfield Osborn, the mammals were very special time-keepers in geology. 'The stages of evolution in animals and plants give us the punctuation points, as it were, or the means of keeping geological time', but unlike the plants, amphibians, fishes and reptiles, the mammals 'are in a state of continuous and incessant change, and what gives them especial chronometric value is that the rate of change or of evolution is the same in many parts of the world at the same time'. Osborn (1910) went on to emphasize the remarkably constant evolution of the multituberculates, in whose teeth changes 'are added with the precision of clockwork', and in the horses, where 'the slow stages in the attainment of perfection in the grinding teeth of the Eocene horses are of great value as time-keepers …' We do not observe any sudden breaks, but a series of minute gradations, always in the direction of adaptation, because it appears that these changes in the teeth, Osborn's 'rectigradations', may be of the same kind as those to which Waagen applied the term 'mutations' in observing shells at successive geological levels.

But the real importance of the mammals resided in the fact that 'in Great Britain, in France, in Switzerland, in the Rocky Mountains, in short, wherever these inconspicuous but important "Rectigradations" are appearing, they arise at approximately the same rate and approximately in the same order even among animals which are widely separated geographically'. Close geological synchrony, moreover, requires a comparison of the entire fauna and entire flora. The survival of a few primitive or arrested types may mislead, as in Australia, for example. Among the mammals as well as among the plants there is a constant progression which is, on the whole, a guide or index to synchroneity. This potential of correlation did not preclude such broad statements as the following: that the general faunal aspect of modern Africa resembles that of Pliocene Europe.

The central problem driving Osborn's research programme was the comparison of mammalian developments in the Old and the New Worlds. His tests were sixfold: (i) *Presence of similar species* – percentages in common reveal an alternating convergence/divergence of Old–New World faunas. Only during the periods of faunal resemblance can we employ test (ii): *Similar stages of evolution* – based on the similarity in the stages of development of like 'phyla' [= clades], ' … as expressed in the detailed changes in the grinding teeth … , in the numerical reduction of the digits, etc. For example, the different transformations of the premolars, or anterior grinding teeth in the horses, rhinoceroses, and tapirs during the Eocene and Oligocene Epochs afford very exact data for correlation purposes.' The next three tests were biogeographic: (iii) *simultaneous appearance or introduction of new mammals* – cryptic appearances due to coincident immigrations from regions unknown, and of great correlational value; (iv) *immigration periods*; (v) *predominance of certain kinds of mammals*; (vi) *extinction periods of certain mammals*.

Now, I remain uncertain as to what constituted Osborn's vision of *stage of evolution*. Did he mean that evolution is progress toward ever better adaptation, in which case each innovation will spread very rapidly in its superiority to the limits of the species' geographic range? Some of the above could support that reading, which seems reasonable enough almost a century later so far as inheritance is concerned, if not the ultra-adaptationist mode. Or was Osborn referring to some clock-like succession of evolutionary innovations that is built into lineages, so that in response to an internal force driving toward a future superior adaptation, the fossils record parallel steps at the same rate in independent lineages? If so, then we have here in rectigradation a foreshadowing of a so-called 'law' of evolution – aristogenesis – which dates (1934) from a period of several such proposals among palaeontologists (Rensch, 1983) and for which Osborn is remembered, perhaps at some expense to the memory of his real palaeontology. Perhaps there is a good reason (if others have been as uncertain as I) for the source of Lindsay's (1989) complaint:

> Stage of evolution has always been well known and widely applied by vertebrate paleontologists, but it has frequently been misunderstood and/or attacked by neontologists, invertebrate paleontologists, and

others, without evoking a response from the vertebrate paleontology community. One might say that 'stage of evolution' has undeservedly received a bad following in the press, especially when its importance in age assignment of terrestrial mammal faunas is acknowledged.

In agreement with the critics, however, it must be emphasized that age determinations based only on 'stage of evolution' is hazardous at best. On the other hand, where mammal faunas have been well studied, as is generally true in Europe, with several lineages (e.g., horses, primates, theridomyids, cricetids, and gomphotheres) well known and widely distributed, the 'evolutionary grade' or 'stage of evolution' within each of these lineages is generally corroborative and age assignment of a particular faunal assemblage is usually straightforward. In those instances, most vertebrate paleontologists would place as much confidence in relative age assignment based on 'stage of evolution' as they would on stratigraphic superposition.

But for excellent examples of 'stage of evolution' Lindsay pointed to evolutionary trends in dental morphology as revealed in recent studies in Neogene rodent biochronology. Fejfar and Heinrich (1989) made the point, by now familiar, that the rapid evolution of many rodents has produced short-ranged species that are good index fossils in the Neogene. Although that is undoubtedly true, it is less clear that real species actually are being used biostratigraphically. For one thing, their revised biochronology demonstrated that the rodent super-zones were diagnosed on genera, not species; and even though the zones themselves are diagnosed on species' ranges and concurrences, a range chart of selected genera (Fig. 4.31) shows that these genera can be used to define every one of the zones. Again, Fejfar and Heinrich pointed to the differences in taxonomic concept among students of the Quaternary vole *Microtus*, ranging from 'highly sophisticated' to 'conservative', presumably meaning splitters and lumpers, respectively. However, their most telling point is the reconstruction of dental morphological changes in the muroid rodent lineage of *Promimomys–Mimomys–Arvicola*. The lineage is illustrated here in Figure 4.32. Note how the well-preserved dental characters come in successively. It is clear in this ex-ample that 'stage of evolution' refers to a within-lineage phenomenon (as recon-structed, at any rate), not to any between-lineage characteristic upon which a model of lock-step parallelism and hence correlation might be constructed. Even so, Fejfar and Heinrich generalized more broadly about the (cricetid) mesodont-lophodont molar condition, which during the Astaracian–Vallesian '*appeared independently from different lineages* ...', and later, in the late Turolian, 'include representatives of different provenance such as invaders from the

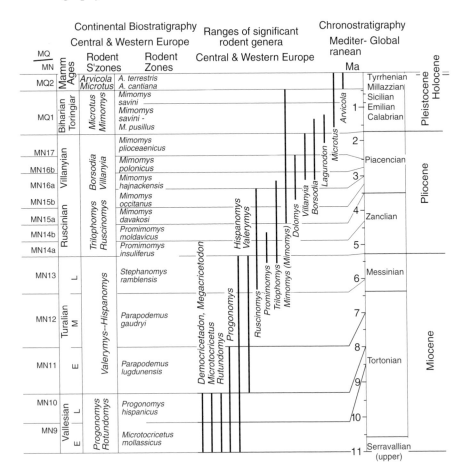

Figure 4.31 Muroid rodent biochronology, late Neogene, central and western Europe (Fejfar and Heinrich, 1989, Fig. 3, with permission).

east ... as well as advanced members of endemic lineages ... ' (emphasis added). It would seem that 'stage of evolution' is still somewhat elusive as a phylogenetic notion; biostratigraphically highly valuable as it is, it retains more than a trace of polyphyly.

Fejfar and Heinrich took a conservative line on the taxonomy of their indices, stressing that we know little as yet about the geographic spread of the extraordinary dental (molar) polypmorphy in these small rodents. Enlightening as their discussion of stage of evolution and biochronology is, at its heart it lacks a confrontation with the question of the recognition of fossil phena and taxa within samples. Their example of a 'highly sophisticated' taxonomy of a highly polymorphic group was Rabeder's (1986) study of *Microtus*, to which we now turn. Rabeder painstakingly sorted the variation

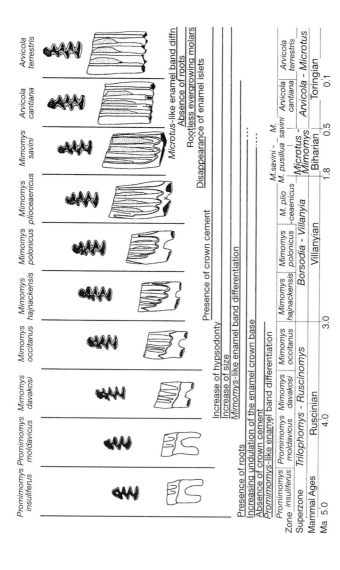

Figure 4.32 'Stage of evolution' illustrated by changes in dental morphology in the rodent lineage of *Promimomys–Mimomys–Arvicola*, Late Neogene, Europe (Fejfar and Heinrich 1989, Fig. 5, with permission). Shown is the first left lower molar in occlusal (upper) and buccal (lower) views. Heavy line shows increasing undulation of enamel crown base.

151

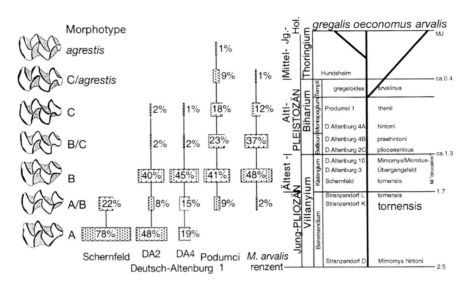

Figure 4.33 Morphotypes and gradual phyletic change in the origin and early evolution of the Quaternary vole *Microtus* from *Mimomys* (Rabeder, 1986). Left, spectrum of morphotypes (molar M$_2$ sinistral) to which specimens from five populations could be assigned, giving percentages. Right, lineage, from late Pliocene to Recent, with localities of collections against age; hatching shows the *Mimomys–Microtus* transition.

especially of the surfaces of the molars into morphological digrams (*Morphodynamisches Schema*; *Morphotypenspektrum*) which can then be quantified, so that a sample can be characterized as percentages of contained morphotypes (Fig. 4.33). The dominant impression from this work is of rapid but gradual change through time: note the cladogenetic succession. The figure emphasizes Rabeder's careful distinction between morphotypes and species.

The tensions in biostratigraphy: can the species survive?

Stripped down to its core, palaeontology consists of three parts. In the first place it maps the history of life, including the great innovations, the spectacular novelties, as well as the emergence of organic diversity and the problems of grappling with that diversity in taxonomy and classification. The problems of progress and extinction, the questions of variation versus *Bauplan*, the very existence of macroevolution vis-à-vis microevolution and tiering – they all belong here. The second part of palaeontology is the reverse of the fact – the truism – of the adaptation of an organism to its environment: if we can work out how a given species lived, then we can use its presence in a rock to deduce the environment of the rock's accumulation. From this relationship flow the great generalizations of biofacies, of transgression and regression and the rise and fall of the sea or the fall and rise of the land, and changes in global climate. And the third part of palaeontology is, of course, the use of fossils to put geological and biological records and inferred events in their right chronological order.

In Chapter 1 I referred to the first tension in this palaeontological triad – the need to disentangle the two signals of fossils as age determinants and fossils as environmental indicators. A second tension, alluded to in the present chapter, resides in this triple role of fossils as age determinants, evolutionary documents and biofacies components. The problems are greater than the mere sorting out of two converging signals and so are the implications. The needs of biostratigraphy diverge so far from those of palaeobiology, in the measured opinions of some, that a separate taxonomy and nomenclature are required. Whereas I have depicted some taxonomy of microfossils as a pragmatic but primitive muddling along, surviving on transformational constructs and virtually avoiding taxic confrontations altogether, whilst investing displacement activity in nomenclatural formalism, there are now proposals to move much further away from biosystematics and toward a modern, automated stamp collecting. These proposals do not include the dismantling of the Linnaean system for classifying the modern biota, but some of them do abjure the use of that system in palaeontology. Philosophically, they differ little from the pragmatic typology of the various taxonomic species concepts, as distinct from the biological and clade concepts (Chapter 8). The proposals represent a return to nominalism, the medieval belief system that there are individuals but not – in biosystematics – species or higher taxa. Nominalism survived longer in botany than in zoology but its adherents were reacting to essentialism and had nowhere else to go. Now, with the establishment of the biological species concept, nominalism can be allowed to wither away at last – or so Mayr (1982) would have it. But as Mayr briefly acknowledged, there are problems in the biospheric record that do not yield easily to a theory of species based in higher animals. In palaeontology there are two categories in particular – organ-taxa, such as spores and pollen grains whose affinities among other organ-taxa based in leaves, say, may not be apparent; and parataxa, which are based on something less than the whole organism. That can be said for all taxa and especially for taxa of fossils such as the teeth of micromammals, but the meaning applies particularly to such groups as the conodonts, whose fossil record is rich and diverse whilst the conodont animal is barely known. Concomitant with this bias in the record is the drive in an applied direction, usually toward biostratigraphy and basin analysis or palaeoceanography. Frustration with the Linnaean system can be real and deep. I refer here to some treatments of the problem.

For Young (1960) there had been an existential split within biostratigraphy for a century. The split was between evolution, never doubted, and Darwinism, and it was only recently, thanks to the modern synthesis, that Darwinism has become acceptable. Where biostratigraphy arose and flourished under the two prevailing concepts of catastrophism and of immutable and specially created species, we can now, at last, classify a rock continuum using a biological continuum – but using biological and stratigraphic nomenclatural frameworks inherited from the earlier milieu. That species vary in time as they do in space changes the entire nomenclatural basis of biostratigraphy, said Young. Acknowledging the classics of lineage correlation, Young asked, somewhat rhetorically, how many existing biostratigraphic techniques involve in their application the basic concept of evolution? A brief review demonstrated, convincingly enough: not a lot. The other techniques include using simple symbols, statistical correlation, assemblage correlation, appearances and disappearances, transition zones between chronospecies, individual occurrences, guide fossils, and correlating segments of the continuum. More important than evolution in most of this lot is superposition – superposition in the local succession and, homotaxially, in testing and confirmation elsewhere. To reinforce his point that Darwin and evolution have not really reached into biostratigraphy, Young cited several works including the 'extreme catastrophism' in the planktonic foraminiferal biostratigraphy of Loeblich and Tappan (1957a, b). He concluded that biostratigraphy had 'lost the battle of the species' (to evolutionary palaeobiology), and that biostratigraphers cannot continue to use the taxa of the 'new palaeontology' for smaller biostratigraphical units because they are not sufficiently refined to meet the need for detail in modern and future stratigraphy.

The clearest case of an analysis of species in nektonic/planktonic organisms as a reason for abandoning them altogether is Shaw's (1969), illustrated by an analysis of a Devonian lineage of conodonts. Shaw showed us six morphological entities, end-members in a continuum, and well-known morphologically and stratigraphically. Those entities can be interpreted differently by two specialists in opposite hemispheres, each well steeped in notions of variation, populations, evolution and biospecies, and by a third specialist, also no philistine but interested in the evolution of conodonts as morphological entities (or transformational evolution, as it came to be known later). Shaw's central point was that the problem is not right or wrong in species' determination, but the impossibility of *any* species determination that does not bury all the important information beyond retrieval. Species cannot communicate ranges, hence correlations, evolution, or morphological differences. Shaw compared us to the alchemists: 'Systematic biology is the only scientific endeavour I know that still insists on treating its subject synthetically rather than analytically. A synthetic science is one that treats the objects of its study as entities, while an analytic science is one that treats the *constituent elements* of its entities separately.' Analytic palaeontology is 'the detailed study of the exact form of each morphological unit, its precise stratigraphic distribution, and the observed combinations with other morphologic units', whereas synthesis into species is disastrous in its destruction of clarity, the collapse of communication, and the onset of chaos. The species must be replaced by a language of entities based in morphological characters and their precise stratigraphic ranges. This, of course, is transformational evolution of a familiar stripe, but formalized and overt.

Terrestrial palynology was the stimulus for the most extensive subverting of orthodox biostratigraphy; indeed, the very word itself is superfluous, along with 'chronostratigraphy' and several others in the '-stratigraphy' group, 'geochronology', and the varieties of 'biozone'. For Hughes (1989) his discipline had now become so constipated by the old order that an extreme purge was the only remedy. He diagnosed an acute difficulty in species-level systematics and nomenclature. In the 'cluster approach' to species, one begins with the holotype and other specimens in the originally studied series – a technique which works in neontology, but which leads to trouble in palaeontology as others identify material from elsewhere, thus extending the species in space and time. This is 'inflation', and it leads to a 'balloon taxon', a taxon which has grown greatly in scope through this repeated attribution. Since species and other taxa are clusters of information, they grow vaguer and more diffuse; and since biozones are clusters of information, they too become vague and ill-defined. And since the normally laudable human traits of caution and parsimony are in play, ballooning is exacerbated. Ironically, the 'iniquitous splitting' which we usually perceive as choking the literature is actually the better way to go, in Hughes's opinion, because we can always lump oversplit taxa in later data manipulation but the reverse is not true: lumped taxa cannot be split without going back to the original material. In an excursion into historical diagnosis, Hughes found three unfortunate consequences of the 'accidentally blinkered taxonomy' of the first – nineteenth-century – phase of universal exploration of the wonders of nature past, when so much newly discovered biodiversity had to be described. One consequence was lumping, which does not overwhelm the data-handling facility of the human brain. A second is the persistent belief that species are natural entities whose true characters must be sought by observation. A third consequence is that because we derive ages from fossils, no taxon can actually be characterized by its known range. This comment seems to echo the warnings of Arkell and Moore that total ranges are highly elusive (Chapter 1). And so, as we pass from the second phase of palaeontology with its detailed exploration and increasing specialization (in the early to mid twentieth century), to the third phase of interpretation and synthesis in systematics greatly assisted by automated data handling (in the late twentieth century and beyond), we have to struggle with this archaic and excess baggage of systematics and nomenclature.

As if this were not enough, Hughes saw the situation in stratigraphy as being little better than in biosystematics. We lack clear and clearly articulated purposes in our research. We have permitted a great superfluity of jargon, weakly defined, to clutter and stifle the discipline. We are timid and probably self-serving in our reluctance even to discuss the possibility of changing our procedures in stratigraphy. Stratigraphy is losing its hold as a reputable and rigorous discipline, and numerical techniques – the 'cloud of mathematical juggling and justification' – far from rescuing it, have deepened the gloom. To kick us over this threshold towards recovery, Hughes proposed that the usual complicated and jargon-laden flow charts of stratigraphic and adjunct activities be replaced by just four activities. (i) There is rock description, including mapping, lithology, palaeontology, palaeomagnetic and radiometric records. (ii) There is construction of the global stratigraphic scale using (i). (iii) Direct correlation of rock successions also uses (i). Or, (iv) there is correlation of an out-succession against the scale developed in (iii).

There is little new in Hughes's stratigraphy as such. The new term 'bracket correlation' refers to a hallowed tactic of constraining ages by two correlations, one above and one below. However, Hughes's taxonomy and species concept are different. Consider his species. As he told it, the current procedure is to build out from the primary types, groping one's way into a miasma of doubt. There are several warnings that a species is not a natural entity, and a major recurrent difficulty is that many authors 'contrarily still believe that "range of a species" is a natural attribute to be discovered only by observation; ... Another way of putting this might be to suggest that there are really two distint species concepts, (a) a simple descriptive outline of some specimens and (b) an interpretative exercise in geologic time, and that the two are not being distinguished.' The taxon does not 'exist' as a part of nature carved at the joints; it is any group of organisms with a similarity as stated by the systematist. Thus, all the recent discussion about the reality of species and species-as-individuals simply is not relevant (Chapter 8). Again, there is the question of the nature of evolution in Hughes' universe (Fig. 4.34). This sketch seems to be saying that evolution is anagenetic and ultra-gradual in its mode, and it is accompanied by the warning that 'it is important to appreciate that in writing a description (definition) round the character-state B, character limits for R and S are automatically laid down by the nature of the species variation concept concerned and accepted; the stratigraphic range becomes a description character and thus is already decided whatever its author may believe.'

Central to Hughes's advocated new regime is *period classification*, a new classification confined to records of fossils from a named geological period;

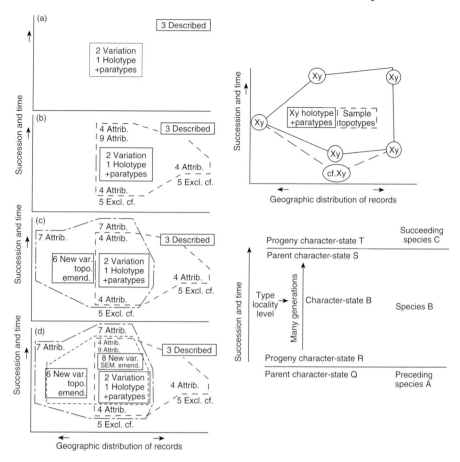

Figure 4.34 Changes in definition and perception of a fossil palynospecies as different authors encounter it down the years, from Hughes (1989, with permission). *Right, top*: The practice of species definition. Only the types and perhaps later topotypes are surely in species Xy. All the others from different places and stratigraphic levels are by attribution and extend the species' meaning, which attenuates further as we encounter doubtful records (determined as cf. Xy). *Left* (a)–(d): a scenario of the above. (a) Original description with types 1, variation 2, and assignment of specimens from sample 3. (b) Next worker attributes 4 to this species based on 2 and 3; 5 is doubtful. By now, there is data for a time–space envelope of the species' distribution (dashed line). (c) Emendation by a third author: study of more topotypes from original locality 6, plus new attributions 7 and exclusion of 3, gives a new envelope of understanding (dash-dot). (d) A second emendation results from SEM study of topotypes 8 and a narrower definition of the species. With new attributions 9 and exclusions due to revision, there is still another envelope (dotted). But (wrote Hughes) it is rare to specify exclusions. *Right, bottom*: 'Limits of fossil species in time. The "species" of fossils B is commonly based on a description of the character-state B as observed at its type locality. It would be more accurately represented by a described range from character-state R to character-state S, with no mention of an irrelevantly placed holotype at B. It cannot be described in terms of a difference from character-state Q to R (or from S to T) because these time-boundaries are from parent to progeny only' (Hughes, 1989).

in *PDHC*, palaeontologic data-handling code, in which a fossil has a formal, three-element name, the *trinomial*. Thus: *palaeobiogroup* (a major morphological non-hierarchical grouping of types of fossil) – *timeslot* (a new substitute for the genus taxon, consisting only of the name of a division of the timescale) – *palaeotaxon* (a new form of immutable base-taxon of fossils).

There is an abundance of illustration of this approach in Hughes's book and a summary here is not feasible. What can be said is that the difficulties encountered in the handling of terrestrial palynomorphs seem to have been surmounted in past years in the study of marine microplankton. None of the problems of taxonomy, species concept, taxic or transformational evolution, diachrony or isochrony of datums, calibration with physical events and radiometric estimates, and so on would justify the draconian treatment advocated here. Indeed, one could make a strong case for precisely the opposite strategy – one could suggest that the classical biostratigraphic foundations of marine micropalaeontology are now sufficiently strong to encourage more classical taxonomy and paleobiology, not less. For skeletonized microplankton have an unrivalled fossil record which can offer far more to the burgeoning field of macroevolution than it has so far. The adoption of Hughes's system in a major fossil group just might assist its application in applied geology, but that group would likely be lost to palaeobiology for ever.

Concluding comments: evolution and biochronology

Several dialectical swards intersect where we consider the evolutionary events used in biocorrelation. Here are six intersects: (i) typological species and morphospecies grading to biospecies; (ii) pseudospecies, pseudospeciations and pseudoextinctions (anagenesis) as against 'real' species, speciations and extinctions (cladogenesis); (iii) punctuated patterns of lineage change against gradualist patterns; (iv) punctuated patterns of community change against gradualist patterns; (v) stratophenetics against cladistics; (vi) allopatric, peripatric and depth-parapatric speciation against sympatric speciation.

Pearson (1998b) discussed the tensions between morphospecies and biospecies and came to a much more congenial conclusion than did Hughes about the evolutionary biology of the entities used most intensively in biostratigraphy. Pearson proposed an evolutionary classification for biohorizons that leaves mostly intact the paraphernalia of kinds of biostratigraphic zone (Chapter 2) and concentrates instead on the nature of the boundary criteria. There are four categories of biohorizon defined on the end-points of the stratigraphical ranges of morphospecies, i.e. two first appearances and two last appearances: dispersal biohorizon, extinction biohorizon, pseudospeciation

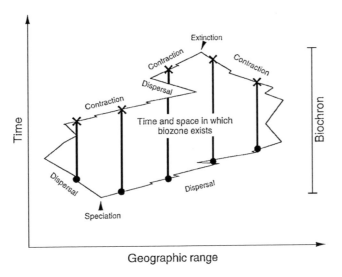

Figure 4.35 Anatomy of a taxon-range biozone illustrating the interaction of evolution and biogeography in determining the range of the zone during the time of the biochron (Pearson, 1998b). In this as in other discussions of the biozone (e.g. Loutit *et al.*, 1988; see Figure 2.14) there is no serious consideration of within-species' morphotypic change of the kind christened gradualism, and to that extent 'biostratigraphers are all punctuationists' (Gould and Eldredge, 1977) although Hughes apparently took gradualism for granted (Fig. 4.34).

biohorizon, and pseudoextinction biohorizon. Dispersal and contraction, and speciation and extinction are shown in a taxon-range biozone (Fig. 4.35).

i.   *Dispersal biohorizon.* If planktonic speciations happen peripatrically or depth-parapatrically and are to be biostratigraphically useful in due course (i.e. are not cryptic, but signalled phenotypically instead), then there has to be dispersal from the point-source of splitting or budding. Dispersal will be rapid or essentially instantaneous, giving an isochronous boundary, or diachronous, or stepped-allochronous, or Lazarus-like. The dispersal biohorizon then is more biogeographic in its essence than speciational (Chapter 5) and will display a relatively cleancut incoming in relatively complete sections. The emergence of *Globanomalina pseudomenardii* and *Orbulina universa* are examples as discussed above. The dispersal biohorizon is objective in the sense that different workers can be expected to identify the same horizon given the same data.

ii. *Extinction biohorizon.* The termination of a lineage without any directly descendent species will be sharp, locally and regionally, but has to be

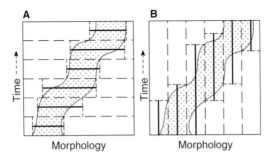

Figure 4.36 Morphotypes in a lineage: options for subdividing a chronocline into pseudospecies (Pearson, 1998b). As discussed above for the *Orbulina* bioseries (Fig. 4.1, 4.17) the options are to recognize anagenetic population shifts through time in chronospecies (A) or to erect typological morphospecies (B). In both cases: heavy line, the 'average' of the taxon; dashed line, boundary or 'grid' line between taxa. B is the inevitable outcome of the typological tactic illustrated for palynology in Figure 4.34.

tested empirically and biogeographically, much as a mirror-reflection of the dispersal biohorizon. Thus, extinction can be expected to display diachronous, stepped-allochronous and Lazarus patterns. The extinction biohorizon too is 'objective'.

iii.   *Pseudospeciation biohorizon.* In a relatively continuous and closely sampled succession, morphotypes are added anagenetically to give a morphocline which can be either a geographic cline or a chronocline, the latter being the more relevant here. Observable phenotypic shifts become morphospecies. This is a 'subjective' event in so far as different workers may disagree on the somewhat arbitrary taxic division of the morphocline, or on which new, transformational character is the most useful and reliable.

iv.   *Pseudoextinction biohorizon.* Morphocharacters may disappear anagenetically in the same way that they appear. The taxa are similarly 'subjective'.

Pearson estimated that of the 58 biohorizons used by Berggren *et al.* (1995a) for Cenozoic zone-boundary criteria (in planktonic foraminifera), as many as 31 are probably gradual transitions. He asserted too that pseudoextinctions and pseudospeciations as gradual anagenetic transitions are likely to be less clear at high resolution than are speciations and extinctions. He instanced the GSSP for the Palaeogene–Neogene boundary (Chapter 7). The *Globorotalia kugleri* group comprises three pseudospeciations: *pseudokugleri* (25.9 Ma), *mendacis* (23.8 Ma), and *kugleri* (23.8 Ma) (datum calibrations from Berggren *et al.*, 1995a; but see Chapter 7). Although Blow (1969) warned of potential confusion between the

morphotypes	alpha taxonomy	datums	zones
α α ε ε ε ε	old species α plus		
α α ε ε ε ε			
α α ε ε ε	new species ε		VI
α α α ε ε ε	[cladogenesis]		
α α δ δ ε ε ε	successional subspecies αδε	LAD δ	
α δ δ ε ε ε			V
α δ δ ε ε		FAD ε	
α α α δ δ δ	successional subspecies αδ		
α α α δ δ δ			IV
α α α α δ δ		FAD δ	
α α α α α	successional species α		
α α α α α			III
α α α α			
α α β β	successional subspecies αβ	LAD β	
α α β β β			II
α α α β		FAD β	
α α α α	successional species α		
α α α α			I
α α α			

Figure 4.37 Morphotypes, typology and populations: a thought experiment in shifts through time. Morphotypes α to ε are distributed as shown, but actually the sample at every level is continuously variable. There is cladogenetic change where morphotype δ drops out, closely comparable in this to the evolution of *Orbulina* after the extinction of the intermediates but the survival of the ancestral *Globigerinoides* (Figs. 4.1 and 4.17). 'Practical' considerations of applied palaeontology would tend to emphasize the FADs and LADs and range morphospecies accordingly. However, zones I to VI are precisely the same in recognition and resolution whether one uses morphospecies datums β to δ or successional taxa α to αδε.

three morphotypes (two erected by himself), Berggren *et al.* (1985a) found that *kugleri* was distinct and relatively easy to identify consistently. However, Pearson (1998b) worried that this pseudospeciation was too fragile to carry the weight of a system-boundary GSSP, for it will by its very nature be temporally variable at 10^4 years scale.

We ask again the question raised in Wade's analysis of *Orbulina* (Fig. 4.1): how to anatomize anagenetic populations in a lineage? The options are horizontal cutting to diagnose populational chronospecies and vertical cutting to diagnose typological morphospecies (Figs. 4.36, 4.37). Whilst 'practical' considerations have been cited frequently to justify the typological viewpoint and procedure, the figures argue that biochronological resolution is not lessened in adopting the populational ideation. Morphotypes, their distribution in space and time and their phylogenetic relationships are at the heart of the patterns exploited

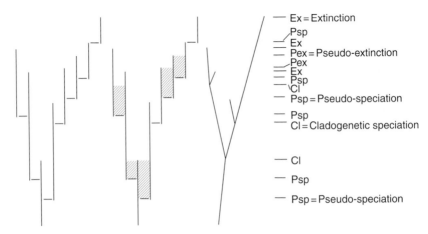

Figure 4.38 Morphotypes and phylogeny: typological species and evolutionary lineages (adapted from Pearson, 1998b, Fig. 5.7, with permission). Left, dendrogram of morphospecies as an evolutionary bush. Middle, acknowledgement of intermediates at some times (shading). Right, lineage phylogeny identifying the actual times of cladogenesis. 'Note that cladogeneses may correspond to pseudospeciations, pseudoextinctions, or even may occur midway in the stratigraphical ranges of two morphospecies. Pseudoextinctions may or may not correspond to extinctions, and anagenetic transitions may be gradual or sudden.' Far right: horizons of these four categories in these diagrams.

and investigated by biostratigraphic zonation and correlation. It is useful to remember that biostratigraphy in this classical sense is but one of the scientific problems to which the large and relatively complete databases of skeletonized microfossils can be put. Pearson makes this point implicitly but vividly in three depictions of a clade, variously morphotypic stratophenetics, more realistic acknowledgement of phenotypic transitions within-sample, and lineage phylogeny (Fig. 4.38).

For three decades Gould advocated the Eldredge–Gould theory of punctuated equilibrium (Eldredge and Gould, 1972; Gould and Eldredge, 1977, 1993), culminating in his magisterial book (2002). One of his cogently recurring themes has been, when in doubt, trust the practitioner – the working hack whose correlations and age determinations have to withstand economic and managerial scrutiny in exploration and development as well as the sceptical colleagues. He asserted that a geologically abrupt origin of morphotypic species followed by prolonged stasis has been 'common knowledge', 'tacitly shared knowledge' – a pattern that has always been recognized by the working palaeontologist. However, in the glare of organic evolution, 'the' great intellectual novelty of the later nineteenth century, our predecessors were 'cowed or puzzled' by their data showing no change. I prefer to recall an analogy from earlier in the century, the point made by W. J. Arkell (Chapter 8) about the systematic field-geological mappers after Smith and Cuvier, who were too busy mapping (and testing and fleshing out the geological succession and timescale) to bother about the intellectual frills. Ever the dialectical advocate, Gould crystallized the situation thus: if most species in the fossil record had changed gradually at geological time scales, then maximal resolution would have been achieved by the *stage of evolution* method. Since, however, most morphospecies have been treated in practice as stable entities, biostratigraphic resolution has been achieved by two other strategies – *index fossils* (short ranges, wide distributions) and *overlap range* zones (narrow overlaps of upper and lower ranges).

This chapter has cited evidence, *pace* Gould, for biostratigraphic practice in the Cenozoic fossil record across the range that he polarized.

5

Systemic stratigraphy: beyond classical biostratigraphy

Summary

There is an unsullied lineage in biostratigraphy from 19C zones to the ISSC *Guide*, then to the integrated geo-magneto-bio-chrono-logical scale based on the ordination and first-order correlation of irreversible events. However, a shift in worldview away from Lyellian gradualism has encouraged systemic stratigraphy, a Quaternary-type systems approach to the use of regional signals of global environmental changes at third-order and higher frequencies. Thus the mainstream of biostratigraphy is now conjoined with another: ecostratigraphy and sequence biostratigraphy, where the reversible events of biofacies shifts and chemostratigraphy together with depositional surfaces can be constrained by classical zones or datums. Sequence biostratigraphy has three aspects: basin analysis, integration with the main scale, and as a template for Cenozoic palaeobiology. A plausible global model of third-order marginal sequences and δ^{18}O-based glaciations can be tested to some degree (and successfully) by neritic biofacies studies.

Systemic stratigraphy

Converting microplanktonic stratigraphic ranges into evolutionary ranges, and tying those fundamental bioevents to physical events such as the geomagnetic chronology, was central to Chapter 3. We are going beyond that here. Correlation is opportunistic in the broadest possible chronological sense in that any and all signals in the exogenic system can be invoked. The global ocean is controlled 'endogenically' by crustal processes which control the volume of ocean basins and thereby sealevel, or the spillage of water across the continental margins, and by the intensity of cycling and exchange into the crust through the ridge systems. The ocean thereby regulates the 'exogenic' system consisting of the hydrosphere, atmosphere, reactive lithosphere (or zone of subaerial weathering and hydrothermal alteration), and biosphere. The most important factors, elusive and not at all easy to extract from the records of times past, are sea level and climate. As a first-order generalization, we are much more conscious than we were a few decades ago that the interpretation of local outcrop and subsurface observations needs to take account of regional and global influences. To use a clichéd term and some polysyllabic jargon, we adopt a systems approach to biogeohistory – to the interpretation of biostrati-graphic, chemostratigraphic and climatostratigraphic signals. There is a name for this: Berger and Vincent (1981) christened as 'systemic stratigraphy' the systems approach to extracting and interpreting those signals in terms of signal input, modulation of the signal within the system, and signal output. Table 5.1 summarizes the elements of systemic stratigraphy. One needs to understand the mechanisms that produce the signal – that is *process* science. One needs too to pick out the global and regional signals from local 'noise' – that is the *chronicling* operations of ordination and correlation. Biogeohistory will not thrive and flourish by concentrating on either process or chronicle to the neglect of the other. Much of this chapter discusses biostratigraphic events and possibilities in a systemic–stratigraphic context. The concept is now deeply embedded in palaeoceanography and palaeobiology but the term itself has not caught on; even so, it is relevantly in the title of this chapter.

Table 5.1 *The elements of systemic stratigraphy (Berger and Vincent, 1981)*

	Elements of systemic stratigraphy
Definition	Determination and correlation of global climate-related trends, cycles and events of the exogenic system (ocean, atmosphere, reactive lithosphere, biosphere) as recorded in lithological and palaeontological sequences.
Basic rules	i. A change in any part of the system produces changes in the others, hence all systemic stratigraphic signals are correlated. ii. Every regional signal has a systemic component, which may be amplified or obscured through regional factors.
Basic driving functions	Fluctuations of irradiation and sea level.
Primary effects through	Changes in temperature and evaporation–precipitation patterns, ice-water balance, fertility in the ocean and on land, carbonate saturation, CO_2 pressure in air.
Secondary effects	Biogeography, ecology, evolution and chemical facies distributions.

Biostratigraphy of reversible events and high-resolution biostratigraphy

Quaternary biostratigraphy

Late Quaternary studies diverged long ago from 'Tertiary' and Cenozoic studies for reasons – very good reasons – of scale. In its duration the Quaternary was comparable to a Cenozoic biozone, and there simply were too few irreversible bioevents to compile a geochronology adequate for the rich stratigraphic and geomorphic record and the fast-moving complexities of later Quaternary biogeohistory. Third-order chronicling is eminently useful for the Palaeogene and most of the Neogene, not so for the Quaternary whose marine faunas and floras are essentially modern. Even so, there is a high-resolution biochronology of calcareous plankton (Berggren *et al.*, 1995b).

Another way has been to relate 'reversible' fossil data to isotopic stratigraphy. Emiliani (1969) found, in continuous time series in oceanic cores (mixing by bioturbation was considered to be minor), that 'a very fine relationhip' existed between isotopic palaeotemperatures and a wealth of foraminiferal parameters, the latter including reversible and irreversible evolutionary (morphotypic and taxic) changes and reversible (likewise) changes induced by the environment. Since he studied only a few species and only a few characters in a 'pallid example' of what is possible, Emiliani concluded that the analysis of a full suite of characters in the whole planktonic assemblage should allow placement of an unknown sample to within a few thousand years in a previously established regional stratigraphic section. This method of integrated, quantitative, morphological analysis, unhampered by nomenclatorial formalism and typology, was to be 'a new paleontology', aimed specifically at establishing an accurate chronometer for geological time.

A more traditional strategy was to employ the response of species and communities to the watermass shifts that are part of global climatic change (Fig. 5.1). Together with chemostratigraphy, this reversible biostratigraphic pattern is climatostratigraphic and can be extended in principle from the marine realm by cross-checking with pollen spectra. Ericson and Wollin (1956, 1968; Ericson, 1961) developed the *Globorotalia menardii* stratigraphy by which to recognize glacial cycles in the Atlantic Ocean and Caribbean Sea. Figure 5.2 demonstrates matches between percentage abundance of *Menardella [Globorotalia] menardii* in the Caribbean and East Pacific and benthic oxygen-isotopic stratigraphies – a good example of *consilience* between independent methods and datasets (Chapter 8). Figure 5.2 suggests that the Indian Ocean was haven for *M. menardii* which resurged in the Caribbean–Atlantic and East Pacific more or less during

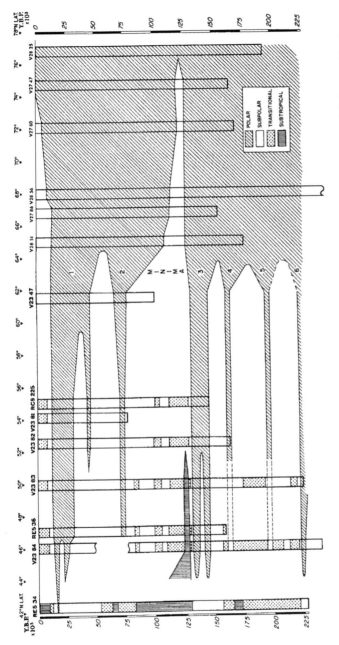

Figure 5.1 Climatostratigraphic correlation in planktonic foraminiferal assemblages, Late Pleistocene and Holocene, North Atlantic Ocean and Norwegian Sea (Kellogg, 1975, with permission). This compilation at mid to high latitudes demonstrates that the four climatically controlled 'faunas' migrate in a coherent way across tens of degrees' latitude in rapid response to climatic and oceanographic shifts at this 10^4 years scale. The very rapid warming immediately after the cold minimum #3 at ~125 000 years has been abundantly corroborated in the oceanic, neritic and terrestrial realms, biogeographically and geochemically, in the decades since this synthesis.

167

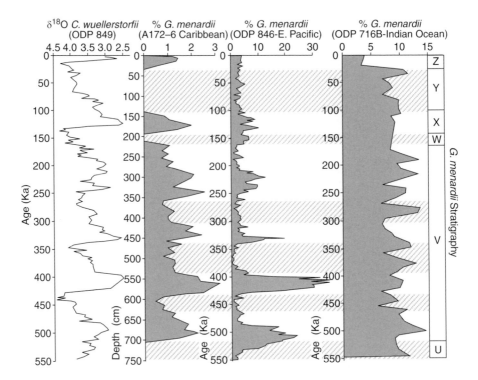

Figure 5.2 Comparison of *G. menardii* abundances in three oceans with Pleistocene oxygen-isotopic events (Norris, 1999, with permission). *G. menardii* stratigraphy, zones U to Z, from Ericson and Wollin (1968). The grey shaded bands indicate *G. menardii* minima in the Caribbean and East Pacific, where the species is more abundant at times of low $\delta^{18}O$ and reduced ice volume. In the Indian Ocean *G. menardii* is more abundant overall and still more so in some of the glacials. Norris used this pattern to argue that *G. menardii* was excluded at times from the tropical Atlantic by unfavourable hydrographic conditions, not by an inability to disperse.

warmings; the species fluctuated during the glacial cycles but not so neatly there (Norris, 1999).

The climatostratigraphic model of Stainforth et al. (1975)

This was a thought experiment intended to show the importance of rapid, far-reaching environmental change described by isotherms; the scale was unspecified but large enough to embrace 8–22 °C isotherms (Fig. 5.3). Species' assemblages (assemblages AB … XYZ) at seven stations (1–7) display disjunct ranges, truncated ranges, and recurrences of key species. Backtracking from the micropalaeontological logs reveals both total ranges and climatic envelopes. Cold and warm peaks permit, in principal at any rate, locking of terrestrial palynomorph fluctuations into a marine framework. This discussion

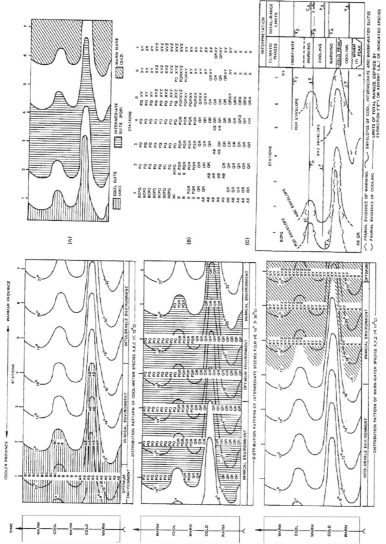

Figure 5.3 Climatostratigraphic model: effects of climatic fluctuations on temperature-sensitive planktonic foraminifera (Stainforth *et al.*, 1975, with permission). (Left half of figure) Three patterns generated as environmental shifts (optimum-inimical, bottom of each panel) affect cool-water, intermediate and warm-water species respectively. As species come and go in response to these hydrographic shifts, examples of locally truncated ranges, disjunct distributions and recurrences of key species are generated. (Right half of figure) A, a digest of the pattern at (Left). B, Species distribution logs for the seven stations shown at (Left). C, interpretation of the grouped data in (B) with envelopes, interpreted climatic phases and species' total ranges.

anticipated sequence-biostratigraphic analysis in several respects (below). Figure 5.4 displays a late Miocene–Pleistocene example of a climatostratigraphic configuration – Ingle (1973) could trace assemblages across tens of degrees' latitude.

High-resolution biostratigraphy

As established in foregoing discussions, several phenomena have comparable durations in the low-single-digit millions of years – the marine microplanktonic zones of the Cenozoic, geomagnetic chrons, Exxon third-order cycles, average species' durations. Also of comparable duration are the oxygen-isotopic cycles interpreted as the Mi glacial cycles of the Miocene, and a succession of planktonic foraminiferal assemblages in the Miocene in southeastern Australia (Li and McGowran, 2000). To compare and contrast these phenomena requires as rigorous a chronology as possible, and 'rigorous' here means both accuracy and especially resolution – one really needs to see through geological time at finer resolution than the phenomena under scrutiny. If the *focal level* on time and phenomena is at the third order, then we need to be aware of both second and fourth orders (Salthe, 1985). There are three modern approaches to high-resolution stratigraphy including biostratigraphy. One is the subdivision of the oceanic record to tease out the isotopic and biotic signals of rapid changes in watermass and climate. Another is based in the especially good stratigraphic record of the Cretaceous continental floodings (Kauffman, et al., 1991), and the third consists of extracting an astrochronology from the rocks (Chapter 3).

A study of mid-Miocene isotope stratigraphy by Woodruff and Savin (1991) provides an oceanic case history. Analysing the dramatic changes at that time required high-resolution correlation – higher than can be provided in the framework of standard biostratigraphic zones, i.e. at the 10^6 years scale, for (to anticipate their conclusions) they could recognize periodic or quasi-periodic fluctuations in $\delta^{13}C$ values at 10^5 years scale. In a more integrated chronology using other time series, there are problems in the oceanic records. In this part of the timescale the strontium isotopic curve is too flat. In biostratigraphically rich sections, such as the equatorial region where high sedimentation rates are common, the record of stable isotopes may be good but the palaeomagnetic signal often is too weak to be used; whereas strong palaeomagnetic signals tend to be found in sections more condensed and with more unconformities, so that a strong signal is ambiguous as to its identity.

The first and most obvious strategy for achieving resolution in correlation is to use multiple biostratigraphic systems. Woodruff and Savin distilled the records from sixteen DSDP and ODP sites in four oceans; species identifications

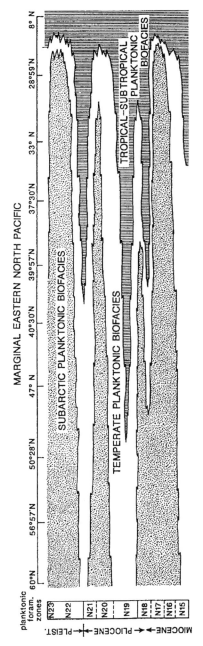

Figure 5.4 Late Cenozoic oscillations in planktonic foraminiferal biofacies in the California Current System and related Alaskan Current gyre (McGowran, 1986, after Ingle, 1973, Fig. 4). This pattern of climatically driven oscillations over tens of degrees' latitude has much in common with Figure 5.1, the major difference being the timescales – 10^6 years or third-order here; 10^4 years or fifth-order in the other.

171

and ranges were by numerous authors and the resolution of the sections is variable, but 'aberrant occurrences' could be recognized during repeated interpolation and were weeded out. The 'absolute' (i.e., numerical) ages of the selected events were tabulated (Table 5.2, Fig. 5.5) – almost eighty events spanning ∼13.2 myr. That potential resolution, impressive as it is, is not down to six events per million years in reality because there are provincial limits on the species. Some events are repeated, implying allochronous interprovincial occurrences. The procedure not only goes beyond the third-order planktonic zonation (because it has to) but it results in the abandonment of that framework in the discussion and all the figures: 'In our experience it is preferable to assign ages, where possible, using multiple biostratigraphic datum levels rather than single zonal boundaries in studies of this sort in order to facilitate recognition of stratigraphic inconsistencies' (p. 770).

Woodruff and Savin went on to establish paired δ^{13}C–δ^{18}O profiles for each section and identify δ^{13}C maxima and δ^{18}O events. For each site they prepared an age-depth diagram plotting biodatums and other relevant features. They used equations of these segments to calculate the age of each sample at each site, the equation being of the form: *calculated age (Ma) = slope x depth (mbsf) + intercept.* Through iterations among the sites the straight line segments were adjusted to maximize the correlations among the isotopic features on the curves of δ^{18}O versus age and δ^{13}C versus age. This exercise in correlation and age determination began with the ordination of a large number of biostratigraphic events and it concluded with highly resolved estimates of the ages of carbon maxima and

Table 5.2 *Table of ages of about 80 events (range tops and bottoms) in four groups of oceanic planktonic microfossils – planktonic foraminifera, radiolarians, calcareous nannofossils, diatoms – and one benthic foraminifer in ∼13.2 myr*

Age (Ma)	Fossil Zone	Datum level	Taxon
8.9	N	T	*Catinaster coalitus*
10.0	N	B	*Discoaster hamatus*
10.4	R*	T	*Actinoma golownini*
11.1	N	B	*Catinaster coalitus*
11.1	D*	T	*Denticulopsis praedimorpha*
11.3	D*	T	*Nitzschia denticuloides*
11.4	R	T	*Cyrtocapsella cornuta*
11.5	F	B	*Globorotalia menardii*
11.5	R***	B	*Diartus petterssoni ZB*
11.6	N****	*T	*Cyclococcolithus floridanus*
11.8	F	T	*Globorotalia fohsi robusta*

Table 5.2 (cont.)

Age (Ma)	Fossil Zone	Datum level	Taxon
11.85	F	B	*Sphaeroidinella subdehiscens* ZB, N12–N13
11.9	D*	B	*Denticulopsis dimorpha*
12.2	N	B	*Discoaster kugleri* ZB, NN6–NN7
12.2	D	T	*Crucidenticula nicobarica*
12.3	R*	B	*Cycladophora spongothorax*
12.3	D*	T	*Actinocyclus ingens nodus*
12.4	R*	B	*Dendrospyris megalocephalis*
12.6	D	B	*Denticulopsis praedimorpha*
12.8	D***	T	*Coscinodiscus lewisianus* ZB
13.0	F	B	*Globorotalia fohsi lobata*
13.4	R*	B	*Actinoma golownini*
13.5	F	B	*Globorotalia fohsi fohsi* ZB, N11–N12
13.5	D*	B	*Nitzschia denticuloides*
13.6	D**	B	*Denticulopsis hustedtii*
13.6	N**	T	*Cyclococcolithus floridanus*
13.8	R***	B	*Didymocyrtis laticonus*
14.1	D***	T	*Cestodiscus peplum* ZB
14.0	N	T	*Sphenolithus heteromorphus* ZB, NN5–NN6
14.0	F	B	*Globorotalia fohsi praefohsi* ZB, N10–N11
14.2	D*	B	*Denticulopsis hustedtii*
14.2	R*	B	*Cycladophora humerus*
14.3	R**	T	*Carpocanopsis bramlettei*
14.3	F	T	*Globorotalia archaeomenardii*
14.4	D*	T	*Denticulopsis maccollumii*
14.6	F	B	*Globorotalia peripheroacuta* ZB, N9–N10
14.6	R***	T	*Calocyletta costata*
14.9	D*	B	*Actinocyclus ingens nodus*
15.2	F***	B	*Orbulina suturalis* ZB, N8–N9
15.2	R	T	*Dorcadospyris dentata*
15.2	D*	B	*Actinocyclus ingens*
15.3	D*	B	*Nitzschia grosspunctata*
15.3	BF	B	*Cibicidoides wuellerstorfi* group
15.5	R***	B	*Dorcadospyris alata*
15.5	D	B	*Actinocyclus ingens*
15.5	F	T	*Globigerinoides diminutus*
15.6	F	B	*Globorotalia archaeomenardii*
15.6	R**	B	*Litheropera renzae*

Table 5.2 (cont.)

Age (Ma)	Fossil Zone	Datum level	Taxon
15.6	D*	T	*Nitzschia maleinterpretaria*
15.7	N	B	*Calcidiscus macintyrei*
15.8	R	T	*Didymocyrtis prismatica*
15.9	N***	B	*Discoaster exilis* ZB, NN4–NN5
15.9	R	T	*Lychnaeonoma elongata*
16.0	F	B	*Globigerinoides mitra*
16.0	N	T	*Helicosphaera ampliaperta*
16.2	R	T	*Carpocanopsis favosa*
16.2	D	T	*Thalassiosira fraga*
16.3	D***	B	*Cestodiscus peplum* ZB
16.35	F	B	*Globigerinoides bisphaericus*
16.4	F	B	*Globigerinoides sicanus* ZB, N7–N8
16.5	F	B	*Globorotalia peripheroronda*
16.7	D*	B	*Denticulopsis maccollumii*
16.8	F*	T	*Globorotalia zelandica*
17.25	R*	B	*Eucyridium punctatum*
17.3	F	T	*Globorotalia zelandica incognita*
17.3	R***	B	*Calocyletta costata*
17.35	R***	B	*Dorcadospyris dentata*
17.4	N***	T	*Sphenolithus belemnos* ZB, NN3–NN4
17.5	N***	B	*Sphenolithus heteromorphus* ZB
17.6	F***	T	*Catapsydrax dissimilis* ZB, N6–N7
17.65	F	B	*Globorotalia zelandica*
17.8	D***	B	*Crucidenticula nicobarica* ZB
18.1	F***	B	*Globigerinatella insueta*
18.2	F	T	*Globigerina binaiensis*
18.2	N***	B	*Sphenolithus belemnos*
18.7	N***	T	*Triquetrorhabdulus carinatus* ZB
18.85	F	B	*Globorotalia praescitula*
22.1	N	B	*Discoaster druggi*

fossil groups: F, planktonic foraminifer; D, diatom; R, radiolarian; N, nannofossil;
BF, benthic foraminifer
T, top of species range; B, bottom of species range
*, Antarctic range
**, tropical range
***, direct palaeomagnetic correlation
****, North Atlantic range
ZB, zonal boundary
(Woodruff and Savin, 1991)

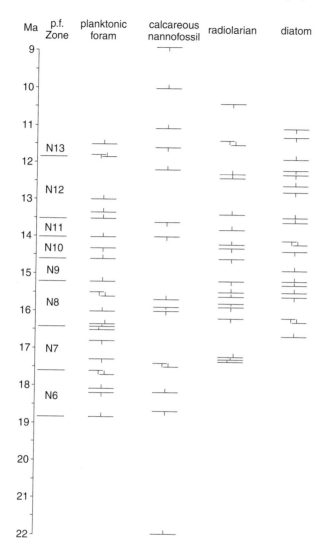

Figure 5.5 A visual display of the events in Table 5.2. Species names are omitted. Being able to ordinate events among four major taxonomic groups, in phyto- and zooplankton with calcite and opal shells, not only increases temporal resolution but permits correlations between different oceanic facies realms and biogeographic regions.

oxygen events found in some or most of nine holes. For these events, ranging in age from 11.49 Ma to 16.64 Ma, the standard deviations of the average ages are better than 160 thousand years in every case but one, and commonly are better than 100 thousand years. An important outcome of such work is the integration of stable isotopic events into Cenozoic geochronology (Miller *et al.*, 1991; Hodell,

1994; Hodell and Woodruff, 1994; Flower and Kennett, 1995). Thus reversible events, controlled and calibrated by independent evidence, are now a powerful tool in the pre-Pleistocene.

Sequence biostratigraphy

Integrated geochronology revisited

Biostratigraphy has made two major advances in recent years and it is of great significance that both have been in thoroughgoing collaboration with other scientific disciplines; neither arose out of a *selfcontained* intellectual break-through in our insights into fossil patterns in space and time. One advance has been the development of integrated systems culminating in the IMBS and in the selection and characterization of boundary stratotypes for the classical stages, as outlined in Chapters 3 and 7. The other advance has been the provision of a new physical-stratigraphic framework within which 'sample-based disciplines', such as biostratigraphy and geochemistry, can be evaluated (Loutit *et al.*, 1988). The framework developed when seismic stratigraphy arose in the 1970s and led to the recognition of depositional sequences, 'fundamental stratal units' that are bounded by unconformities and their correlative surfaces and are recognized and correlated in the new discipline, sequence stratigraphy. Loutit *et al.* (1988) put this advance into context (Fig. 5.6) and I emphasize its powerful unifying force in Figure 5.7 – for biofacies and correlation are tied together more strongly than they have been for a long time. And not only fossils: the integration of all disciplines has increased (Fig. 5.8).

There were three roots to the development of sequence stratigraphy. (Wilson (1998) listed nineteen events, from Hutton and Lyell onwards, comprising the foundations.) One, the most comprehensive in its influence, was the rise of seismic stratigraphy and its demonstration that bedding planes or bundles of bedding planes could be traced from one facies to another, and that they pinch out by onlapping in the advance (including transgression) and downlapping in the retreat (including regression). Seismic stratigraphy 'has initiated a revolution in stratigraphic analysis as profound as that caused by plate tectonics' (Cross and Lessenger, 1988), not least because sedimentary distributions could be predicted in favourable circumstances from the gross geometry of seismic patterns, obviously of interest in basin analysis and sedimentary geometries and processes (Fig. 5.9)

A second root was to take unconformity-bounded stratigraphic bodies (allo-stratigraphic units) seriously as something more than a frustratingly imperfect stratigraphic record. Instead of unconformities being apologized for as the main

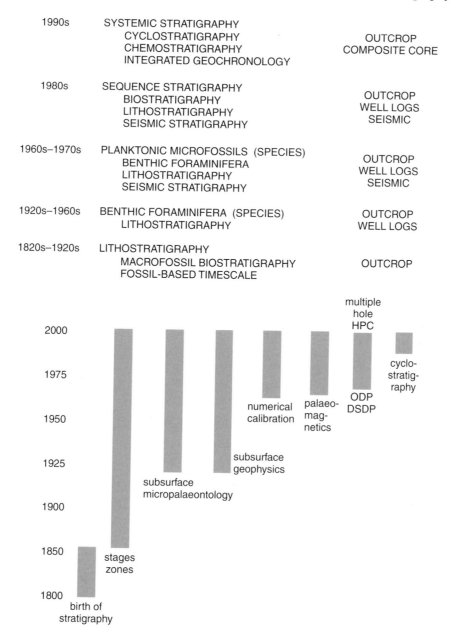

Figure 5.6 (*Upper*) Major periods in the development of stratigraphy and stratigraphic disciplines and their data sources. (*Lower*) The timing of major events in the development of stratigraphy. (Both adapted from Loutit *et al.*, 1988.)

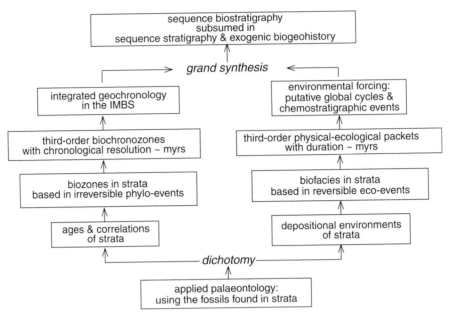

Figure 5.7 Sequence biostratigraphy – the uniting of the two major signals from the fossil record, namely, age and environment. Left pathway, the development of classical biostratigraphy with its ranges and zones based on key species, enfolded into the integrated geochronology of recent times. Right pathway, fossils and biofacies in ecostratigraphy, yielding packets of strata and fossils, cycles and sequences. Both pathways are essential to progress in sequence biostratigraphy. (McGowran and Li, 2002, Fig. 3, with permission)

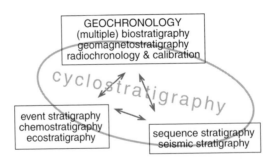

Figure 5.8 Interaction of the major stratigraphic disciplines in modern biogeohistory, to emphasize that cyclostratigraphy actually or potentially pervades the three major areas – the 'classical' geochronology culminating (in the Cenozoic) in the IMBS, the various events, physical and biological, and seismic and sequence stratigraphy.

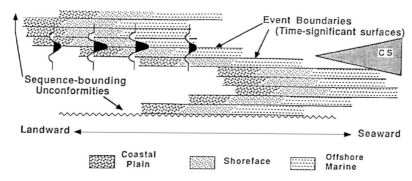

Landward ◄──────────────────────────────► Seaward

▨ Coastal Plain   ▨ Shoreface   ▨ Offshore Marine

Figure 5.9 Basis for seismic stratigraphy (Cross and Lessenger, 1988, Fig. 4, with permission). Their caption reads: Depiction of hierarchical stacking patterns of progradational units within the context of seismic sequences. Above a sequence-bounding unconformity, progradational units step progressively seaward, then become vertically stacked then step landward. The condensed section (labelled CS) corresponds in time to the landward-stepping events, and its duration expands in a seaward direction. Subsequent progradational units are vertically stacked and begin a seaward-stepping phase prior to the development of the upper sequence-bounding unconformity. Durations of unconformities expand in a landward direction. Schematic zero-phase couplets ... show that reflections are generated at the time-significant surfaces bounding the progradational units. The amplitude or phase of the wavelet may vary along those surfaces as a result of difference in impedance contrasts generated by superposition of different sedimentary facies across event boundaries.

manifestation of an all-too-imperfect geological record, they are information-rich and to be cherished, especially since Sloss *et al.* (1949) and Sloss (1963) proposed sequences and demonstrated cratonic sequences at the Phanerozoic scale separated by continent-wide breaks. Allostratigraphic units also emerge from the deep as oceanic sedimentary sections are becoming recognized as more hiatus-ridden than many have assumed hitherto (Aubry, 1995).

A third sequence-stratigraphic root was the sedimentary cycle (Israelsky, 1949; Fig. 5.10) and the rising, systemic-stratigraphic belief that cycles of transgression and regression might be regional signals of a global eustatic rhythm that has been modified but not blotted out by local sedimentary swamping or regional tectonism. Eustasy, the nineteenth-century concept of Eduard Suess (Datt, 1992; Hallam, 1993), has a new lease of life. Understanding 'accommodation space', wherein sediments can accumulate, requires the disentangling of eustasy from isostasy. A marine transgression can be very clearly signalled in the strata – but was it a rise in sea level or a regional crustal subsidence that let the sea in? An accommodation model yields a model of coastal onlap and, in its turn, a curve of inferred eustatic sea level. The 'global curve' (Haq *et al.*, 1987, 1988; Hardenbol *et al.*, 1998) has been controversial (Miall, 1997, 2004; Miall and Miall, 2001, 2002, 2004)

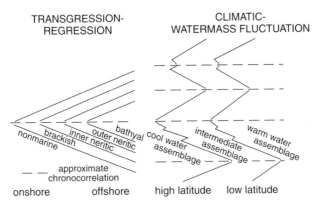

Figure 5.10 Potential chronological employment of reversible events in event correlation (McGowran, 1986a). Left, the Israelsky oscillation of transgression–regression, as in many texts (e.g. Eicher, 1976 , Fig. 4–5). Right, 'acme events' in the cycles of translatitudinal migrations by planktonic protistan communities (based on Haq, 1982).

but it is important to affirm that inferred eustatic configurations are not the core of sequence stratigraphy which will survive any amount of eustatic/isostatic controversy.

What is sequence biostratigraphy?

Biostratigraphy and the configuration of biotas in depositional sequences can be termed sequence biostratigraphy, which has three major components each focusing on a central problem.

i.   In applying micropalaeontology to depositional sequences in sedimentary basins one objective is to identify the components of the sequence as biofacies, which might be termed *sequence biostratigraphy in basin analysis*. The main concerns are intrabasinal correlations and facies and environments, especially palaeodepths of deposition.

ii.   The chronological correlation of sequences to construct and test a global configuration, the *geochronology of sequence biostratigraphy*, is commonly but not necessarily synonymized with timing the shifts in a putatively global sea-level curve. The main concerns are the linking of regional insights to global – especially oceanic – scenarios of geochronology and environmental change.

iii. The *ecostratigraphy and palaeobiology of sequence biostratigraphy* is a rather more strongly palaeontological approach than merely the identification of palaeoenvironments and ages. It exploits the fact in (ii) that there is now available for the first time, for studying the distribution and

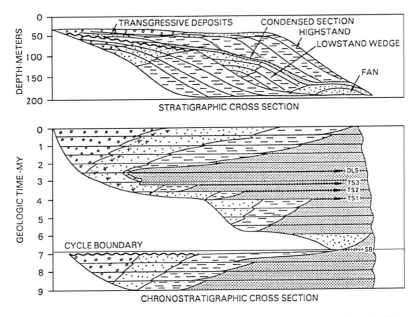

Figure 5.11 The Exxon clinoform model and three major surfaces (Loutit *et al.*, 1988, Fig. 9, with permission). The wedge of sediment in a depth-distance diagram (upper) is exploded into a time-distance diagram (lower). The three major physical-stratigraphic surfaces are the sequence boundary (SB), transgressive surface (TS1) and downlap surface (DLS).

evolution of ancient (late Phanerozoic) communities, a reasonably cogent if not consensual physical scenario of sequences together with proxies for temperature, nutrient levels and sea levels at the second and third orders. Its concerns are such questions as environmental forcing of evolution and coherence of communities over long timescales.

Depositional sequences are made up of parasequences – sedimentary cycles sandwiched by flooding surfaces and commonly displaying such classical features as upward-shallowing and Waltherian facies relationships (gradational, with lateral changes matching vertical changes, as in the textbooks). Parasequences come in sets which extend intra-parasequence trends to display prograding, aggrading and retrograding – outcomes of differing balances between sediment supply and accommodation. Tracts of parasequences are the lowstand, transgressive and highstand systems tracts of the depositional sequence (Fig. 5.11). The three key physical surfaces of the depositional sequence (*sequence* for short – a highly useful general word most unfortunately pilfered for this more specific meaning) are the sequence boundary, the transgressive surface and the downlap surface. The sequence boundary is the most

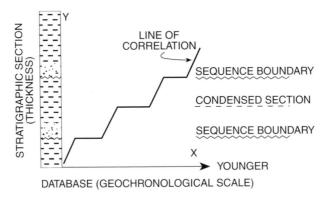

Figure 5.12 Sequences show marked variations in accumulation rates when subjected to a thickness–time plot. Highly schematic (no scales; no indication of chronological controls or resolution), this diagram displays stratigraphic discontinuities at sequence boundaries and the condensed section at the downlap surface. From Carney and Pierce (1995, Fig. 2) with permission.

widespread, recognizable from non-marine to oceanic environments and defined by stratal onlap or truncation. Marine transgression forms a surface as the high-energy zone advances inland, tending to bring a zone of sediment starvation behind it; this is the marine flooding surface. At the time of most extensive sediment starvation (the maximum flooding surface), the turnaround begins and prograding or seaward-downlapping commences. In a highly simplified graphic plot of thickness–time (Fig. 5.12), sequences are shown as alternating in their accumulation rates between relatively high and very low, the latter characterizing the boundary and the maximum flooding surface. As to age ranges: parasequences seem to fall mostly in the 10^4–10^5 years band and sequences mostly in the 10^6 (range 10^5–10^7 years). Wilson (1998) suggested that in greenhouse times third-order sea-level change produces sequences at 10^{6-7} years, whereas in icehouse times fourth- to fifth-order change produces sequences at 10^{4-5} years.

Olsson (1988) discussed foraminiferal modelling of sea-level change. Holland (2000) summarized and predicted from modelled fossil distributions in sequence stratigraphy, the testing of stratigraphic ranges (first and last appearances), species abundances, polyspecific abundances (biofacies), and morphological changes in lineages. The stratigraphic range of a fossil species in the rocks, hence first and last local appearances, are affected by four important biases – sampling, unconformity, facies and condensation (Figs. 5.13, 5.14). Sampling bias causes an abrupt event to be smeared into a gradual pattern in a local succession (the Signor–Lipps effect). Unconformity bias conflates events that may have been spaced through the hiatus of erosion and non-deposition, clustering them in the stratigraphic section. Zonal boundaries in the rocks will tend to fall at sequence

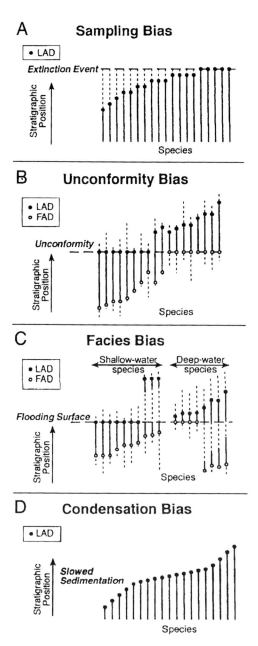

Figure 5.13 Four significant biases of stratigraphic ranges recognized in sequence-stratigraphic models of the fossil record (Holland, 2000, Fig. 4, with permission). Sampling bias spaces out LADs that actually were clustered; it decreases with denser sampling. Unconformity bias clusters LADs and FADs that actually were spaced. Facies bias clusters events at abrupt facies changes, such as a major flooding surface. Condensation bias bunches up events in condensed sections.

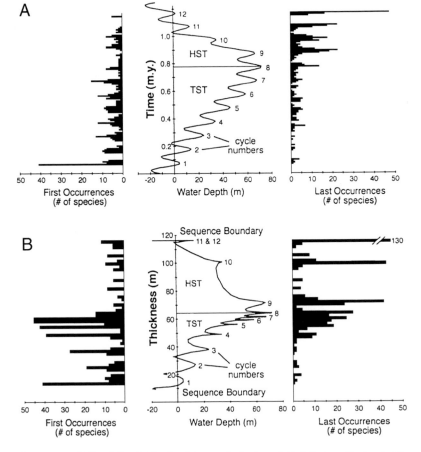

Figure 5.14 Holland (2000, Fig. 5, with permission) modelled first and last appearances through a depositional cycle (A, duration 1.2 myr equivalent to B, sediment thickness ~120 m) consisting of a dozen smaller-scale cycles. Against the timescale, first and last appearances bundle respectively near the base and the top of the sequence. Against the thickness scale, clustering increases where cycles are thinned by condensation or non-deposition.

boundaries (see below). Likewise with facies bias – ranges are truncated by abrupt facies shifts such as flooding surfaces or rapid regressions. Both will occur together at the sequence boundary. Again, condensation bias will cluster tops and bottoms especially at the maximum flooding surface. Holland fleshed out these patterns by modelling abundances of individual species and biofacies (Fig. 5.15). Not surprisingly, the maximum flooding surface and sequence boundary loom large, and the effect is enhanced by the likelihood of coincident apparent shifts in different clades, so that biofacies shifts also are abrupt. As to morphological change: a lineage ranging over more than one sequence might be

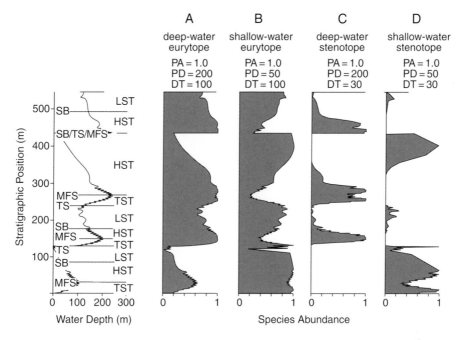

Figure 5.15 Holland (2000, Fig. 7, with permission) modelled changes in species' abundances through depositional sequences. (Modelling parameters: PA, peak abundance; PD, preferred depth; DT, depth tolerance. Depth is binary – shallow and deep; tolerance is binary – eurytopic and stenotopic.) Abrupt changes in species abundance tend to occur near SBs and MFSs and to occur synchronously, giving abrupt changes in biofacies. More gradual changes tend to occur in LSTs and HSTs.

chopped into disjunct segments in its record in the section, and changes through time will give the appearance of distinct successional morphotypes – probably the source of systematic oversplitting in the past and perhaps of interpreting iterative evolution instead of succession (Fig. 5.16).

On the craton, depositional sequences tend to lack the lowstand tract so that the sequence boundary and transgressive surface are together; also, the transgressive tract may comprise several flooding surfaces, closely spaced if sediment accumulation rates are low (more characteristic of the TST than the HST). Taken together, these configurations predict that the TST plus SB is the locale of most turnover. Holland concluded from the modelling that the fossil record was highly episodic, even though the actual biohistory was characterized by stability in rates of origination and extinction and in ecological and community structures. Shifting environments not only influence (control?) species and communities – they also shape the fossil record. In investigating such questions as the effect of environmental impact on ecology and evolution we first have to deconstruct the stratigraphic impact on the fossil record.

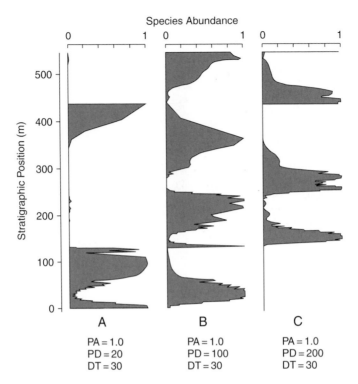

Figure 5.16 Holland (2000, Fig. 9, with permission) generated an artefactual pattern of iterative evolution. (Parameters, water-depth and sequence architecture as in Fig. 5.15. Only preferred depth differs among these three species – C deepest, A shallowest.) The changes are pronounced and successional but not symmetrical as in an Israelsky diagram (Fig. 5.10) – note the abrupt replacement of A by C. Nor is this pattern due to iterative evolution; Holland suggested that small changes in, e.g., C could tempt oversplitting. These spikes could model epiboles (Chapter 6).

Sequence biostratigraphy in basin analysis

This is the application of micropalaeontology to depositional sequences, clarifying the intra-sequence and inter-sequence architecture of a sedimentary prism or basin fill (Simmons and Williams, 1992; Martin *et al.*, 1993). Thus integration with sedimentary analysis, seismic profiles and down-hole logs is paramount. This facet of sequence biostratigraphy is basinwide in scope, using rapidly determined ratios such as the planktonic–benthic ratio, which broadly is higher at the transgressive surface and at or near the max-imum flooding surface than in the other parts of the third order sequence. More generally, this technique develops microfossil signatures of the major parts of the sedimentary sequence. The age-component of a fossil signal helps

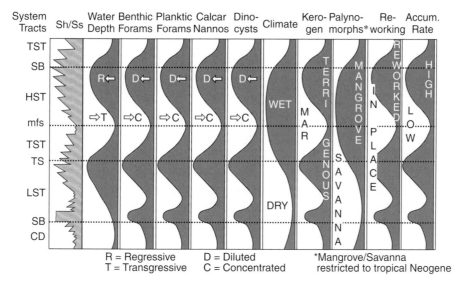

Figure 5.17 Armentrout (1996, Fig. 17, with permission from the Geological Society) produced a simplified, smoothed, schematic model of microfossils in depositional systems tracts in the interplay of water depth, climate, and sedimentary accumulation. The patterns will be spiky and details will vary along spike, of course.

identify the sequence; the environment-component contributes to biofacies, palaeoenvironment, palaeodepths. Together with shale–sand ratios, kerogen types, reworking, and climatic generalizations, microfossils in clastic wedges were sketched by Armentrout (Fig. 5.17). All of these indicators are rapidly determined and cost-effective in industrial palaeontology. Martin *et al.* (1993) demonstrated how high-resolution biostratigraphy could be used operationally and industrially in a sequence–stratigraphic content in the Gulf of Mexico; Wakefield and Manteil (2002), likewise in the Indus Basin, Pakistan.

Marine microfossils are expected to reach their higher abundances in fully marine facies in conditions of sediment starvation (but not nutrient starvation), i.e. in condensed sections. These conditions are found at marine flooding surfaces, especially at the maximum flooding surface, where benthics will indicate most deepening. Conversely, trends upwards towards a sequence boundary will include biofacies evidence for shallowing together with decreasing specimen numbers. In addition, the condensed sections will have higher infaunal numbers and faunas higher in their tolerance to lowered oxygen levels, in keeping with increased organic content and authigenic minerals especially glauconite. Anticipating Holland's modelling, Vail *et al.* (1991) had already generalized that the TST is the characteristic zone for fossil change, faunal and extinction events and stage boundaries, and that fossil abundance and diversity both peak in the

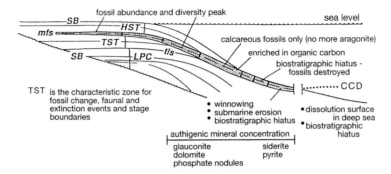

Figure 5.18 Generalizations, including fossil, mostly on the maximum flooding surface extending from shallow-neritic to deepwater settings (Vail *et al.*, 1991, Fig. 20). With permission.

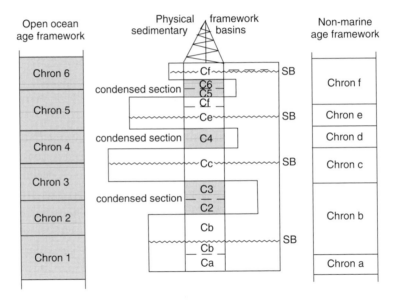

Figure 5.19 Loutit *et al.* (1997, Fig. 2, with permission) sketched a section displaying guide fossils from the pelagic realm interleaved with guide fossils from the terrestrial realm – a highly valued configuration for two centuries' biostratigraphy, with the recent addition of sequence boundaries strengthening the integrated geochronology.

MFS (Fig. 5.18). The MFS extends into the deep ocean where dissolution of aragonite then calcite increase, and hiatuses are clustered (see below).

A central problem in basin analysis is the crossover at the strand – the physical link between marine and non-marine frameworks, whereby two independent ordinations have to be turned into one. Loutit *et al.* (1997) emphasized the value of the sequence boundary in providing this link (Fig. 5.19).

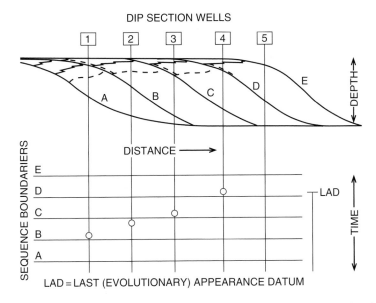

Figure 5.20 A schematic section plotted against both depth and time showing the distribution of microfossils in prograding sequences (from Loutit 1992, unpublished, with permission). Compare with the diachronous 'climbing' of the *Heterostegina* biofacies in the Gulf Coast (Chapter 1). The LAD is found only at one of the four species tops logged – the highest.

The geochronology of sequence biostratigraphy

This is the use of standard or classical biostratigraphic events to relate depositional sequences to an integrated global geochronology. The problem is this: How do we project the highly integrated geochronology, developed in oceanic facies and embodied in the IMBS, into stratigraphic sections on continental margins? (See Fig. 5.19.) This biostratigraphic problem has two aspects: a 'proximal' problem which is the distribution of species within the tracts of the depositional sequence, and a 'distal' problem of how fossils might be employed to integrate depositional sequences into a putative global model (including a eustatic sea-level curve).

Two sketches illustrate the proximal problem (Fig. 5.20). Within a depositional sequence, species distribution may display a strong variation across the clinoform in responding to rapid environmental shifts. Between prograding depositional sequences, the topmost occurrence of a given index fossil will 'climb' in the basinwards direction, forcefully recalling Lowman's demonstration of *Heterostegina* climbing basinwards (Fig. 1.13). The pattern is ecostratigraphic with the environmental controls linked to sequence patterns.

The distal problem in the geochronology of sequence biostratigraphy is the refining of biostratigraphic events – datums, evolutionary first and last appearances refined in open-oceanic facies – against marginal sequences as well as against geomagnetochronology and the IMBS, and against the ^{87}Sr–^{86}Sr curve (Hodell, 1994). The correlation and calibration of the putative global second- and third-order sequences on continental margins and plateaus at the same time tests the actual existence and chronological position of 'the' sequence boundary and 'the' maximum flooding surface. To the extent that this testing can be undertaken and the physical surfaces can be pinned down chronostratigraphically, so too can their identification become an integral part of correlation and age determination in neritic-continental stratigraphy. This exercise is global in scope. The controversial aspects of 'global sea level' are three-fold: (i) Does it exist – is there such a thing as 'the' sea-level curve, beyond a bundle of regional curves, each isostatically distorted? (ii) How many and how old are the sequence boundaries (and other surfaces)? (iii) What were the actual amplitudes in meters of the putative eustatic cycles? The latter is a forbiddingly difficult task, for to achieve it we must disentangle the triad of processes controlling the stratigraphic record on the continental margin: namely, eustasy, tectonics (including thermal subsidence, isostasy, compaction, flexure), and sediment supply (Miller, 1994).

Vail *et al.* (1991) sketched their notion of a great Neogene sedimentary wedge – a stratigraphic signature, implicitly global, based on dates and amplitudes in Haq *et al.* (1987, 1988) (Fig. 5.21). Thus the second-order bundling of third-order packages was and is essentially correct, notwithstanding underestimates of the third-order packages in the Neogene (e.g. Martin and Fletcher, 1995) largely corrected in the (Hardenbol *et al.*, 1998) version.

The last in this series of conceptual models of the world of sequence stratigraphy suggests the location of the two most common chemostratigraphic profiles (Fig. 5.22). It shows the case of oxygen and carbon varying through time as approximate mirror images. This is a plausible situation – δ^{18}O peaks early in the lowstand at a time of cooling, perhaps with the glaciation associated with the sequence boundary, perhaps due to maximum continentality. At this time there is minimal fixing of organic (light) carbon in the neritic and terrestrial realms and maximum erosion of previously fixed carbon, and maximum recycling by vigorous circulation. Early in the highstand there is maximum neritic space, both for trapping heat and organic production, and highs in humidity and forestation around the shore; there is also less vigorous circulation and recycling. In this diagram Abreu and Haddad commit very clearly to unconformities recording coeval hiatuses in the neritic and abyssal environments – erosional hiatuses linked climatically. The alternative possibility is that

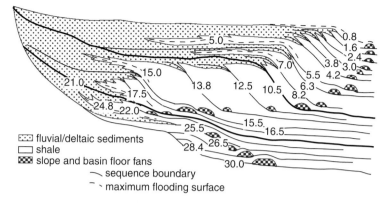

Figure 5.21 Vail's stratigraphic signature for the Neogene, modified from Vail *et al.* (1991, Fig. 12, with permission). It shows an idealized section across a continental margin with terrigenous sediments supplied from the left. Advances basinwards by fluvio-deltaic sediments represent marine retreats, alternating with marine advances (transgressions each culminating at a maximum flooding surface). Each advance – retreat comprise a third-order package of strata or cycle; these units are packaged in turn into three supercycles (heavy lines) spanning the past 30 myr. The two carbon-isotopic excursions in Fig. 8.4 match the first and second supercycles. The Miocene optimum peaks at the 15.0 Ma peak; chill III (Chapter 6) is at the 13.8 boundary and the lowpoint is at the 10.5 Ma boundary. The Pliocene reversal accompanied the 5.0 Ma flooding. The number and ages of the Neogene sequences has been revised (Hardenbol *et al.*, 1998) but the general anatomy and principles shown here remain intact. With permission.

oceanic and neritic hiatuses alternate through time, the oceanic being associated with sediment starvation and more aggressive carbonate dissolution at the maximum flooding surface (McGowran, 1986b). This possibility is clearly illustrated in Figure 5.25. Like all reconstructed or modelled patterns in stratigraphy, chronological resolution and correlation would settle this matter.

Fossils and sequences in the early Palaeogene: the Wilcox in the eastern Gulf Coast

Recent studies of stratal successions that are mature in the sense of having a long history of stratigraphic and micropalaeontological investigation include Olsson and Wise (1997), Olsson (1991) and Olsson *et al.* (2002) on the New Jersey margin and Mancini and Tew (1991, 1995) on the eastern Gulf Coast. We look at the latter. The physical stratigraphy displays several packages of marine and marginal facies separated by unconformities which mostly coincide with the boundaries of planktonic foraminiferal biozones (Fig. 5.23). Note that in this siliciclastic regime, the shelly horizons are on the flooding surfaces and lignites are in the highstands. Mancini and Tew used these sequences to distinguish varieties of sequence symmetry (Fig. 5.24) which go some way towards explaining

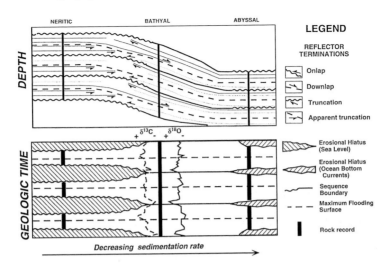

Figure 5.22 A sequence stratigraphic model with hypothetical profiles of carbon and oxygen isotopes (Abreu and Haddad, 1998, with permission). 'Conceptually', maxima in δ^{18}O are slightly younger than the sequence boundary and minima slightly younger than the maximum flooding surface. δ^{13}C is considered to covary with δ^{18}O. Note that deep-ocean erosional hiatuses are coeval with neritic hiatuses, unlike the starvation and dissolutional hiatuses which are chronologically offset from the SB by falling at the MFS instead (see, e.g., Figs. 5.9, 5.21, 5.25).

previous observations as to the position of the biozone boundary. They went further in demonstrating the relationship of the biozone to the biochronozone, the latter falling at the boundaries of the stacked condensed sections in the basin – but offering no evidence for such a neat configuration (Fig. 5.24). However, Holland (2000) observed that the confluence of depositional sequence boundaries and biozonal boundaries in this example is consistent with his modelling (above).

The ecostratigraphy and palaeobiology of sequence biostratigraphy

This is the taxic- and biofacies-profiling of second order and third order sequences, a programme to relate neritic fossil events and assemblages to sequences and to draw such problems as recurrent biofacies and communities and their hierarchical structure (Chapter 6) into a physical framework based on stratigraphic sequences. As Holland (1995) expressed it, any fundamental change in stratigraphic thought should stimulate an examination of palaeontological thought. This field begins at the regional or basinal level but is potentially global in its scope. Put biofacies or recurrent fossil assemblages into sequences and you have this component of sequence biostratigraphy. Later (1999) Holland christened the *New Stratigraphy* which mostly comprises sequence stratigraphy, in which the rock record is divided into genetically

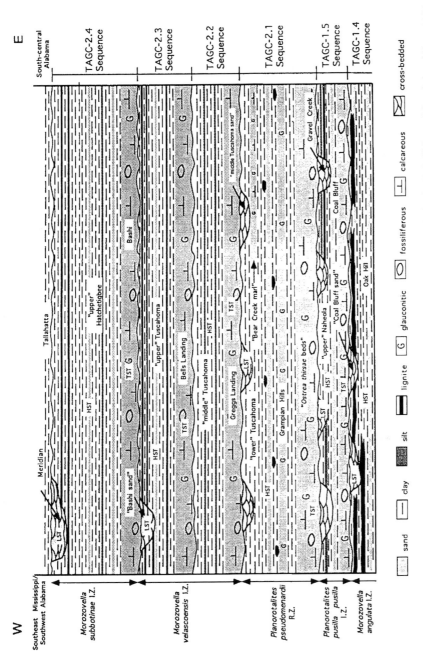

Figure 5.23 Stratigraphy, biostratigraphy and facies of the Wilcox Group, Alabama–Mississippi, US Gulf Coast (Mancini and Tew, 1995, Fig. 4, with permission) displaying unconformity-bounded packets of strata which are sequences. (I.Z., R.Z., interval and range zones.) Note that the zonal boundaries (left) are all at the unconformities identified as sequence boundaries and bearing pockets of cross-bedded sediment designated as lowstands. The calcareous, glauconitic and fossiliferous beds are transgressive and the lignites are in the highstands (most are interrupted by the cut-and-fill beginning the next sequence).

193

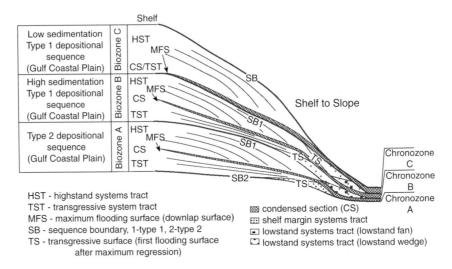

HST - highstand systems tract
TST - transgressive system tract
MFS - maximum flooding surface (downlap surface)
SB - sequence boundary, 1-type 1, 2-type 2
TS - transgressive surface (first flooding surface
       after maximum regression)

▨ condensed section (CS)
▦ shelf margin systems tract
▣ lowstand systems tract (lowstand fan)
▢ lowstand systems tract (lowstand wedge)

Figure 5.24 Schematic section of depositional sequences in a marginal transect showing biozones and chronozones in tight stratigraphic relationship to physical surfaces and sequence architecture (Mancini and Tew, 1995, Fig. 6, with permission). The sequences vary in distal (outer) facies and asymmetry (position of MFS).

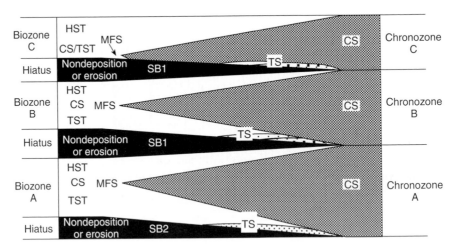

Figure 5.25 Schematic section of depositional sequences in a marginal transect showing biozones and chronozones as in Fig. 5.24, but with a time axis making unconformities into hiatuses (Mancini and Tew, 1995, Fig. 7, with permission). The boundary hiatus expands landward whilst the condensed section of the MFS expands (time-wise) seawards. A competing view is shown in Figure 5.22.

related packages bounded by unconformities. It asks the questions of eustasy and tectonics that were (wrote Holland) stifled (with exceptions) by the traditional bookkeeping of stratigraphy and its nomenclature. Among other virtues, the New Stratigraphy provides a physical framework for interpreting the fossil record palaeobiologically. Holland urged that palaeobiology add an alliance with the New Stratigraphy to its fruitful collaborations with such disciplines as geochemistry, geochronology and ecology.

In his crisp summary of four models of recurrent fossil assemblages, W. Miller (1993) contrasted the two *top-down* approaches – one is Boucot's, discussed in Chapter 6 – with the two *bottom-up* approaches. The former strategies subdivide whilst the latter build. In the Cenozoic we can take a third way, beginning with a stable physical scenario of sequences and proxies for temperature, nutrient and sea level at the second and third orders. Thus we can approach *from the side*, as it were – from the physical framework sketched at the correct timescales for assessing biofacies and faunal change. It is this strategy that give insights into 'external' environmental forcing vis-à-vis 'internal' or self-organizing community dynamics in chronofaunas (Chapter 6).

This sequence-biostratigraphic approach begins with the configuration of biotas in neritic sequences. I present two examples from southern Australia, one in the late Palaeogene and the other in the early Neogene. I carry the discussion further in Chapter 6, for these questions of stratigraphy and palaeobiology merge seamlessly.

Foraminiferal biofacies across the Eocene–Oligocene boundary: St Vincent Basin, southern Australia

The pivotal boundary horizon is a regional downcut and backfill followed by a transgression. Microfossil correlations are consistent with this event being the local manifestation of glaciation Oi1, which 'ought' to be at the Eocene–Oligocene (E–O) boundary (Chapter 7). There is strong correlation of this section with the sequences of the Bartonian, Priabonian and Rupelian Stages (Hardenbol *et al.*, 1998). The Tortachilla Limestone with *Acarinina collactea* is below SB Pr1; the Blanche Point Formation with *Isthmolithus recurvus* is above SB Pr2; Pr1 and Pr2 are buried in multiple hardgrounds at the contact. The downcut unconformity is bracketed between top *Globigerapsis index* (uppermost Eocene) and base *Cassigerinella chipolensis* (lowermost Oligocene); this uncontestably is SB Pr4–Ru1. Therefore, the base of the Tuit Member, a sequence boundary, 'must' be SB Pr3. Thus we have a moderately well-controlled succession (given the lack of geomagnetics) across a major global event and also a highly contrasting succession, from grey-green-almost-black, opal-rich sediments, below, to brown-yellow quartz-bryozoan-rich neritic sediments above (Fig. 5.26).

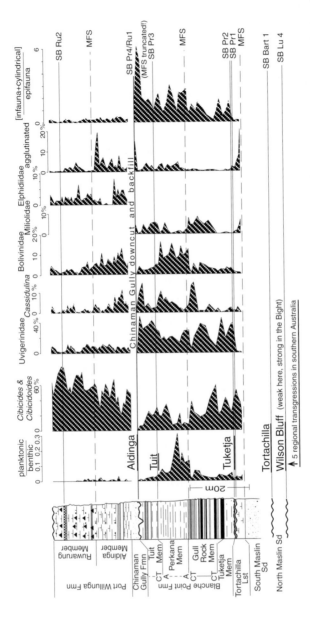

Figure 5.26 Biofacies across the Eocene–Oligocene boundary, St Vincent Basin, South Australia (Moss and McGowran, 2003). Note the unambiguous identification of the Hardenbol *et al.* (1998) sequence nomenclature. The basin was restricted by Kangaroo Island (a Caledonian massif) lying across its opening to the still-narrow Southern Ocean. The spectacular changeover between the essentially epifaunal Cibicididae and infaunal Uvigerinidae across the E–O boundary (together with outgoing Miliolidae and incoming Elphidiidae) parallels a change from grey-green opaline chemofacies to typically inner neritic bryozoan-quartz facies – a contrast reinforced by the infaunal index at far right. The 'Chinaman Gully downcut' is a downcut of some 50 m in the St Vincent Basin infilled by marginal marine siliciclastics succeeded by the upward-deepening Aldinga Member. Within the (Priabonian) opaline chemofacies, opal-CT is in the TST and opal-A in the HST. Note the pronounced biofacies changes, amounting to tens of percent in relative abundance, paralleling the opal-CT-A switches – Uvigerinidae, *Cassidulina*, Bolivinidae, and Miliolidae. These clades track environmental perturbations in this nutrient-rich environment more succinctly than do the clades in the early Oligocene. Similarities to modelled biofacies in sequences are apparent (Fig. 5.15).

The major biofacies contrasts across the E–O boundary are seen in the dominant families, the epifaunal Cibicididae and infaunal Uvigerinidae. There are three outstanding features: the decrease in infauna, the dampened amplitude of swings, and the most pronounced taxic overturn through the Priabonian and Rupelian (not shown) Thus there is a very strong local-neritic signal to one of the more significant global transformations during the Cenozoic Era – the onset of well-established ice sheets and the development of the psychrosphere.

There were several well-marked mineralogical changes in the late Eocene opal-rich section. The abrupt change fom opal-CT to opal-A is at the top of the darkest sediments richest in infaunal gastropods (*Spirocolpus*) and traces (*Thalassinoides*), exactly at a diagenetic change from hard–soft couplets to soft spicular marls and securely identified as the MFS (Gull Rock–Perkana boundary). This is the strongest change in biofacies, especially in the relay from Uvigerinidae to Bolivinidae in the dominant infauna, which is reversed at the abrupt reversal fom opal-A to opal-CT and reversion to hard–soft couplets, identified as the next SB and the Tuit transgression. The environment was restricted in circulation and planktonic numbers were very low. The changes at the E–O boundary are comprehensive – opaline to quartzose, infaunal-to epifaunal-dominated, grey-green to yellow-brown, sponge-rich to bryozoan-rich; broadly, from somewhat poorly aerated to well aerated (but planktonic numbers remaining low). The SBs are well marked but hiatuses are brief and constrained. It seems likely that the patterns shown are not greatly affected by the biases on the fossil record imposed by sequence-stratigraphic architecture, as outlined above. However, a taxic comparison with an open-neritic section (benthic diversities higher, planktonic numbers much higher; no significant opaline content or spicularites in the Eocene) shows several parallels (Fig. 5.27): incomings high above SB Pr2, no evidence of condensation at the MFS, highest outgoings in the topmost Eocene, incomings rise above Pr4–Ru1, faunal similarity highest immediately at the recolonization in the earliest Oligocene before provincial tendencies recur. We infer from these patterns that sequence-stratigraphic bias is not strong.

Figure 5.28 displays a three-way chronological relationship between (i) five regional, neritic, marine transgressions (Wilson Bluff to Aldinga), (ii) the third-order Exxon sequences TA3.5 to TA4.4 (Hardenbol et al., 1998) and, (iii) the five benthic positive spikes (implying sharp coolings?) in a $\delta^{18}O$ composite profile of bottom and surface waters (striped envelope, adapted from Shackleton, 1986) that alternate chronologically with the transgressions and cycles. These approximate correlations of regional marine transgressions, third-order sequences, and composite $\delta^{18}O$ profiles suggest, plausibly if not compellingly, that: (i) the positive, presumably cool spikes in the deep water profile could easily fit

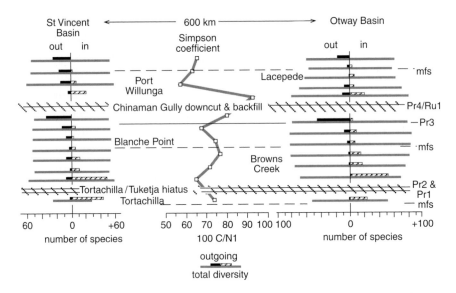

Figure 5.27 Taxic shifts across the Chinaman Gully event at the Eocene-Oligocene boundary (Moss and McGowran, 2003). Distribution of first and last appearances against species numbers across the Eocene–Oligocene boundary in open-neritic (right) and restricted-neritic (left) facies. The Simpson coefficient measures faunal similarity, which maximizes at the major environmental change, suggesting a temporary homogenizing tendency. The pattern of outgoings and incomings resembles a modelled pattern (Fig. 5.15) and presumably is due to a mix of change in nutrient regime (emphasized in Fig. 5.26) and unconformity bias (see Fig. 5.14).

chronologically between the marine transgressions on the southern Australian margin; and therefore that (ii) the three-way fit between deep-benthic cool spikes, marine transgressions and global third-order sequences TA3.5 to TA4.4 is good enough to corroborate the hypothesis that the following generalizations hold good at the *third* order:

> *sequence boundary* = *cool* = *regression* [= *contracted trophic resource continuum*]
> and *maximum flooding surface* = *warm* = *peak transgression* [= *expanded trophic resource continuum*]

Foraminiferal biofacies in the Miocene in East Gippsland, southeastern Australia

An oil-mine section in Gippsland samples the Miocene in an extratropical neritic environment which was exposed to two major and counterpointing influences, the East Australian Current (strengthening during global warming) and the Subtropical Convergence (strengthening during global cooling). Although it lacks geomagnetic and modern downhole logging (it was a 1940s wartime oil-exploration shaft) the section was well sampled before cementing-in.

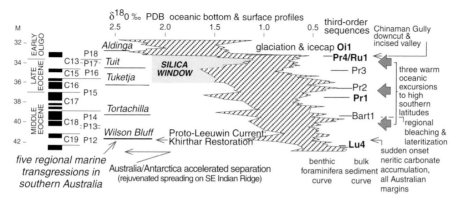

Figure 5.28 Third-order anatomy of the Khirthar restoration establishing a three-way chronological relationship between (i) five regional, neritic, marine transgressions (named Wilson Bluff to Aldinga), (ii) the six third-order sequences Lu4 to Pr4-Ru1 (nomenclature of Hardenbol *et al.*, 1998) and (iii) the five benthic cool spikes in a δ18O composite profile of bottom and surface waters (striped envelope, adapted from Shackleton, 1986) that alternate chronologically with the transgressions and cycles. Three arrows at right: three warming events in the Indian Ocean, inferred from warm, shallow-intermediate water penetrating to high southern latitudes (Zachos *et al.*, 1992). Silica window: a timeslice of neritic and oceanic opaline facies in the Southern Hemisphere including the Blanche Point Formation (Fig. 5.26). Diagram modified from McGowran *et al.* (1997a).

The foraminiferal profiles display reversible, quantitative characteristics reflecting depth, temperature, and nutrient. Our main conclusion here is that the patterns can be correlated plausibly with third-order global sequences. More 'community' aspects of this section are discussed in Chapter 6.

Interaction between the dominant planktonic groups is expressed in the cancellate/spinose ratio of which the main component is the *woodi–bulloides* ratio (Fig. 5.29). Broadly, the cancellate-spinose ratio is an oligotrophic–eutrophic ratio. Planktonic factor 1 reinforces the patterns (Fig. 5.30) (but includes on the higher nutrient side the microperforates, also a monophyletic group). The notion of the trophic resource continuum implies that there should be a broader range of environmental options at the Miocene climatic optimum (Chapter 6) – confirmed amply in the increased amplitudes in this timeslice in Fig. 5.29. This implies a component of temperature but of nutrient levels as well – increased numbers of the *Globigerina bulloides* group may indicate a cooler watermass or a more fertile watermass. To distinguish cooling from fertility in the water column, benthic profiles can act as an outgroup. Conversely, we have the benthic problem: if increased numbers of infauna imply increased buried food supplies, then is this due to increased productivity or to increased

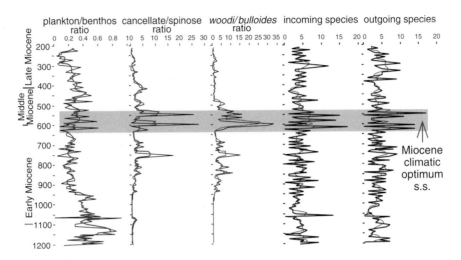

Figure 5.29 Plankton in the Lakes Entrance Miocene section: measures used to profile the planktonic foraminiferal succession (McGowran and Li, 1996, Fig. 8). The stippled interval corresponds to the Miocene climatic optimum at the zeniths of the various global curves (see Figure 8.4) and divides the succession into three parts. Light lines, sample-by-sample plots; heavy lines, three-point moving averages. P–B, plankton/benthos ratio; *woodi* s.l.–*bulloides* s.l. a subset of the cancellate–spinose ratio. Incoming and outgoing species: these plot the reversible appearances and disappearances of species, as distinct from the first and last appearances of conventional range charts.

preservation? – and the plankton can act as an outgroup. Benthic factor 1 expresses the interplay of two groups which happen to be the dominant epifaunal and infaunal benthic groups respectively (Li and McGowran, 1994, 1995). The former are the Family Cibicididae and the latter the Families Buliminidae plus Bolivinidae, all well founded taxonomically in robust clades. The parallels between factored plankton and benthos in Figure 5.30 are such that the warmer, relatively oligotrophic planktonic group (cancellate–spinose especially the *woodi* group) varies together with the benthic epifaunal group especially the Cibicididae; and the plankton flourishing in cooler or upwelling conditions (spinose globigerinids and microperforates) varies together with the benthic infaunal Bolivinidae and Buliminidae. In this way we distinguished several third-order episodes of upwelling in the section, essentially at reversals in the trend towards increasingly dominant epifauna (Li and McGowran, 1994). Towards the other end of the TRC, epifaunal dominance includes larger species which were (and those extant still are) photosymbiotic. Since those species are also good indicators of warming, we used them to indicate six third-order warm intervals. The same benthic data were rearranged to derive a palaeodepth curve

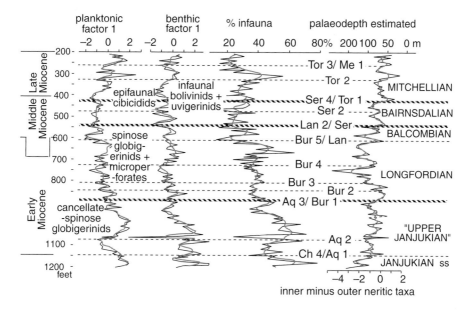

Figure 5.30 The Miocene foraminiferal-biofacies record at Lakes Entrance in east Gippsland, southeastern Australia (modified from McGowran and Li, 1996, Fig. 9). In the factored succession (two profiles at left), parallel scores for planktonic factor 1 and benthic factor 1 suggest intergroup associations, suggesting in turn the same environmental forcing factor – warmer, lower-nutrient waters versus cooler, higher-nutrient waters – thus identifying heightened production rather than enhanced preservation of C_{org} as the stimulus for raising the infaunal numbers. The epifaunal–infaunal ratio (shown as % epifauna with a three-point running average) trends towards epifaunal dominance at the top but with several reversals signalling upwellings. The palaeodepth curve based on biofacies encourages estimates of where sequence boundaries fall and of correlation with the Mi glacials (Fig. 6.33).

by estimating the relative abundance of inner, middle and outer shelf species (Li and McGowran, 1994, 1995). They are different ways of scrutinizing overlapping parts of the benthic dataset.

All four curves in Figure 5.30 show a similar second-order trend. The *woodi* group increases in the plankton, epifaunal benthics increase over infaunal, and the Lakes Entrance Platform evidently shallows. Regional stages, erected on lithological and molluscan grounds, and subsequently characterized biostratigraphically (Chapter 7), turn out to comprise stratal packages between natural breaks.

Predicting third-order sequences

Third-order variations are rather less clear. It is one thing to show various third-order variations including inferred upwelling and warming–cooling; it is another thing to relate such changes to a physical-stratigraphic framework. In the absence of outgroup control such as seismic or physical-stratigraphic interpretation, the main control is the inferred palaeodepth curve in Figure 5.30. Third-order biofacies changes should permit predictions of sequences – an aspect of sequence biostratigraphy. SBs should be at shallowing events and epifaunal peaks, whereas infaunal peaks should indicate burial of organic carbon, i.e. quieter and usually deeper water as well as upwelling. A simple one-to-one fit is not expected because there can be strong parasequence effects additional to third-order sequence rhythms. There should be changes at the MFS in both the plankton, responding to watermass shifts, and the benthos as the TRC changes with the sedimentary shift from condensation to prograding. Inspection of Figure 5.30 reveals abundant evidence of third-order change, but its patterns are more coherent on the broader second-order template than on the third-order.

There are sufficient planktonic foraminiferal biostratigraphic events to make an approximate match with a global scenario of third-order sequences and glaciations, the absence of isotopic or GPTS controls notwithstanding. The correlation is plausible in many respects, if not entirely compelling. There are biofacies signals of shallowing where they ought to be and there is more evidence for hiatus in the Lakes Entrance section after the main ice growth (Mi3) than before.

Such correlations are a promising beginning to the regional task of identifying regional stratigraphic packages and testing the timing of a 'global' curve. It may well be that third-order patterns, notorious for falling into the shadow zone between orbital patterns and tectonic patterns (Table 5.3) will be clarified by the use of templates and fine tuning, as developed for fourth to fifth order phenomena in the late Neogene and now rapidly pervading the Palaeogene and pre-Cenozoic. Strasser *et al.* (1999, 2000) developed arguments based on carbonate-dominated sections in the Oxfordian and Berriasian–Valanginian in western Europe. They found reason to correlate 'many' cyclostratigraphically recognized packages with the third-order packages of sequence stratigraphy. Their arguments turned on two points or possibilities: (i) If low-amplitude changes in insolation can be translated into low-amplitude fluctuations in sea level, then there is reason to relate nodes in the 400-kyr eccentricity cycles to the generation of third-order sequence boundaries. (ii) The authors found evidence to support this proposition – a long-term (*second*-order) rise in sea level will favour the enhancement of the *third*-order maximum flooding surfaces and attenuation of sequence boundaries; the corresponding fall in sea level will favour the reverse – attenuation of the MFSs and enhancement of the SBs. Thus, accommodation space might not be sufficiently appressed at one critical point in the 400-kyr eccentricity cycle during a fall to generate the facies

Table 5.3 *Hierarchy of sequences or cycles, based on Strasser* et al. *(2000), to emphasize that the third order is both the most familiar, being the one seen in outcrop, and the most difficult, being the meeting place of tectono-eustatic and glacio-eustatic causes*

Hierarchy cycles/packages	Usual suspects	Strasser *et al.* (2000) Mesozoic, neritic, carbonate-dominated successions
First-order	Tectonic and tectono-eustatic changes: Oceans open and close; supercontinents make and break	
Second-order	Tectonic and tectono-eustatic changes: seafloor spreading changes leading to MOR volume changes	Long-term second-order sea-level rise: third-order MFSs enhanced, SBs attenuated?
		Long-term, second-order sea-level fall: third-order MFSs attenuated, SBs enhanced?
Third-order	Combined tectonic and eustatic changes? Intra-plate stresses? Tectonic rifting and convergence superimposed on second-order MOR volume changes?	Large-scale: *some* 400-kyr cycles correlated with third-order sequence packages – low-amplitude insolation translated into low-amplitude sea-level fluctuation?
Fourth-order	Climatically controlled in Milankovitch frequency band	Medium-scale packages: ~400-kyr eccentricity
Fifth-order	Climatically controlled in Milankovitch frequency band	Small-scale packages: ~100-kyr eccentricity
Sixth-order	Climatically controlled in Milankovitch frequency band	Elementary packages: ~20-kyr precession

MOR, mid-ocean ridge; SB, sequence boundary; MFS, maximum flooding surface.

contrasts needed to justify the attribution of a sequence boundary, whereas that might happen the next time around. There are implications here to resonate with parallel cyclostratigraphic arguments from the pelagic realm, where the oxygen-isotopic proxy for a third-order glaciation (Mi1 at the Oligocene–Miocene = Palaeogene–Neogene boundary) falls at the confluence of low-amplitude eccentricity with low-amplitude variability in obliquity – unlike the nodes before and after in the 400-kyr eccentricity cycle (Zachos *et al.*, 2001b).

We do not yet have a Theory of Everything, tying together orbital dynamics, oceanic stable-isotopic signals of climate and productivity, and allostratigraphic sequences, but progress is discernible. The most sustained regional testing of putatively global patterns and their explanations is on the western North Atlantic margin (Miller and Kent, 1987; Miller, 1994; Miller *et al.*, 1987, 1991, 1998, 2003; Kominz *et al.*, 1998; see also Miller *et al.*, 1993). These studies, strictly stratigraphic in the most integrative sense, have delineated allostratigraphic packages (composite New Jersey sequences) separated by brief hiatuses, and shown good matches with the Exxon third-order global sequences and putatively global curve (although the estimated amplitudes of the latter in the 1987 version were too high). They have also shown good matches with oceanic oxygen-isotopic signals of third-order glaciations, implying glacio-eustasy not only in the Neogene with its uncontroversially large icecaps but in the Palaeogene and late Cretaceous. Sea-level amplitude changes of >25m in <1 myr would seem to indicate glacial punctuation of a 'greenhouse' world.

6

Biostratigraphy and biohistorical theory II: carving Nature at the joints

Summary

Geological timescales and phenomena in time series, biological classifications and assemblages as remnants of ancient ecologies – all of these constructions and reconstructions are hierarchical. There is a strong third-order parallel through time between depositional sequences and biozones based on speciations and extinctions. In the same window of 10^6 years there is also a rank in biofacies and community, i.e. the entities of ecostratigraphy. A 'Palaeozoic' notion of fossil assemblages coherently shifting in response to environmental shifts (e.g. sea-level changes) is not useful in Cenozoic biostratigraphy, where assemblage zones were abandoned long since. Although there are natural events in the sense of real ranges and real extinctions and speciations, there are not natural biostratigraphic units beyond that; integrated geochronology is splendidly opportunistic. Nature is carved at the joints more readily in phylogeny than in community. This is because the components of a phylogenetic tree are robust individuals tightly constructed, whereas communities are looser and less robust. Even so, there is a rhythm in biochronological resolution which can be related to rhythms in Earth history, perhaps responding more cogently to environmental notions such as the trophic resource continuum than to macro-evolutionary notions such as the effect hypothesis.

Hierarchy, tiering and the scales of time and life

The *genealogical hierarchy* comprises genome, deme, species, and mono-phyletic taxon. It is stable, it supplies the players, and it records the differential results, namely the outcome of the game of life. But the game of life is actually played out in the *economic hierarchy* – the economic aspects of organisms, avatars, and local and regional ecosystems (Eldredge, 1989). This point recalls G. Evelyn Hutchinson's evocative *Environmental Theatre and Evolutionary Play*. The economic hierarchy is a looser, less stable, and less consensual categorization than the genealogical hierarchy, but ecological clumping and cohering are the subjects of recent lively discussion. How loose are the units of the economic hierarchy? Do the community types persist because the species persist, or do the species persist because the economic system persists? (Eldredge, 1989.) DiMichele (1994) recalled an old confrontation between 'Eltonian' and 'Gleasonian' notions of 'community' – two end-member worldviews labelled after prominent ecologists. (a) In an 'Eltonian' world, communities are exclusive associations of interdependent and coevolving species each with its own special role or niche. Communities that are coevolved multispecies assemblages have emergent properties beyond properties of the constituent parts. One implication might be that a community will retreat in the face of environmental adversity and return when the good times return. (b) In a 'Gleasonian' world in contrast, communities are ephemeral or even happenstance associations of species with similar resource requirements – samples of a regional species pool that just happen to be there and able to cope: it could have been other species instead, and there is great ecological redundancy. In this case communities might be looser. The latter *individualistic dynamics* dominate in biotic responses to buffeting by Quaternary-scale climatic fluctuation, at least in terrestrial environments (DiMichele, 1994; see also W. Miller, 1993). Further back in time, on the other hand, there is persistence and recurrence in *ecosytem structure* and taxonomic composition for up to millions of years. Perhaps this is a scaling effect – perhaps we see individualistic dynamics up close whereas the organizational patterns are seen only in deep time. Or, on the contrary, perhaps we see a contrast between times of stability and times of disruption (Darwin's example of environmental forcing, the waxing and waning ice ages of the Quaternary, being the latter). Individualistic dynamics then would be the mode of post-crisis opportunists before a new, ordered system is established, and that mode is not typical of most of geological time. This suggestion recalls Kauffman's (1987) *uniformitarian albatross*: today's relatively resilient species being quite untypical of the fragile biotas characterizing most of Phanerozoic time (cf. McGowran, 1991). Boucot (1990a) called the contrast the 'Pleistocene paradox', to explain which he invoked both the different timescales and the changed global gradients brought on by late Neogene cooling.

Community ecology has shifted in the Gleasonian direction on this spectrum just as ecologists are learning to take the long view, discovering that neontology cannot by itself predict major patterns in the biosphere at 10^5–10^8 years scales and must rely on palaeontology. Eltonian notions have emerged mostly from studying the faunas in the neritic realm of the lower Palaeozoic, like the recurrent biofacies model advocated by Boucot (1990a,b, 1994) and the 'coordinated stasis' pattern in which a fauna can remain relatively stable both morphologically and taxonomically for a long period before abruptly changing (Brett and Baird, 1995; Holland, 1995).

Table 6.1 presents an ecological hierarchy based on a discussion of hierarchy by DiMichele (1994). DiMichele emphasized that such schemas are far from settled and, likewise, some superfluity is apparent here: e.g. the need for four levels in the benthic invertebrate column is not clear. There is an approximate comparison between living and fossil hierarchies, and a still more tenuous comparison with the hierarchy of eustatic cycles (Haq *et al.*, 1987; Hardenbol *et al.*, 1998b). The tabulation makes at least the point that we have some way to go to attain clarity and consensus. Table 6.2 makes the point that it is no easy matter to take an ecological process operating in ecological time and extrapolate it to geological time. The example is ecological succession (Gili *et al.*, 1995) which ostensibly is biotically driven on a level playing field, but whose patterns may look similar to an environmentally driven biotic succession at the longer timescales.

Table 6.3 presents a temporal hierarchy as the tiers of time and life – hierarchies and the concept of tiering in Earth history and the history of its biosphere. Ascending tiers I to IV at successive orders of magnitude of time identify something more than the mere accretion of events and phenomena from lower levels. The essence of 'tiering' in temporal hierarchy (Gould, 1985, 2002; Bennett, 1990, 1997) is in the perceived explanatory disjunctions between levels – the notion, for example, that mass extinction is different in kind from background extinction, or species selection from natural (organismal) selection. *Level I* is the domain of biology and geology in the modern environment and the action of natural selection as evolutionary agent. *Level II* is the domain of the great climatic swings of the Pleistocene ice ages, of high resolution stratigraphy (Kauffman *et al.*, 1991; Woodruff and Savin, 1991), orbital tuning (Hilgen *et al.*, 1993), and of cyclostratigraphy and Milankovitch cycles in greenhouse times (Fischer, 1986). Orbital forcing is the evolutionary agent (Bennett, 1990, 1997). *Level III* is simultaneously where much of the action is and where there has been most neglect; and it is on this level that the present study of biostratigraphy focuses. It is the time band of the *third-order cycles* of sea-level change, the level actually seen in geological outcrops, the level of stratigraphic packaging most used by the working stratigrapher (Haq *et al.*, 1987; Vail *et al.*, 1991). It is the time band of the *Mi* glacials which modulate at 10^6 years the rise and fall of climate and sea level in the Miocene oscillation (Wright *et al.*, 1992). Raup (1991) flirted

Table 6.1 A preliminary attempt at an ecological hierarchy inspired by and based on a discussion of hierarchy by DiMichele (1994). At right is the sequence-stratigraphic hierarchy (Vail et al., 1991); there is not a direct lateral equivalence from ecological units to sequences, because the former are said to persist and recur (as indicated for several). However, there is some rough equivalence between sequences and the actual occurrence of these eco-units.

Ecological analogue	Fossil marine benthic invertebrate	Fossil vertebrate	Fossil plant	Sequence-stratigraphic chrono-equivalent(?)
Province	Province	Province	Floristic Province: regional-continental assembly of biomes	First-order sequence and megacycle in sea-level oscillation
Biome	Ecological-evolutionary unit: an assembly of regional faunas or community groups, which have same overall duration as EEUs: perhaps 5 to 20 million years	Chronofauna: conceptually very similar to marine community groups, but (for mammals, at any rate) of shorter duration	Biome: assemblages of landscapes	Second-order sequence and supercycle: 3 to 50 million years
Landscape	Ecological-evolutionary subunit: groups of biofacies ~the regional fauna; the community groups manifest in stratal sections at this level	North American Land Mammal Age: which is a biochron	Multicommunity landscape: comprises species assemblages; persists 2–3 million years in Carboniferous	Third-order sequence and cycle
Community	Biofacies or recurrent assemblage: habitat-specific species group ~the community; persist 5–6 myr	[chronofaunas and LMAs overlap]	Species assemblage: habitat-specific; persists 5–6 million years	0.5 to 3 million years
Guild	Guild: functional group of species; filled during millions years by same species group		Ecomorph: ~ guild; persists for millions of years	Milankovitch band fourth-, fifth- and sixth-order sequences: 0.03 to 0.5 myr

Table 6.2 *Ecological succession versus sedimentary succession, from Gili* et al. *(1995). Assemblages a to d have a direct causal relationship in an ecological succession, which seems to be matched in a sedimentary succession. Here, however, sequence a to d is the outcome of changed conditions from A to D with no interactive biotic relationship between a and b, and so on. Although taphonomic feedback may operate between a and b, etc. (Kidwell and Jablonski, 1983), an ecological succession and a sedimentary succession are still largely attributable to different causes, the latter being more visible and analysable in the stratigraphic record. The example used by Gili* et al. *is the succession of coral-to-rudist assemblages in the Tethyan Cretaceous, which have been attributed to biotically driven processes, i.e. were an ecological succession, but which Gili* et al. *argued was forced by environmental changes, probably increases in sediment flux.*

	Ecological succession				
Communities	a→	b→	c→	d→	etc.
	Sedimentary succession				
Sedimentary succession	A→	B→	C→	D→	etc.
	↓	↓	↓	↓	
Communities	a	b	c	d	

briefly with the notion that extraterrestrial impacts not only caused mass extinctions at level IV, but also caused the events at level III that define what we call biozones. It is here that we have hypotheses on the impact of environment on evolution, as in the turnover-pulse hypothesis, in which climatic change is at once the forcing factor in evolution and the link between plants and animals, terrestrial and marine (Vrba, 1980, 1985; Vrba *et al.*, 1995). And it is here that we have the *biozones* and their chronological equivalents, the biochrons, and the defining events of speciation and extinction, or datums. Speciation by isolation and cladogenesis is the evolutionary agent (Bennett, 1990, 1997). At *level IV* we have the theory of polytaxic oceans interrupted every 30-odd million years by the oligotaxic state (Fischer and Arthur, 1977) (Fig. 6.1), anticipating the theory of cyclical mass extinction (Raup, 1991; but see Boucot, 1994). Mass extinction via species sorting is the evolutionary agent. At this level there is a case for inferring some cause-and-effect from matching oxygen-isotopic cycles and putatively eustatic supercycles with the evolutionary radiations in the planktonic foraminifera (Fig. 6.2). Fischer's sweeping theory of polytaxy/oligotaxy is corroborated very well by the more recent compilations in the latter figure. This too is the time band of the big chunks in the fossil record, the chronofaunas (Olson, 1952, 1983). Bakker (1986) perceived four successional megadynasties in the large land herbivores since late Carboniferous – finbacks, proto-mammals, archosaurs and mammals.

Table 6.3 *The tiers of time and life: hierarchies and the concept of tiering in Earth history and the history of its biosphere (McGowran and Li, 1996). Ascending tiers I to IV at successive orders of magnitude of time identify something more than the accretion of events and phenomena at lower levels. At the emphasized level III we have the coincidence of fundamental biological phenomena with the critically important third-order cycles. It is the actual pattern of fossils vis-à-vis cycles that is not documented rigorously enough to test the turnover pulse of environmentally forced evolution at this tier (Vrba, 1985), or to explore the notion that pulsed extinctions mark not only the major boundaries of the geological timescale but also stage boundaries and probably some zonal boundaries (Raup, 1991). And it is this pattern that marine microfossils in neritic–oceanic transects are pre-eminently qualified to deliver.*

	Hierarchies of natural units in biogeohistory	
Tier	Biological	Physical-environmental
'deep time' – 'macroscale'	(i) theory of polytaxic oceanic states[1]	(i) second-order cycles in sea-level oscillation[6] [3 to 50 Ma], e.g. the *Miocene oscillation* due mainly to thermotectonic subsidence and plate-tectonic reorganization
level IV 10^7 to 10^6 years	(ii) theory of cyclical mass extinction[2] (iii) *chronofaunas*[3], *ecological-evolutionary units*[4]; the successional *megadynasties* of tetrapods[5]	(ii) impacts on planet[2]
'deep time' – 'mesoscale'	(i) *palaeobiological*: species replacement – evolution; dispersal; biofacies shifts; *chronoecograms*[11] *coordinated stasis*; *ecological-evolutionary subunits*; *turnover pulse hypothesis*[7]	(i) third-order cycles in sea-level oscillation[6] [0.5 to 3 Ma] due to eustasy of unknown type, but possibly including orbital amplification.
level III	(ii) *biostratigraphic and geochronological*: biozones, biochrons, datum spacing[8]	(ii) the *Oi* and *Mi* glacial cycles of the Oligo-Miocene[9]

10⁶ to 10⁵ years	(iii) average species duration is about 4 Ma	(iii) impacts on planet[2]
'Q-time' – 'microscale' level II 10⁵ to 10³ years	*Milankovitch time*: communities ['biofacies'] track (or fail to track) spatiotemporal shifts in environment; *epiboles and outages –* *above this line: palaeontologists' cladogenesis and stasis macroecological breakpoint?*	cyclostratigraphy, orbital tuning[10]: fourth-order cycles [0.08 to 0.5 Ma] fifth-order cycles [0.03 to 0.08 Ma] sixth-order cycles [0.01 to 0.03 Ma] due to main solar-orbital frequencies – *Milankovitch perturbations –* which force environmental change (e.g. via glacio-eustasy)
⋯⋯⋯⋯⋯	⋯⋯⋯⋯⋯	⋯⋯⋯⋯⋯
'real time' level I up to 10³ years	*below this line: ecologists' gradual and continuous community change – anagenesis* *ecological time*: populations, natural selection, microevolution	actualistic exemplars of geological processes – current, folkloric, human-historical – e.g. 'little ice age'; human-induced greenhouse, desertification, extinctions, plagues

[1] Fischer and Arthur 1977, [2] Raup 1991, [3] Olson 1952, [4] Boucot 1990a, [5] Bakker 1986, [6] Vail *et al.* 1991, [7] Vrba 1985, [8] McGowran, this Chapter,
[9] Wright *et al.* 1992, [10] Hilgen *et al.* 1993, [11] van Harten 1988

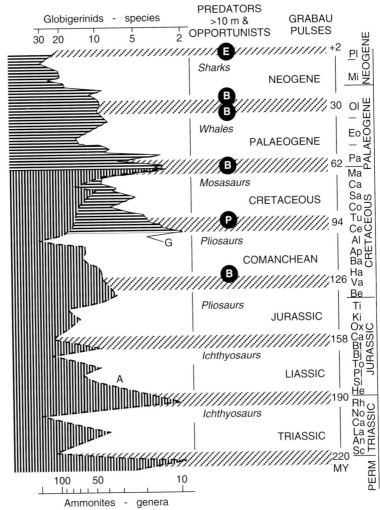

Figure 6.1 A theory of natural divisions in biogeohistory (Fischer and Arthur, 1977, Fig. 1, with permission; Fischer, 1981, Fig.1). Cyclical episodes or pulses of diversification and ecological expansiveness culminating in marine superpredators (*polytaxy*) were separated by crises of moderate to high intensity marked by pelagic blooms of opportunists or disaster species in a mode of ecological contraction (*oligotaxy*). B, *Braarudosphaera* (coccolith), P, *Pithonella* (problematicum), E, *Ethmodiscus* (diatom). These biotic pulses essentially coincide with the transgression pulses recognized by Grabau, as named.

Although Lyell and Darwin were correct to insist that theories at this high level of geological time must not violate the laws of nature as seen to operate at the lower levels, the history of the biosphere and especially the discipline of macroevolution and the fossil record (evolution at and above the species level) is

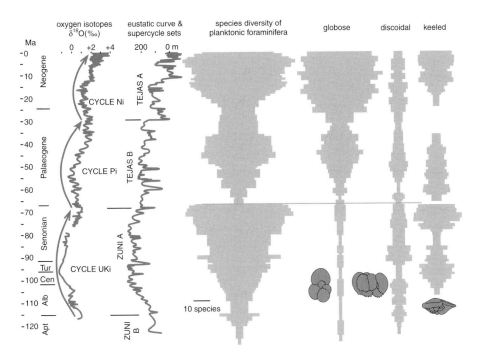

Figure 6.2 Three radiations congruent with three natural environmental divisions of the past 100 myr of biogeohistory (Abreu *et al.*, 1998, Fig. 2; Norris 1991a, Fig. 1.3, with permission). The three-part division of the physical environment is seen independently in oceanic oxygen isotopes as a proxy for water-temperature change, and a eustatic curve based on oceanic spillage across the continental margins. The Senonian, Palaeogene and Neogene radiations in planktonic foraminifera (Cifelli, 1969), long known as the basis for taxonomic revision (McGowran, 1968a), are expressed here simply on species diversity, also broken down to recurring gross test forms, globose, discoidal and keeled (the latter morphogroup is clearly quadripartite through time).

an autonomous field of enquiry beyond the mere upwards extrapolation of neontological theory – evolutionary genetics, ecology, etc. – into phenomena visible at geological timescales (Gayon, 1990). That assertion will stand regardless of the possibility that macroevolution-as-process is but an epiphenomenon riding on microevolution (e.g. Smith, 1994). It will stand simply because the history of the biosphere as revealed at geological timescales cannot be predicted by theories of evolution and inheritance developed at laboratory and field time-scales – macroevolutionary theory cannot be derived a priori from microevolutionary knowledge (Ayala, 1983). Likewise, palaeontologists have been unable to apply the principles of ecological succession to the fossil record because the two processes and patterns are separated by orders of magnitude of time (Chapter 8). Miller (1986) accordingly distinguished three basic levels in the ecological hierarchy: *ecological succession, community replacement* (the sometimes abrupt succession seen in the fossil record), and *community evolution* (origin of new types of communities). Theory and phenomena at the first level do not predict or subsume theory or phenomena at the second or the third levels. Later, Miller (1993) expanded this notion for mostly ecological processes over different timescales in the development of reefs, thus (at ascending timescales): *competitive interaction/substratum colonization/organism growth/community response/secondary succession/ primary succession/community replacement/regional dynamics/community evolution.* (Valentine and May (1996) argued that these processes do not aggregate upwards: community for example is not a collection of processes at lower levels. This sequence of phenomena accordingly does not form ranks and is not a hierarchy.)

We want to find the turning points, the natural caesuras in biogeohistory – in *Plato's metaphor,* we hope to *carve nature at the joints* (Hull, 1984). Questions of the fossil record from our biostratigraphic viewpoint include:

   i.   Are there natural biostratigraphic units?

   ii.   Why are good index fossils, good index fossils?

   iii.   As to the durations of biozones and biochrons: are there intrinsic limits to resolution in biostratigraphic zonation and correlation?

Are there natural biostratigraphic units?

The first of these questions turns on 'natural', by which I mean some kind of package to be discovered and recognized – well-bounded, well-defined assemblages of fossils recognized by clusters of biostratigraphic events. Natural units will reflect strong pulsing in the biosphere but whether coevally through all environments or not is to be discovered, not assumed. At the highest level there are three successional chunks of biospheric history demarcated by the inventions in turn of the procaryotic, eucaryotic and multicelled levels of biological organization. At the next level down there have indeed been breaks in the record, so that the fossil-based Palaeozoic, Mesozoic, and Cenozoic which were introduced by Phillips (1840, 1861) have not been blurred by subsequent advances in knowledge, and today's close scrutiny of mass extinctions have, if anything, sharpened our perceptions of the Ediacaran–Cambrian, Permian–Triassic, and Cretaceous–Palaeogene boundaries. As Fischer (1984) showed, there is a case for perceiving a natural, two-part Phanerozoic–Palaeozoic and Neozoic – but that does not reduce the sharpness of the biotic discontinuity between the Maastrichtian and Danian. Such matters are biostratigraphic, but they concern high-level carvings of the record. Of the same order are the three 'evolutionary [marine] faunas' of the Phanerozoic (Sepkoski, 1978) and the divisions of the terrestrial vertebrate succession and the fossil record of land plants. Important as they are as perceptions of biohistory, they are not biostratigraphic in the same sense as the fossil-based eras are biostratigraphic.

At a lower level, designated level IV in Table 6.3, we find several attempts to carve up the fossil succession. There is the chronofauna of Olson (1952, 1983) which recognized the persistence through time of community types (in Permo-Triassic terrestrial vertebrate communities). There are networks of species holding together through local variants of the overall environment and those species can be replaced by others in more or less the same role without destroying the structure of the chronofauna. In his study of the Clarendonian (late Miocene) chronofauna, Webb (1984) described the development of an essentially self-contained entity, based in large ungulates, held together by an elaborate set of positive and negative feedback loops, all in response to a climatic tendency toward cooling and drying at temperate latitudes in North America, and displaying wholesale evolutionary convergence in detailed resemblances with the modern African ungulate fauna. Webb distinguished between two concepts here. There is the land mammal *age*, primarily biostratigraphic and geological, used to define time spans on the basis of faunal change and requiring narrower and more rigorous definition, and there is the *chronofauna*, capturing palaeobiologically the broader ecogeographic and evolutionary continuities within the succession of faunas, and not requiring comparable rigor in definition (a risky distinction). But Webb saw a clear continuity, persisting in the Clarendonian for about ten million years, sandwiched between discontinuities produced by relatively rapid and disjunctive transformations. In this respect the Clarendonian chronofauna and its biostratigraphic counterpart are natural slices of the record.

In parallel with the development of these notions in the tetrapod record was Krassilov's (1974, 1978) attempt at causal biostratigraphy, in which stratigraphic classification and correlation are based on the recognition of ancient ecosystems, which in turn are the outcome of the interaction of geological events – climatic, tectonic – and organic evolution. Stratigraphic units of lower rank correspond to palaeoecosystems. The 'catena' is a chain of plant communities, inferred from the respective assemblage zones, which has the familiar horizontal and vertical distribution in, say, a cyclothem, where succession predicts vicarious coeval relationship among adjacent communities. Units of higher rank correspond to palaeobiospheres.

> *Recurrent biofacies or communities: 'reconciling d'Orbigny with Darwin' – the thoughts of A. J. Boucot*

Boucot's starting points in his voluminous writing on palaeocommunities were threefold:

i.   A severe loss of data is implicit in three habits among palaeontological research programmes. One is the compiling of taxic data from monographs where ranges are available only down to the level of the Stage (in the Cenozoic, almost all >3 myr' duration). A second habit is our focusing on index fossils for biostratigraphic correlation and age determination – the ordination of irreversible events for geochronology – to the neglect of most of the fossilized community. The third habit is the conflating of data from ecologically unrelated sources, thus randomizing the patterns which actually were non-random in the living state.

ii.   Biofacies are not random – they are neither a homogenized mix nor a unique succession of fossil assemblages. In between those extremes are the *recurrent biofacies* (*community groups*), and the discipline known as *ecostratigraphy* works out the evolutionary, ecologic, biogeographic, biostratigraphic, basin-analytical consequences of the community groups' distribution in space and time.

iii.   The 'prime facts of evolutionary importance' to Boucot were: (a) There is a finite number of biofacies present within each ecological-evolutionary unit (EEU). (b) There is a fixed number of EEUs. The time interval is of the order of magnitude of 5 to 20 myr for the average level-bottom, marine benthic community group. It would be much shorter for a mammalian community group, possibly 1–5 myr. Olson's (1952) vertebrate chronofauna was very similar to Boucot's community group.

Under the heading of 'ecostratigraphy', Boucot used the faunas of the Palaeozoic, neritic, level-bottom communities to promote the recognition of natural associations (Boucot, 1982–1994; with references). His central concept was the *community group*, a stable association of genera whose species' content may change considerably through time, at least in the uncommon to rare species. 'Stability' means survival through geological time, to be distinguished from 'persistence': community groups do not persist in stratigraphic sections – they come and go responding to environmental shifts which they track through space and time. Typically, community groups take 1–2 myr to become established (less in fast evolvers such as the proboscideans and microtine rodents) and are fixed for millions of years, sometimes lingering longer. A further property is that the species of each genus tend to maintain similar abundance levels – species of rare genera remain rare – and it is the species of rare genera that evolve the most rapidly (both anagenetically and cladogenetically). Figure 6.3 begins with Boucot's illustration of a community group as abstracted from Palaeozoic benthic invertebrate data (see below). The concept is developed in Figure 6.4

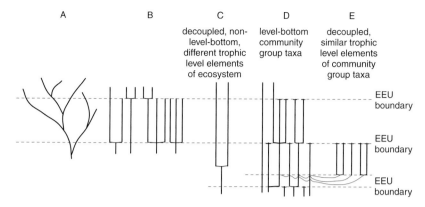

Figure 6.3 Boucot portrayed (A) the standard, random, gradualist model of phylogeny 'which ignores the constraints imposed by what we know about community evolution'. In contrast is the punctuated, synchronized pattern of cladogenesis (B) that emerges when one considers both community evolution and the basics of biostratigraphy – new community groups occur during the adaptive radiation near the beginning of the fundamental (but high level) unit, the ecological-evolutionary unit (EEU) or subunit. That punctuation, seen well developed in the centrally significant level-bottom community (D), is not demanded of another decoupled ecosystem (C), but a breakaway community (E) is still subjected to the environmental stresses whose effects mark the EEU boundary. The accommodated contrast is Boucot's '*reconciling d'Orbigny with Darwin*'. McGowran and Li (1996), with permission.

where these characteristics are displayed in three community groups X, Y and Z. On the left, there has been no environmental change in any of three separate sections (and note again the inverse correlation between abundance and speciation rate); on the right, the disjunct stratigraphic ranges have been produced by the tracking of the environmental changes across the site of accumulation of a single stratigraphic section by the three community groups. To obtain the true ranges, particularly of the rarer (and biostratigraphically more important) genera, one must (wrote Boucot) ensure sampling adequately the preserved beginnings of the 'adaptive radiation', and sampling likewise the record of extinction. And this point introduces the envelope enclosing these shifting community groups – the ecological-evolutionary unit. Figure 6.4 spans one EEU with typical time values for establishment, stable existence, and destruction. There are twelve EEUs in the Phanerozoic, but eight of them are in the Palaeozoic and there is but one for the entire Cenozoic – they are large entities. Of the eleven interfaces of the EEUs, seven are designated as major terminal extinction events.

   Boucot distinguished two types of biostratigraphy. In the style associated with the name of d'Orbigny, there are a relatively small number of marked changes in the marine benthic record especially in the level-bottom biota.

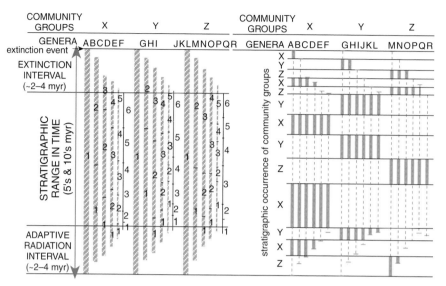

Figure 6.4 Boucot's community group concept (genera A–R, species numbered) expanded to show the generation of recurrent biofacies by the environmentally driven migrations of three community groups X, Y, Z. Left block, the true ranges of taxa, which are easily missed if the right facies is not sampled at the boundaries, giving a known range less than the true range. Right block, succession in a single section constructed to reinforce this point. More interesting here, though, is that although there is speciation and therefore change in each community group, the respective recurrent biofacies maintain their very high integrity. It is not entirely clear, but I infer from his emphasis on rare taxa that Boucot would intend such a stark mutual exclusion among X, Y and Z to be a reasonable digest of the situation among realistic diversities. After McGowran and Li (1996), with permission.

Boucot's EEUs and their divisions are equivalent to d'Orbigny's *étages* and *sous-étages* respectively. The appellation *d'Orbignyan* referred to extinction, adaptive radiation and dispersal. In the biostratigraphic style associated with the name of Oppel (within-*étage*) there are more gradual, within-community, species-to-species changes in each genus in each community. The appellation *Oppelian* referred to the phyletic evolutionary content of the fossil record. By this hierarchical distinction of styles Boucot disentangled the conflated data that give us the standard impression of random cladogenetic patterns (Fig. 6.5). By considering both the pattern of community evolution and 'the basics of biostratigraphy' in which community groups appear near the beginnings of the EEU, the strongly non-random pattern of ranges is constructed. Thus Boucot's most persistent message is that the distribution of fossils in space and time is neither an homogenized mixture nor simply a set of non-repetitive occurrences in which every locality preserves a unique mixture. Instead, there is during each

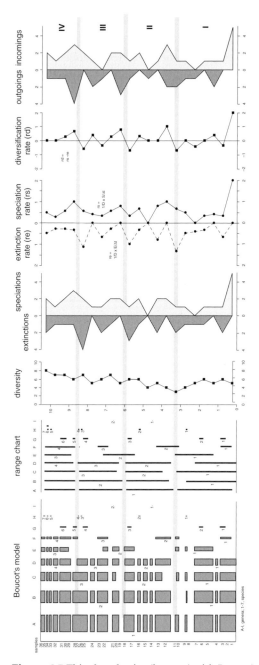

Figure 6.5 This chart begins (bottom) with Boucot's much-published model of a community group (A to I, genera, with relative abundance; numbers, species). We then have a range chart and a series of plots analysing the community group. Taxic counts are converted into rates (D, diversity; S, speciations during time interval Dt; r_e, r_s, r_d, rates of extinction, speciation, diversification). At the top are the

of his EEUs a distinctive, finite number of community groups (in the palaeobiological argot) or biofacies (in the argot of applied palaeontology). 'This fact has been common knowledge among geologists and paleontologists for well over a century' (Boucot, 1990a). However, biofacies have been treated mostly for their value in the reconstruction of physical environments – as an often-indispensable adjunct to stratigraphy and sedimentology. It is the task of ecostratigraphy to broaden the enquiry by tracking these community groups through evolutionary time and from section to section. By distinguishing and sorting them ecostratigraphically we attain a clarity and precision that too often is lacking in two dominant modes of research. One research mode is Oppelian biostratigraphy which in its accumulating of composite ranges from composite stratigraphic sections tends to conflate separate associations, such as level-bottom and non-level-bottom. (The equivalent conflating in the planktonic and oceanic domain would be to ignore the postmortem mixing of stratified pelagic communities that occurs as their shelly residues accumulate on the bottom of the sea.) The other research mode addresses taxic macroevolution and its main strategy is the conflating of data from treatises and monographs into a stage-by-stage pattern of origination and extinction. Boucot claimed no great conceptual advance in all this but rather a necessary focusing on biofacies, so that there is a back-and-forth feedback between environment, evolution, and correlation and age. The biostratigraphic consequences of ecostratigraphy lie in the clarification of precision and reliability. If *precision* refers to the finest, ultimate time division achievable using fossils, and *reliability* is the reproducibility, section to section, of the designated boundaries between fossiliferous units, then, wrote Boucot, the greatest potential for precision goes with a relatively low level of reliability (because of the rarity of the species concerned), and the highest reliability, i.e. reproducibility, is found in moderately abundant taxa, for they have a more continuous record even though their taxic-evolutionary rates are somewhat lower.

The discussion of community groups has avoided causation. Is there evidence of environmental impact generating these parcels of life and fossils? Boucot found only very weak correlations between various geological variables and the boundaries of his EEUs. He took pains to emphasize the decoupling of marine benthic EEUs from packets in other realms, as in terrestrial plants, for example,

Figure 6.5 (*cont.*)
 comings and goings of species-potential data that are smoothed out of the picture in the range chart. Note the fairly non-arbitrary ('natural') division into four successional biotic groups I to IV ('faunas'?). Diagram constructed by Qianyu Li (McGowran and Li, 1996, with permission).

and to point out that there are time lags within the EEUs in the development of tiered communities, such as reefs (Fig. 6.5). But he could suggest only that the controls are more likely to be physical than biological, weak though the evidence is. In his balancing of the changes between the EEUs with the changes within, he invoked the resonating *'reconciling Darwin with d'Orbigny'*, rather than the usual opposing of the two worldviews of the biogeohistorical record, the punctuated and the gradual.

The d'Orbignyan–Oppelian duality arises from the neritic record, and so it is interesting to note a parallel coming from the vast populations of protists in the pelagial: Emiliani's (1982) 'extinctive' and 'competitive' evolution – not opposing worldviews, but modes which combine into a unified model of evolution.

Coordinated stasis

Sheehan (1996) suggested that the last six of Boucot's EEUs (Ordovician–Cenozoic) display a consistent pattern of a brief 'reorganizational-EEU' succeeded by a temporally longer 'stasis-EEU'. Brett and Baird (1995) distinguished within one of the EEUs (Silurian–Devonian) multiple intervals of relatively stable communities separated by brief times of marked community change. Diversities of marine invertebrates were in the range of ~50–300 + species, less than 20% of which were carryovers but a strong majority of which persisted throughout the respective timeslice. There are about ten of these ecological-evolutionary subunits, with durations in the range of 2–3 to 7–8 myr' duration. As mentioned above, this pattern theory has a distinguished pedigree going back to Cuvier's extinction-bounded units and d'Orbigny's stages bounded by intervals of extinction and origination (Chapter 7). Brett and Baird (1997) recognized two hierarchical levels of bioevents – lower level, epiboles and outages (Chapter 5); higher level, evolutionary pulses in community structures giving rise to the EESUs. This higher pattern is the basis for 'coordinated stasis'. It is an extension of the argument for punctuation and stasis in evolutionary lineages – since species in a community are interdependent, it 'stands to reason' that punctuational events occur collectively (Brett and Baird, 1997). The pattern comprises relative stability (meaning several millions of years) in both ecological structures and evolutionary genealogies in fossil faunas (arising in the marine, neritic, Palaeozoic fossil record).

Coordinated stasis was prominent in discourse on (palaeo)community and ecosystem through the 1990s (W. Miller, 1990; Morris *et al.*, 1995; a set of papers (Ivany & Schopf, 1996) introduced by Brett *et al.*, 1996; A.I. Miller, 1997; Patzkowsky and Holland, 1999; Ivany, 1999). Ivany listed several areas of enquiry having to do with the interaction of evolution and ecology in deep time, where coordinated stasis sits.

i.   Do the ecological characteristics of a lineage determine its stability; will that stability be coincident with other lineages in the assemblage; are perceived patterns of change affected by sample size (and ignoring the rare species)?

ii.   Does coordinated stasis manifest in some environments more than others? This question is an extension of questions about species longevity in asymmetrical clades (see below and Fig. 6.16).

iii. In more or less cogent instances of coordinated stasis, do similar assemblages persist intact through the timeslice of stability, or do they reconstitute themselves similarly each time the appropriate habitat reappears? This question is a refinement of the simple Eltonian/Gleasonian opposition. Repeated reassembly is claimed in Pleistocene corals (Pandolfi, 1996). The alternative is the tracking of preferred habitats by faunas, as in the mid-Palaeozoic studies by Brett and his colleagues, this habitat-tracking producing lateral shifts by broad biofacies belts. Tracking, not some evolutionary response to environmental stress, may account for most local faunal changes, for a 'pattern of high-fidelity habitat-tracking is typical of many marine-benthic communities' (Brett, 1998;). However, a scrutiny of Boucot's recurring community groups (Fig. 6.4) does not clearly refute reassembling in favour of tracking, at least to me. Thus, as Ivany noted, these are patterns with different ecological-evolutionary implications.

iv.   Are the clusters of turnover events punctuating seemingly stable associations a function of evolutionary change or biogeographic shift? This either/or is the basis for A. I. Miller's (1997) 'coordinated stasis or coincident relative stability'.

v.   Ivany's last area of necessary enquiry is the relationship between environmental change and faunal change. If there is a punctuated pattern – abrupt and coincident turnover of taxa bounding stable intervals – is this a sign of some threshold arising in the community itself, in response to protracted physical change, or is it a prompt reaction to a threshold shift in the environmental itself?

There are uncontested difficulties in extracting macro-community changes from fossils and strata, even with rigorous statistics (e.g. Bambach and Bennington, 1996; Jackson *et al.*, 1996). Jackson *et al.* observed that examples of coordinated stasis are based on records of species persisting for millions of years then changing in a much shorter time – such stasis and turnover seemingly good evidence for punctuated evolution without necessarily saying much about punctuated ecology.

Punctuation?

Here is a pungent example of the move away from gradualism and from the habit of assigning all disjunctions to the gaps in the record (Krassilov, 1978):

> Darwin prophesied the decline of the 'noble science of geology' (that is, palaeontology and stratigraphy), and it is undeniable, in Rudwick's words (1972, p. 264), 'that palaeontology was withdrawing more and more from the position of intellectual importance that it had held in the public mind earlier in the century'. Darwin contributed to its decline not by demonstrating the discouraging imperfection of the fossil record (which he, in fact, failed to do), but by his long-standing reductionist view of natural selection as competition between organisms against a steady-state geological background. Only recently was it realized that causal explanation of evolution should be sought not at a biotic but at a higher geobiotic level of organization and that the 'survival of the fittest' is a tautology unless the trend of geological development is specified. The gradualistic concept of evolution stemmed from ignorance of such general system properties as homeostasis. Resilience of a system is directly related to its complexity. The more complex a system is, the more discontinuously it evolves, and the layered rocks are a paradigm manifestation of this character of geobiological evolution.

Contrast that statement with the thoroughly Darwinian conclusion demanded by a lack of synchrony and diastrophy (Simpson, 1952, 1965):

> The evidence summarized in this essay is consistent with the view that most of the broad features of vertebrate history might have been much the same if the earth's crust had been static (provided that the surface remained sufficiently varied and with large connected and nearly connected land and sea areas). Crustal movements may have had essential roles only as regards details of timing and of distribution, important details in some cases but still only details. The older and still perhaps more common belief in causal synchronism of periodic world-wide evolutionary and diastrophic episodes is certainly not disproven, but the evidence runs rather more against than for it. The most likely points at which physical events may have had decisive influence are the extinctions of aquatic vertebrates around the end of the Permian and of terrestrial vertebrates around the end of the Cretaceous.

The world does not change in so stately a manner as it did when Simpson was writing magisterially. The Neogene vertebrate record is the source of the turnover-pulse hypothesis (Vrba, 1985, 1995) – a concentration against the geological timescale of events of first and last records. Vrba suggested, to assess the scaling, a concentration of events within 100 000 years preceded and followed by a million years of predominant stasis in the same monophyletic groups. The hypothesis assumes the phenotypic conservatism, and the concentration of phenotypic change in speciation, of punctuated equilibria, with speciation and stasis at the same scales, but with the important addition of synchronism among lineages. The tuning factor would seem to be climatic change under astronomical forcing, working through disruption of geographic range by habitat disruption (ecological vicariance). Vrba argued that a turnover pulse occurred among African mammals (including speciations in the human clade) at ~2.5 Ma in response to the major climatic change (Chill IV herein). The theory does not imply that speciation is determined only by environmental change, but it says that if changes do occur in lineages, they will occur together. It does not demand synchroneity among different biofacies or biogeographic realms. The theory does require (i) that the record is good enough to demonstrate the required pattern among lineages, including some rigorous correlation, and (ii) further cogent correlations with evidence of sharp climatic change. If turnover pulses actually exist, then they are of a scale comparable with biostratigraphic zones (i.e. level III; Table 6.3) and they would indeed be the boundaries of natural units at obvious horizons, with a causal tuning mechanism as an explanatory bonus.

However, the turnover pulse was not corroborated when tested in the Pliocene vertebrate record in east Africa by Behrensmeyer *et al.* (1997). Nor did it get support from the late Miocene Siwaliks of northern Pakistan, as noted below. Bobe and Eck (2001) found marked changes in bovid assemblages and several lines of evidence for marked environmental shifts, 2.8–2.1 Ma, but clear corroboration for neither the turnover pulse nor the prolonged shift detected by Behrensmeyer *et al.* (1997). The Late Pliocene changes in the terrestrial realm as chronicled by Bobe and Eck have remarkable parallels with the pelagic realm in the North Atlantic (see below and Fig. 6.14).

Like chronofaunas, turnover pulses were based conceptually in terrestrial fossil records. Community groups spring from the neritic fossil records of the Palaeozoic. But is there an answer to the question at the head of this section so far as late Phanerozoic microplanktonic biostratigraphy is concerned? On the one hand, Boucot undoubtedly was recognizing natural associations which are biostratigraphic – they pertain to the distribution of fossils in space and time and the consequences of those patterns; likewise with the terrestrial vertebrate patterns. On the other hand, the microplanktonic zones are developed by testing, refinement and extension geographically – ultimately, circumglobally – and acceptance by usage, i.e. consensus. But we are dealing here with defining events and datums, not associations, when we accrete hard evidence of the consistent ordering and palaeomagnetic calibration of the first and last appearances of taxa for biostratigraphic systems. The rise of microplanktonic biostratigraphy of the later Phanerozoic has exacerbated the split between the biostratigraphy of the 'more useful' planktonic protists and the taxonomy/palaeobiology of the 'less useful' benthic invertebrates. Ecostratigraphy and macroevolution undoubtedly will increase our insights synergistically, but that is in the future, when taxic abundances are as well known as taxic ranges. Meanwhile, the answer for the late Phanerozoic microplanktonic zones is, no, they are not natural in the sense of this discussion, because the aim of research is to decouple datums completely from their biofacies context and recouple them anew into the IMBS. It follows that the active consideration of such matters is occurring rather less in microplanktonic biostratigraphy than elsewhere, as among the benthic communities, as observed, and in such realms as the fossil ostracoda, to which we now turn, and the rich macrofossil record of the American Cretaceous, to be invoked in the next section.

The ostracods are mostly benthic organisms but they can be split three ways as to the 'extension' (range) zones based on zonal markers (Colin and Lethiers, 1988). The difference is between endemic benthics, ubiquitous benthics, and pelagics – in that order, geographic spread increases and biozonal boundaries based on members of the respective groups progressively approximate isochrons more closely. Those zones are 'conventional', meaning that they ignore ecological sensitivities, and they are contrasted with 'ecostratigraphic units characteristic of particular depositional environments'. But Colin and Lethiers concluded, reasonably enough, that 'In fact, the so-called ecostratigraphic zonations are nothing more than regional multiple biozonations (topozones or teil-zones, Hedberg, 1976), with different biozonations referring to each main type of environment'. Which is to acknowledge that ecostratigraphy is not so much a new idea as a newly invigorated scrutiny of biofacies configurations in space and time, as for community groups, as mentioned above. But there is more. Colin and Lethiers (after Lethiers, 1983, 1987) went on to consider causal or event biostratigraphy, based on the sigmoidal pattern of species ranges in neritic environments. Inspect Figure 6.6. The central pattern shows the species' ranges arranged to demonstrate the sigmoid of one cycle (no timescale). One part of the interpretation of this pattern is from a suggested correlation with the third-order global eustatic (Exxon) cycles, as shown. In the other part, Lethiers made these correlations: *transgression* ∼ *higher diversity* ∼ *taxonomic stability* ∼ *longer ranges* ∼ *geographic ubiquity* and *regression* ∼ *lower diversity* ∼ *taxonomic volatility* ∼ *shorter ranges* ∼ *geographic endemism*.

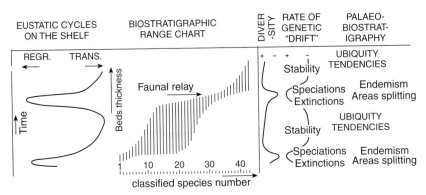

Figure 6.6 Sigmoidal stratigraphic distribution of species of ostracods and the eustatic signature in the neritic realm (Colin and Lethiers, 1988, with permission), the inference being that the latter influences or controls the former. The supra-ecological patterns in Figures 6.6–6.8 were based on ostracod studies, anticipating the discipline and research programmes at timescales well beyond neo-ecology and now known as evolutionary palaeoecology.

The pattern is equivalent to the turnover pulse, it is 'natural', and it has a ready-made causation. The configuration was developed somewhat further in parallel ostracod studies by van Harten and van Hinte (1984) and van Harten (1987) as 'chronoecology', the study of long-term change in the overall environment through geological time. They began with three propositions: ostracods occur in almost all aquatic environments; individual species tend to be highly sensitive to their respective environments; and many species seem to maintain their requirements constantly throughout their time ranges. If a species is inflexible in its needs, then it will survive only so long as the niche is available. Therefore, species longevity becomes a function of environmental stability. Physical stability means longer-lived species and reduced turnover. Environmental change means shorter-lived species and increased turnover. The endemic speciation of the brackish-water genus *Cyprideis* in central Paratethys is a nice example of the sigmoid: rapid turnover and short ranges in the Pannonian; slower turnover and much longer ranges in the Pontian (Fig. 6.7). But the authors introduced a contrast between the sigmoidal pattern of a single section and the triangular signal of a composite, regional succession (Fig. 6.8). To explain this contrast they invoked the dubious belief that physical instability locally is a transient condition but regionally is migratory and prolonged, so that the ranges of the generalist taxa are long. At any rate, that segment of the conclusion accords with orthodoxy, as will be considered below.

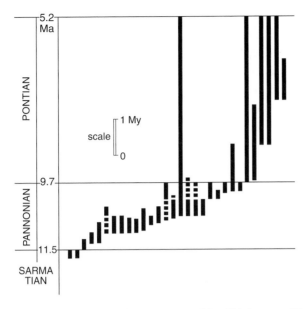

Figure 6.7 The brackish-water ostracod *Cyprideis* in central Paratethys (van Harten, 1988, with permission) – a chronoecological graph: 'Duration pattern reflects asymptotic process of salinity change.'

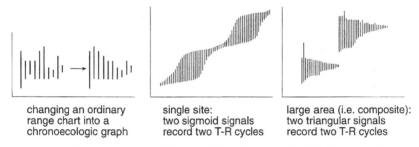

| changing an ordinary range chart into a chronoecologic graph | single site: two sigmoid signals record two T-R cycles | large area (i.e. composite): two triangular signals record two T-R cycles |

Figure 6.8 Chronoecology (van Harten, 1988, with permission). Left, rearranging a range chart into a chronoecological graph. Centre, two sigmoidal cycles imply two transgressive-regressive cycles. Right, in the larger perspective, composite charts show two T–R cycles in triangular distributional patterns.

Cenozoic chronofaunas of Tethyan neritic foraminifera

The chronofauna was christened by Olson (1952, 1983) who recognized the persistence through geological time – far beyond ecological time – of community types (in Permo-Triassic terrestrial vertebrate communities). Networks of species were held together through local variants of the overall environment; those species can be replaced by others in more or less the same role without destroying the structure of the chronofauna. There is stability and coadaptation whilst evolution proceeds. Studying the patterns of ecological organization and change through long periods of time, i.e. the anatomy of chronofaunas, is the discipline of evolutionary palaeoecology (Wing *et al.*, 1992), the logical bridge between ecology and evolutionary biology – and the logical bridge between the tiers of ascending timescales.

The abundance of larger foraminifera in fossil assemblages, often to the point of rock-forming density, reflects living conditions: adequate light and trace-nutrients are not density-limited to the degree that filtering or foraging for food would be. The diversity of larger foraminifera reflects a refined partitioning of niches. More than other physical factors, light intensity and quality would appear to be the prime control: different wavelengths penetrating to different water depths are exploited by different, host-specific symbionts.

Hottinger (1982, 1983) crisply summarized the distribution of the late Phanerozoic larger foraminifera in space and time in terms of three different and – he opined – largely independent sets of processes: (i) *facies specificity*, (ii) *biogeographic limitations*, and (iii) *frequent synchronous replacement*. Tied closely to these three sets of processes is the concept of evolutionary community maturation which is hierarchical with three tiers: ecologic spreading through the photic zone, development of chronofaunal maturity, and innovation between chronofaunas. These matters are discussed together here.

i.   *Facies specificity*. A lateral succession onshore→offshore in morphology has developed repeatedly when the photic zone was extensively colonized by larger foraminifera, as follows: *Conical-agglutinated→discoidal-porcella-nous→fusiform-porcellanous→thickly-lenticular-perforate→flat-lenticular-perforate* or *discoidal-perforate*.

Thus, it has long been known that recurring morphotypes, such as the orbitoidal or the fusiform, record phyletic convergence which can be quite spectacular – e.g. the Senonian, Palaeogene and Neogene iteratively fusiform 'alveolinids', repeating the fusiform morphologies developed in the late Palaeozoic fusulinid radiation. The group 'larger foraminifera' is gloriously polyphyletic, but equally importantly it comprises an evolutionary grade in which morphologic convergence signals ecologic replacement and relay. (See Figure 4.21.)

The assortment of characteristics identifies the large foraminifera as extreme K-strategists – uniquely so among protozoans (Hottinger, 1982, 1996). The combination of high diversity and low productivity requires long-term environmental stability, which depends in turn upon climate and nutrient concentration. The mature K-selected community takes time to develop in full, hence two strong and recurring points in Hottinger's discussions – the concept of community *maturity* and the importance of phyletic *history* in the attainment of that maturity. Hottinger has emphasized that an assemblage of these taxa at a given time depends not only on the environmental factors of the time and place but on the historical background of the constituents.

In regional or local disruption of ecologic balances, maturation is produced by colonization and/or immigration producing at levels of equal maturation a repetition of assemblages. In mature communities depth differentiation is at its most refined in a trade-off between closely related species.

ii. *Biogeography* largely means provinces in this discussion. (We should add from southern Australasia: major climatic and watermass fluctuations extend by tens of degrees latitudinally.) There are three major biogeographic provinces: Caribbean, Mediterranean, Indo-Pacific (Adams 1967, 1983). These provinces are successors to Tethys. However, biogeography figures in Hottinger's maturity model at a lower level, i.e. within-Tethys (see below).

iii. *Frequent synchronous replacement*. The 'worldwide' loss of genetic information stimulates evolutionary processes to produce new communities playing analogous ecological roles. These major replacements are attributed to pollution by excess nutrients: it takes 5–10 myr to recover diversity and full oligotrophic adaptation (Hottinger, 1997). Meanwhile, there is a maturation factor at a still higher level that is preserved in the geological succession of innovations in skeletal wall structure: the agglutinated large foraminifera which appeared in the middle Lias are supplemented successively by porcellanous (miliolid) forms in the Cenomanian and by lamellar-perforates (with hyaline, crystallographically constrained wall structures in the skeleton) in the Santonian, representing higher grades of complexity in biomineralization and wall textures.

The predicted pattern of development for each 'chronofauna' (as we christened them: McGowran and Li, 2000) is: (i) rapid diversification from small-sized and simply-structured, monospecific, cosmopolitan forms; then (ii) stable or slowly decreasing diversity with (often pronounced) size increase and complexity of structure; concomitantly (iii) an increase in provinciality; (iv) meanwhile, the communities spread out through the photic zone, from shallower, brighter and warmer to deeper, dimmer and cooler waters (and not in the reverse direction). This expansion by colonization and diversification is limited by light intensity, nutrient pollution, and the cutoff minimum temperature tolerated by the photosymbionts.

The Early Palaeogene chronofauna in western Tethys

Hottinger (1990) used the Early Palaeogene record of the nummulitids to characterize development through time as an evolutionary cycle. Two 'phyla' or distinct, unified phyletic branches (clades) are mutually distinguished by anatomical complexity: *Assilina* was simpler, *Nummulites* was more complex in its canal systems. *Assilina* begins in the Thanetian Stage, some 8 myr after the end-Cretaceous extinctions. This was a time of smallish, monospecific pioneers in the 'generic' radiation of larger foraminifera – *Discocyclina, Ranikothalia, Glomalveolina, Fallotella, Broeckinella*. During the later Ilerdian Stage only a few of those groups continue to flourish and take over in a 'specific' radiation – *Assilina, Nummulites, Alveolina* (Hottinger, 1997). *Assilina* and *Nummulites* each diversified (Fig. 6.9) but at very different rates. (The early Palaeogene stages in Figure 6.9 keep faith with Hottinger's and others' chronostratigraphic usage in larger-foraminiferal studies.) Diversity is constant in the Cuisian and Lutetian, which was the mature phase during which other indices of K-strategy developed fully, such as the large size of specimens and, especially, the occupancy by larger foraminifera of niches throughout the photic zone – a full occupancy which took ten million years and more since the comparable adaptive range was previously achieved in the Maastrichtian, then extinguished. Hottinger devised a measure of provincial tendency in western Tethys and its surrounds by scoring each species by its presence in one, two or three bioprovinces, the provinces gaining their identity from oceanographic barriers in that tectonically complex region between Africa and Europe, with all-important nutritional barriers marked by upwellings. Provinciality was higher in the more diverse genus, increasing after taxic diversity stabilized and subsequently vanishing among the few late Eocene survivors. Diversity dropped markedly in the Biarrritzian to the few survivors in the Priabonian (Late Eocene), the Biarritzian–Priabonian boundary marking a 'faunal revolution'.

Brasier (1995) used the same Paleocene–Eocene Tethyan record to develop an evolutionary model for these oligotrophic ecosystems. Its stages are incorporated into Figure 6.9: (i) Recovery interval: normal conditions return after a collapse; pioneer species become established. (ii) Radiation interval (non-critical phase): development of benthic communities with symbiosis and interdependence; symbiosis perhaps largely facultative, so that any environmental perturbations at that time did not have a catastrophic impact. (iii) Radiation interval (critical phase): size and diversity reach their maxima, symbiosis inferred to be obligate and interdependence to increase, the ecosystem having by now become

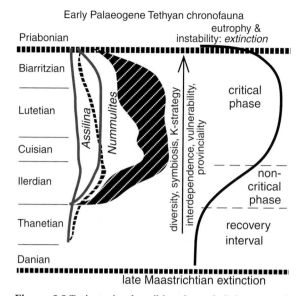

Early Palaeogene Tethyan chronofauna

Figure 6.9 Trajectories describing the early Palaeogene chronofauna (as it is called here) in Tethyan neritic large foraminifera, adapted from Hottinger (1990) and Brasier (1995). Hottinger plotted the diversities of the nummulitids *Assilina* and the anatomically more complex *Nummulites* (solid lines) and devised a formula for west-Tethyan provinciality expressed by the width of the hatched zones. Thus, the genus *Nummulites* both is more provincial than *Assilina* and itself is most provincial in the later stages of the chronofauna, with diversity and provinciality collapsing in the 'faunal revolution' at the Biarritzian–Priabonian (= middle/late Eocene) boundary. The three phases described by Brasier are fitted; the unscaled solid line describes the rise and fall in diversity of photosymbiont species. After McGowran and Li (2000) with permission.

very vulnerable to environmental perturbation; provinciality increases. (iv) Extinction interval: the ecosystem collapses due to eutrophic pollution of an inherently specialized, interdependent and unstable system.

Cenozoic chronofaunas

Hottinger (1982) sketched the r→K trajectories of the main groups of larger foraminifera and listed (1997, Table II) the members of the successive communities in the upper, middle and lower sectors of the neritic photic zone. These ideas are sketched in Figure 4.21. Just as it takes time – in millions of years – to become a fully mature lineage or phylum, in the sense of achieving the ultimate K-mode lifestyle, so too did the communities need time to spread down the shelf (not up-shelf) towards the base of the photic zone by speciation and niche partitioning (it took the early Palaeogene chronofauna more than the entire duration of the Paleocene epoch to achieve full euphotic colonization).

There are four Cenozoic chronofaunas, unequal in duration, separated by faunal revolutions. The shortlived Priabonian (late Eocene) is anomalous among these chronofaunas in that the spread of communities through the photic zone was rapid (as we discuss below, the Priabonian itself was a highly transitional and unusual time in Cenozoic history).

Likewise, the discrete units of evolutionary community maturation marked off by 'biological revolutions' would seem to exemplify the notion of chrono-fauna among the neritic larger foraminifera. The community holds together even as evolution proceeds by speciation and diversification and as ecological trade-offs continue down through the photic zone. It is interesting that the early Palaeogene chronofauna survived two major environmental perturbations before it succumbed to the events in the late Biarritzian. The earlier perturbation is marked by the sharp warming spike and deep-ocean extinction at the end of the Paleocene. The second is the cooling at the end of the early Eocene – 'Chill I'. It may well be that this apparent robustness in the face of environmental impact was a function of the evolutionary immaturity of the chronofauna in its 'non-critical' phase in the early Eocene, in contrast to its Biarritzian (= Bartonian) 'critical' condition of fragility and metastability.

Indo-Pacific larger foraminifera and the East Indies Letter Classification

An extra biogeographical dimension is given by sketching a latitude-time envelope of the Indo-Pacific Letter Classification (Chapter 7, Fig. 7.21), drawing attention to the response of the essentially tropical larger foraminifera to fluctuations in climate and in sea level. It is a theory of pattern and is eminently falsifiable (McGowran, 1986a,b). The elements of the pattern are the rapid, shortlived extratropical excursions and the tentative correlations of the letter stages – shown as less-than-satisfactory defining events – with the planktonic foraminiferal P- and N-zones. The large-foraminiferal record in southern Australasia reveals very little development in the sense of a maturing of the communities. Instead, we receive sporadic samplings of the Indo-Pacific communities via warm currents (McGowran et al., 1997b) without speciation. These warm horizons marked by pantropical immigrants are ecological-biogeographic patterns, not evolutionary-biogeographic as in taxic provinciality.

Correlation of Tethyan, Indo-Pacific and New Zealand patterns

The various configurations are assembled in Figure 6.10. The patterns begin at the Maastrichtian–Paleocene boundary. Whereas there were excursions to high southern latitudes in the warmest time, Paleocene–early Eocene, the other indices do not respond either to that or to Chill I at the end of the early Eocene.

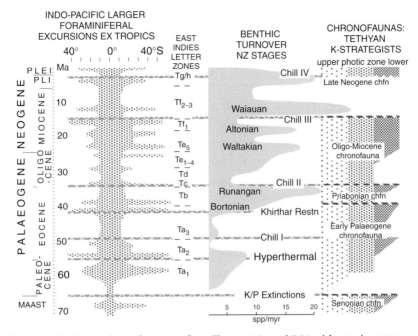

Figure 6.10 Comparison of patterns from Figures 4.21 and 7.21 with a taxic pattern from the New Zealand benthic foraminiferal record of species' incomings and outgoings, regional stage by stage, compiled from Hornibrook *et al.* (1989). (McGowran and Li, 2000, with permission.) Counts are normalized to events/myr (but not taking account of standing diversity) using their geochronology and correlations of the regional stages. Note the cyles of build-up in taxic overturn successively to peaks in the Bortonian, Runangan, Waitakian, Altonian, Waiauan and Waipipian stages. Major environmental events, mostly the four second-order chills, from McGowran *et al.* (1997a). The matches between the three datasets are not comprehensive but there are enough temporal parallels to suggest that evolutionary and biogeograhic changes surely had some environmental forcing.

The late middle Eocene is different. This was the time of increased immigration to western-southern Australia, due mostly to the Leeuwin Current (McGowran *et al.*, 1997b), and of the first (Bortonian) peak in taxic overturn in New Zealand. It is followed immediately by the largest Tethyan faunal revolution of the Palaeogene and a coeval and comparable change in the Indo-Pacific fauna, marked by the Ta_3/Tb boundary. The ensuing Kaiatan-Runangan warm cycle matches exactly the second (Priabonian) Tethyan chronofauna. But the Priabonian seas were too fertile to encourage extratropical migrations (see below). The strongest correlations of the Oligocene–Miocene are between the three successive peaks in turnover – Waitakian, Altonian, and Waiauan – and strong warmings recorded by immigrations of large foraminifera. Between-times, faunal changes marking the Indo-Pacific zones can be ascribed to combinations of regression, cooling, and eutrophication: there is no formulaic, billiard-ball, cause and effect. The most apparent example of a fertility-induced problem is at the Te_5/Tf_1 boundary, where the disappearance of abundant *Eulepidina* and *Spiroclypeus* (but survival of *Nephrolepidina, Cycloclypeus* and *Miogypsina* s.l.) matches very well the onset at ~17.5 Ma of the Monterey carbon positive excursion as seen on southern shelves (Li and McGowran, 1994) and in central Parathethys in the regressive Karpatian Stage (Chapter 7). The previous major carbon excursion, across the Oligo-Miocene boundary (Fig. 8.4), embraces the lower Te/upper Te boundary.

The Oligo-Miocene Tethyan chronofauna was a coherent unit through all this time. But the middle Miocene faunal revolution which is the chronofaunal termination finds several resonances in the Indo-Pacific and southern Australasian regions. Immediately preceding the termination is a 3.5-myr period of instability in several parameters, global and regional, physical and biological (see below). It is simultaneously the time of maximum diversity of large for-aminifera in the Indo-Pacific region (Tf_1) and the most marked warming and immigrations into southern Australia (Batesfordian–Balcombian Stages) and New Zealand (Altonian–Clifdenian Stages) as well as into the Northern Hemisphere (e.g. the Badenian Stage in central Parathethys; Japan). The actual termination is low in Zone N10 at ~14 Ma which is 'Chill III' and the onset of rapid growth of the Antarctic icecaps. In southern Australia it is a transconti-nental surface capping the Batesfordian–Balcombian neritic carbonates, which were at their most extensive; the Nullarbor Plain is this withdrawal surface and so too is the strongest seismic reflector in the Gippsland Basin. Biostratigraphically, this horizon is marked by the biggest loss of Indo-Pacific neritic taxa at the Tf_1/Tf_{2-3} boundary. In all respects, then, we can confirm that (i) the termination of the Tethyan chronofauna is strongly matched in the Indo-Pacific province; (ii) termination was driven by a global environmental event; and (iii) well corroborated is the notion of instability and community fragility in late-stage maturity (cf. Fig. 6.9) prior to the crash, and not just in the neritic Tethys but in the Indo-Pacific and southern temperate regions as well.

As in Tethys, the record then becomes sparse and chronologically poorly constrained. However, the very large Waiauan overturn matches exactly the last major excursion out of the Indo-Pacific during the Miocene (by *Lepidocyclina* and *Cycloclypeus*).

Neritic-pelagic-terrestrial parallels among Palaeogene chronofaunas?

Since nutrient patterns are watermass patterns, changes in the latter should be recorded in parallel responses in two groups with K-strategists in the

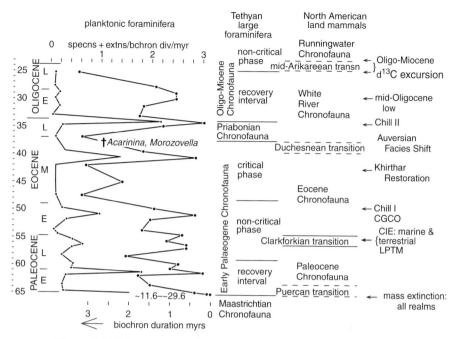

Figure 6.11 Palaeogene correlations across three realms. Planktonic foraminiferal overturn is based on the phylogenetic tree in Pearson (1998b), and taxic end-events were counted for each zone or subzone (Berggren *et al.*, 1995) ignoring differences between speciations and pseudospeciations or extinctions and pseudoextinctions (Chapter 4). Biochron durations, from Figure 5.12. Tethyan chronofaunas and their phases, as in Figure 4.21. North American land mammal chronofaunas, mostly from Webb and Opdyke (1995). From McGowran and Li (2002) with permission.

euphotic zone – larger benthics in the neritic realm and planktonics in the pelagic realm. Hallock *et al.* (1991) demonstrated such parallels through the Paleocene and Eocene in diversities, extinctions, originations and overturn (see also McGowran, 1992; Brasier, 1995). Since some of the most characteristic planktonic genera of the earlier Palaeogene, *Acarinina, Morozovella* and *Planorotalites*, became extinct at the end of the middle Eocene, a planktonic chronofauna must have many of the same benchmarks as the large-benthic chronofauna. These parallels are compared in Figure 6.11. It is clear that planktonic foraminiferal taxic evolution was strongly episodic and that some but not all peaks have parallels in the large benthics. The North American land mammal succession is divided into the land mammal ages, primarily biostratigraphic and geological, used to define time spans on the basis of faunal change and requiring rigorous definition (Chapter 7). However, the succession displays clumped events, both evolutionary and migrational, which can be used to distinguish chronofaunas capturing palaeobiologically the broader ecogeographic and

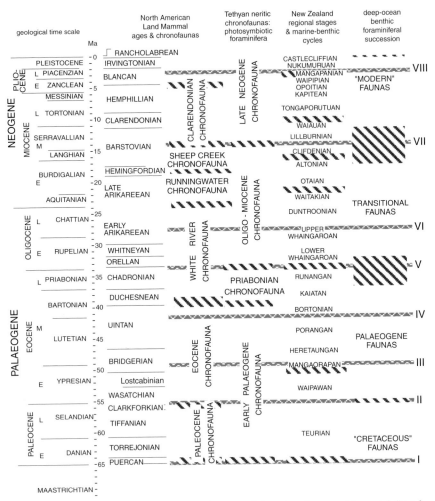

Figure 6.12 North American Land Mammal Ages (northern and terrestrial) and chronofaunas compared to Tethyan chronofaunas (neritic) and New Zealand taxic cycles (southern and marine) taken from Figures 4.21 and 6.10. The two sets of chronofaunas have boundaries in common at end-Maastrichtian, Bartonian, and end-Langhian. Boundaries are drawn with thick cables as a reminder that chronofaunal changeovers are not instantaneous. Geological timescales from Berggren *et al.* (1995). For recent surveys and discussion of the North American succession, see Webb and Opdyke (1995), Woodburne and Swisher (1995), and Prothero (1995). In the deep-ocean benthic foraminiferal succession (Thomas *et al.*, 2000, Fig. 5.11), there is one brief but pronounced turnover when the Cretaceous fauna went down, much later than its counterparts in the planktonic, neritic and terrestrial realms, and two prolonged changeovers in the late Eocene and mid Miocene, both times of greenhouse metastability truncated by major chills with far-reaching hydrographic impacts. Environmental punctuation is shown by eight well-corroborated horizons

evolutionary continuities within the succession of faunas, and not requiring comparable rigour in definition (among others: Webb and Opdyke, 1995; Woodburne and Swisher, 1995; Prothero, 1995, 1998).

Correlations between the neritic, pelagic and terrestrial chronofaunas suggest four parallels: (i) at the Cretaceous–Palaeogene (K/P) mass extinctions; (ii) at the Paleocene–Eocene boundary and LPTM; (iii) within the Auversian facies shift in the later Eocene; and (iv) from just before to within the Miocene optimum (Figs. 6.11, 6.12). The parallels at the LPTM are increased if we acknowledge that the Tethyan, large-foraminiferal, Early Palaeogene chronofauna contains a significant caesura called the larger foraminifera turnover, or LFT (Orue-Etxebarria et al., 2001).

We reviewed the accumulating evidence of global environmental instability beginning at about the end of the Lutetian age (McGowran et al., 1997a). The TRC expanded, with oligotrophic limestones at one end and massive coals at the other; and fluctuations increased in magnitude with marine transgressions interleaving with the first cogent evidence of icecap growth foreshadowing the Oligocene ice ages. There are indications in all biotas in all realms of fundamental responses to the Auversian facies shift. McGowran (1990, 1991) identified the Lutetian–Bartonian boundary as one of the major turning points in early Palaeogene history, citing deep-benthic, planktonic, neritic and terrestrial taxic and biogeographic evidence in support of the claim. Berggren and Prothero (1992) and Prothero (1994a,b) showed that whilst changes happened over the ten million years of the late-middle Eocene to early Oligocene, there were clustered events as well, extinctions occurring especially at the middle–late Eocene transition where diverse biotas of warm times were decimated by climatic change. In the North American land mammal succession, Webb and Opdyke (1995) identified a strong turnover pulse marking the changeover from the Eocene chronofauna (characteristically 'subtropical forest') to the White River chronofauna (characteristically 'woodland savannah'). This pulse was at the end of the Uintan age and into the brief Duchesnean age. The end of the Uintan displays by far the most significant extinctions: >25% land mammals (Berggren and Prothero, 1992) whereas the immigrations giving the White River chronofauna its complexion occurred mostly in the (late Eocene) Chadronian age. This major faunal and environmental shift was coeval with the end of the early Palaeogene chronofauna in the neritic larger foraminifera and the disappearance of the

Figure 6.12 (cont.)

I to VIII – consisting of Chills I–IV (horizons III, V, VII, VIII), two 'restorations', or major returns to warmings and transgressions in the late middle Eocene (Khirthar) and late Oligocene (Chattian) (IV, VI), the end-Cretaceous extinction (I), and the end-Paleocene hyperthermal (II). From McGowran and Li (2002, Fig. 5) with permission.

most prominent elements in the planktonic chronofauna – *Acarinina*, *Morozovella* and *Planorotalites* (Hallock *et al.*, 1991).

The notion of the chronofauna captures ideas of community in the long term – 10^6–10^7 years scale. One such idea is of relatively gradual change due to migration, speciation and extinction punctuated every so often by more rapid and comprehensive turnover. There are two fairly obvious processes whereby such a pattern can occur. In one process, the community itself evolves as its species do what species do – exist, speciate, extinguish; the community matures in a kind of super-ecological succession into a highly specialized but meanwhile highly fragile or metastable state. In the alternative process, communities change in response to external physical change, itself displaying a strongly punctuated pattern in Cenozoic history.

These alternatives are not mutually exclusive. The importance of the historical context, emphasized by Hottinger's model of what we call the early Palaeogene chronofauna, is that physical impacts of various kinds may recur in some detail but the actual effect on the biotas will not. In the early Eocene marked warming is accompanied by a succession of higher-than-usual frequency of third-order cycles before the termination in Chill I. Since the communities were immature and in non-critical phase (Fig. 6.9) there was not a revolutionary outcome in large-foraminiferal communities. The benthic faunas of New Zealand show their first peak overturn in the earliest Eocene (Fig. 6.10). Nor was there a revolutionary outcome in coeval terrestrial floras (Wing and Tiffney, 1987). Although the North American land-mammal succession displays the biggest immigrational wave of the Cenozoic Era in the Clarkforkian-early Wasatchian ages (early Eocene), this resulted from a conjunction of benign climate and land bridges at high latitudes.

In the late-middle Eocene (the Bartonian, Biarritzian, and Auversian Ages–Stages in the various diagrams denote virtually the same time slice) biotic revolution is more apparent in neritic (tropical and extratropical), pelagic and terrestrial communities, and we predict that these parallels will become stronger with more comparative scrutiny. What happened at that time? Hansen (1988) suggested that Bartonian molluscan faunas diversified more strongly than did their forerunners, simply because it took more than 20 myr for the communities to recover from the end-Cretaceous disaster. Did it likewise take so long for large herbivorous mammals to evolve, before brontotheres could occupy the elephant-like niche of opening closed forests into woodlands? In a slightly different but still 'internalist' or progressively 'self-organized' mode (Brasier, 1995), one might infer from Figure 6.9 that oligotrophically highly mature neritic communities were a disaster waiting to happen – a disaster triggered by some physical perturbation.

We suggest that the highly mature phase of the early Palaeogene chronofauna was brought on by the Khirthar transgression which did three things: it increased neritic living space, it encouraged, through warming knock-on effects, latitudinal expansion of tropical or pantropical biotas, and it expanded the TRC in all realms – pelagic, neritic and terrestrial. In its turn the expanded TRC did three things in the neritic: increased physical space was accompanied by steepened trophic gradients encouraging isolation and provincial tendencies, overall diversities increased, and the entire biota was rendered hypersensitive to perturbation. The latter effect is preserved in the record by very strong alternations between more eutrophic and more oligotrophic components of the fossil assemblages. This entire situation is characteristic of a warmer, higher-sea-level world; the outcome of upsetting it was a cooler, lower-sea-level world. According to the Monterey hypothesis the biosphere itself had a hand in this process through cooling via CO_2 drawdown by carbon burial. The notion was first applied to Chill III (Vincent and Berger, 1985) – later tested by Flower and Kannett (1993) – then to Chill I (McGowran, 1989), then briefly suggested for the build-up to Chill II (Thomas, 1992) and Chill IV (Raymo, 1994).

Thus, we suggest that the ultra-mature phase of that chronofauna was forced environmentally and so too was, eventually, the chronofaunal demise.

The arguments sketched in Figure 6.9 depend on strongly K-strategy behaviour evolving iteratively at one end of the trophic resource continuum (TRC) (see below; Fig. 6.28) and Hottinger took pains to emphasize that ecological and evolutionary arguments based on photosymbiosis do not export well to other sectors of the biota – e.g. he was reluctant to make palaeodepth predictions based on non-photosymbiotic benthic foraminifera. However, there are strong chronologic parallels with turnovers in other communities, demonstrating the pervasive impacts of physical change. There is a good match between increased taxic turnover in New Zealand, invigorated migrations ex-tropics, and maturity in the photosymbiotic communities. We find two strong correlations in particular: one correlation is between instabilities in all realms preceding a crash; and the other is between disparate sets of evidence, some with and some without major photosymbionts.

Terrestrial, neritic, and deep-ocean patterns through the Cenozoic Era are displayed in Figure 6.12.

Neritic – pelagic – terrestrial parallels among Neogene chronofaunas?

Figure 6.13 compares evolutionary studies in the pelagic and terrestrial realms. Within the pelagial there can be seen some counterpoint between two ecologically complementary clades of planktonic foraminifera, the globigerinids and the globorotaliids, with some reason to suppose that five global-environmental events – two of the four Cenozoic 'chills' (III and IV), two carbon shifts (Monterey and Messinian) and a sea-level minimum (the lowest for a quarter-billion years) – had some impact on episodic changes in the planktonic communities. The taxic changes in the horses were argued to another agenda as shown: the classical pattern of an evolutionary burst comprising radiation, steady state and reduction (before the renewed speciation at ~2.5 myr) as the successional outcomes of the balance between speciation and extinction. It is likely that the Equinae marched to the same environmental drum as the planktonic foraminifera. If so, then we should expect to find parallel patterns in other biotas as well.

The late Miocene mammal record has been extensively studied in the Siwaliks of Pakistan (Barry *et al.*, 2002, with references and synthesis). On a very large database of specimens, localities and diversity (115 mammalian species or lineages) Barry *et al.* found taxic turnover through five million years with three pulses at 10.3, 7.8 and 7.3–7.0 Ma. They inferred abrupt changes in community ecology and found good matches with the stepped appearance then dominance in C_4 vegetation (spread of grasslands, as inferred from carbon-isotopic shifts in soil carbonates). Thus there is punctuation with a climatic explanation for the two latest Miocene pulses. However, the authors concluded too that the Siwalik record supports neither coordinated stasis nor turnover pulse, for there was too much background turnover for stasis, and declining species richness and abrupt, uncoordinated changes in diversity.

Recent studies have been revealing strong patterns of the chronofaunal type in the late Pliocene at and after the major climatic shift labelled Chill IV in Figure 6.13. '*The modern era began* with the extinction of large proportions of the marine and terrestrial biota between 2 and 1 Ma' (Jackson and Johnson, 2000, emphasis added). Jackson *et al.* (1996) found strong increases in rates of origination and extinction in Caribbean molluscs and corals, in contrast to the intervals before and after. They made the jump from a strong taxic pattern to a palaeo-community pattern by using (on the corals) an ordination of samples by global non-metric non-dimensional scaling, revealing striking evidence for punctuated community change coinciding with the late Pliocene taxic turnover. Thus: Late Miocene and Early Pliocene communities are 'entirely distinct' from Early Pleistocene to Recent communities; in both intervals the communities display relative constancy in species composition; and the Late Pliocene overlaps extensively with the intervals before and after but does not show their relative constancy. The taxic change does indeed match the community change. The community membership was more stable in the Late Neogene than expected. This is not the Eltonian notion of tightly integrated communities ('simply wrong'); nor though is it strongly Gleasonian in the way usually ascribed to individual species tracking and happenstance association in the late Neogene

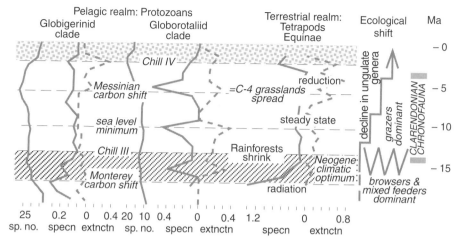

Figure 6.13 Comparison of episodic taxic overturns in the terrestrial and pelagic realms through the late Neogene: horses, including the radiation, steady state and reduction phases of a radiation (before renewed speciation at Chill IV), from Hulburt (1993); planktonic foraminifera as two major clades, from Stanley *et al.* (1988). The two foraminiferal clades, globigerinids and globorotaliids, have ecological strategies that overlap but are broadly distinguishable (Hemleben *et al.*, 1989) and respond differently to environmental events. The Clarendonian chronofauna extended from the early Miocene warm period, through the extreme vegetational changes forced by Chill III into the late Miocene, terminated only by the early Pliocene warming. The decline of the ungulates whilst dominance switched from browsers and mixed feeders to grazers ~15–12 Ma, from Webb and Opdyke (1995). From McGowran and Li (2002, Fig. 7) with permission.

terrestrial realm – this pattern in coral faunas is one of repeated reassembly from a limited species pool (point (iii) in Ivany's (1999) areas of enquiry, above).

Unsurprisingly, pelagic fossils in oceanic facies offer more detail and chronological constraint through this Late Pliocene timeslice. Recent studies have focused on the variance biogeography, speciation and extriction of members of a clode (*Menardella*) of globorotaliid planktonic foraminifera caught, as it were, between the closure of the tropical Atlantic–Pacific oceanic connection and the expansion of northern ice sheets (Norris, 1999; Chaisson, 2003). We look at this time of change through Chapman's (2000) study of ODP holes in the North Atlantic. Shown is a presently subtropical site (~18°N, 21°W) in which abundances vary by tens of percent (Fig. 6.14). The cooling controlled the threefold planktonic foraminiferal pattern, thus: (i) 3.2–2.7 Ma, relatively stable, warm conditions; (ii) 2.7–2.5 Ma, a transition phase; (iii) 2.5–1.9 Ma, pronounced climatic variability. The deterioration is seen most strongly in the abundances of the cold-water taxa. Note the relative shift at ~2.7 Ma from less-cold

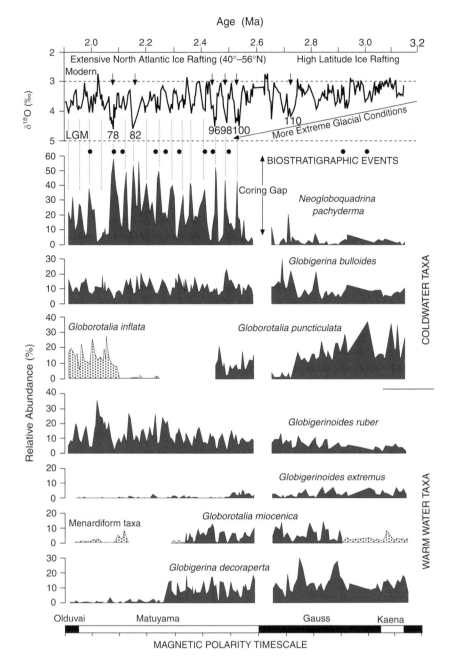

Figure 6.14 Chapman (2000, Fig. 6.3, with permission) profiled the strong planktonic foraminiferal response to the late Pliocene climatic deterioration of Chill IV at a subtropical site in the North Atlantic. There is a threefold pattern – (i) 3.2–2.7 Ma, relatively stable, warm conditions; (ii) 2.7–2.5 Ma, a transition phase; (iii) 2.5–1.9 Ma, pronounced climatic variability. Relative abundances are grouped as cold and warm water taxa. In land-mammal terms this marked and comprehensive faunal turnover would qualify as an excellent example of a geologically rapid transition (~200 000 years) between two chronofaunas.

Globorotalia puncticulata to more-cold *Globigerina bulloides*, then the takeover by subpolar *Neogloboquadrina pachyderma*. Note too the congruence of these biotic events with the physical evidence of ice rafting and bottom-water cooling in the benthic oxygen-isotopic profile.

All components of the planktonic foraminiferal fauna were affected – cold- and warm-adapted species, surface- and deeper-dwelling species, and numerically abundant and rare species. There is a marked faunal turnover. The extinct component of the Late Pliocene fauna decreased from ~40% at 2.8 Ma to ~2% at 2.0 Ma. Warm-water taxa were strongly affected. Chapman observed that the keeled globorotaliids – the menardiform group – lacked the calcite crusts that form below the thermocline in their modern counterparts, implying that they were restricted to tropical waters with a deep semipermanent thermocline and that cooling and steepening gradients outpaced their ecological tolerances. Recolonization took about 500 kyr, leaving something of a faunal gap between the last extinction at ~2.3 Ma and a modern fauna by ~1.8 Ma. The end-Pliocene fauna was very similar to the late Pleistocene–Holocene; there have been very few speciations in the interim; the keeled globorotaliids were adapted to deep- dwelling (as in Fig. 4.8); and species such as *Globorotalia truncatulinoides* and *G. hirsuta* expanded biogeographically into the mid latitudes.

There is a remarkable chronological and macroecological parallel here between the North Atlantic pelagic microfauna and the Caribbean neritic macro- fauna. In this sense of 'natural' chronofaunal stability punctuated by turnover, the birth of the modern in both cases is towards the end of the Pliocene.

Indications of chronofaunal pattern are seen in the plankton, deep-ocean benthics, and large-neritic benthics. There are not comparable patterns in the rich record of small benthic-foraminifera in the neritic. The most sustained studies of palaeocommunities, on the embayments of the North American Atlantic margin, are by Buzas and Culver (1984, 1989, 1998; Culver and Buzas, 2000). Aside from a tepid suggestion of possible coordinated stasis, there is little in the way of pattern and their conclusions seem to be strongly Gleasonian – each species has a unique distribution in space and time; there is no movement or structure as a species group; environmental regime influences community com- position but there is no apparent correlation with physical-environmental shifts; and biotic interactions and hierarchical structure are explanatorily superfluous.

There are strong chronological parallels with turnovers in other commu- nities, demonstrating the pervasive impacts of physical change. There is a good match between increased taxic turnover in New Zealand, invigorated migrations ex tropics, and maturity in the photosymbiotic communities. All of this resonates very strongly with the North American land-mammal ages (NALMA) succession. The authors cited above agree in the sense of a build-up in warmth, diversity, and metastability – thus the shortlived Runningwater and Sheep Creek chronofaunas interpolated beween the longer-lived (inherently more stable?) White River and Clarendonian chronofaunas. At the most general level we find two strong correlations in particular: one is among signs of instability in all realms preceding a crash (i.e. Chill III); the other is between disparate sets of evidence, terrestrial, neritic, pelagic, in the neritic, some with, some without major photosymbionts.

Do chronofaunas exist? – divergent views

Alroy (1992, 1998b; Alroy et al. 2000 was unimpressed by the NALMA chronofaunas – but then he was unimpressed by the Ages themselves and regarded all such divisions of the biostratigraphic record as superfluous (see Chapter 7). Tang and Battjer (1996) found stasis without coordination.

A recent, extensive review convinced DiMichele et al. (2004) of long-term stasis in ecological assemblages (plant and animal, marine and non-marine), probably due to long-term tracking of the physical environment. Bonuso et al., (2002) could not corroborate coordinated stasis on its home territory. Similar outcomes about turnover pulse are mentioned above.

Gould (2002) discussed coordinated stasis and turnover pulse in terms of the power and relevance of punctuated equilibrium (Chapter 8). He pointed to the distinction between the two hypotheses, namely that Brett and Baird (1995) assessed stability in the former as due partly to ecological dynamics whilst collapse was due to rapid environmental turnover; in the latter, Vrba (1985, 1995) attributed both stasis and abrupt replacement to environmental stabilities and vicissitudes. Gould distinguished two debates. One is about whether the patterns exist (he was inclined to agree that they do) and he pointed to the difficulty of comparing the robustness in species and clades (evolutionary individuality) with a lesser robustness in more 'leaky' ecological units. The second debate asks what forces hold biotas together with such intensity for so long. There are two propositions on this one. One is 'conservative', perceiving a passive response to overriding extrinsic (environmental) events. External forces impose coincident endings and beginnings defining the temporal packages that are the units of coordinated stasis and turnover. The second proposition is that active causal mechanisms are intrinsic – e.g. ecological locking and incumbency.

On the durations of Cenozoic biozones and biochrons

Since W. A Berggren began his calibrations of Cenozoic chronologies (Berggren, 1969a, 1971b) it has been apparent that the zones are not only unequal in duration but are clustered, so that chunks of well-resolved time contrast with poorly resolved chunks. That effect has persisted whilst calibration has progressed, and it was soon apparent too in the first comprehensive attempt to correlate an extratropical biostratigraphic succession with the calibrated tropical standard (McGowran *et al.*, 1971).

Presumably the zones that have been erected and tested have evolved pragmatically in the heat of trial and error, although some are older and have had less practitioners down the years than have others. If zones are based on evolutionary events of speciation and extinction, is there a relationship between the rate of evolution, as expressed in taxic overturn, and biostratigraphic resolution, meaning the brevity of the acceptably defined, utilitarian biochronological interval? Can that link if it exists remain visible through the thicket of other criteria for zonation and correlation, such as ease and consistency in recognition of the defining taxa, and their geographic spread?

I address that question by comparing the zones of biostratigraphic systems for the late Cretaceous and the Cenozoic. Moore and Romine (1981) compared zonations in three major groups of oceanic microfossils in 1969 (pre-Deep Sea Drilling Project) and 1975 (Fig. 6.15). Resolution, expressed as boundaries per million years, varies through time. Resolution improved especially in the Cretaceous and also in the Neogene. However, the variation in resolution through time did not change greatly. During the next two decades zonation and correlation improved and the stock of magnetostratigraphically calibrated datums increased (Berggren *et al.*, 1995a). I acknowledge this in two sets of histograms in Figure 6.16 and compare these oceanic phenomena with the patterns of marginal third-order stratigraphic sequences as they were in 1987 and 1998. The most striking feature of these comparisons is the subdued biotic and eustatic activity in the middle Eocene, an effect enhanced still further by adding the radiolarian 1975 pattern in Fig. 6.15.

Another way of presenting biochrons (Fig. 6.17) is simply to graph biochron durations against age. Again we see reduced biochronological resolution in the middle Eocene – in both oceanic protists and terrestrial mammals. This plot is unsmoothed, but smoothed plots of the same kind are discussed below.

A plot of biochron numbers against biochron durations (Fig. 6.18) seems to contrast two groups of oceanic plankton with the terrestrial NALMA divisions, the latter record not yielding the same overall density of usable events.

Spencer–Cervato *et al.* (1994) assembled data on Neogene oceanic events (see also Chapter 3) and compared theoretical versus actual resolution. The ~400 planktonic events detected in the 24 myr of the Neogene should yield an average age resolution of 400:24 = 0.06 myr. Using only the 124 geomagnetically

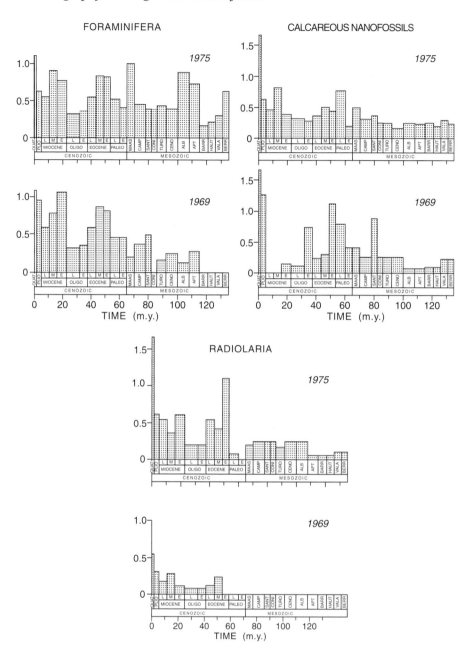

Figure 6.15 Oceanic biostratigraphy after a half-decade of deep-ocean drilling. The number of zonal boundaries per million years, averaged over subseries (Cenozoic) and stages (Cretaceous). Each of three major microfossil groups was compiled for 1969 and 1975 (Moore and Romine, 1981). The main changes were in the improvement in resolution of the Cretaceous section; the Neogene was also improved. Resolution varied strongly through time and that did not change greatly. With permission.

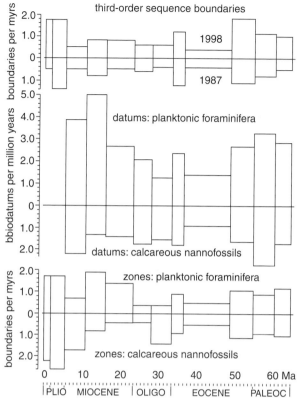

Figure 6.16 Histograms of boundaries per million years, averaged over each subepoch. Bottom, two sets of zone-defining calcareous microfossil first and last occurrences using calibrations in Berggren *et al.* (1995). Middle, same, for calcareous microfossil datums calibrated magnetostratigraphically, same source (the seemingly richer planktonic foraminiferal record probably reflects more study). Top, third-order sequence boundaries, comparing the pattern in 1987 (Haq *et al.*, 1987;) with the pattern in 1998 (Hardenbol *et al.*, 1998). Note (i) variation through time in all datasets, (ii) some strong parallels, especially in subdued biotic *and* eustatic activity in the middle Eocene and to a lesser extent in the early Oligocene (except in nannofossil zones). Time scale at bottom is for all histograms.

calibrated events gives an average resolution of 0.194 myr. However, events are unevenly distributed as to time and taxic group: the number available per myr decreases with age (Fig. 6.19). This was due not to an increase in taxic evolutionary rates through the Neogene but to an increase in number of magnetostratigraphically controlled sections (Spencer-Cervato *et al.*, 1994, Fig. 4). This trend can be expressed by standard deviation. For the Neogene as a whole, the average resolution possible of 0.194 myr (above) is exceeded threefold by an SD of 0.598 myr. The convergence of the two trends in Fig. 6.19 is a measure of

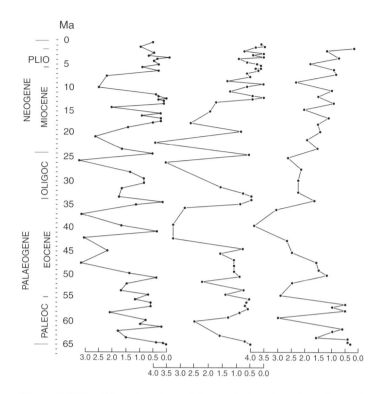

Figure 6.17 Zonal durations through Cenozoic time, ignoring subzones by treating them as zones. Curves are unsmoothed. Dates of boundaries, planktonic foraminifera (left) and calcareous nannofossils (centre), from Berggren *et al.*, 1995; for North American Land Mammal 'Ages' (right), from Woodburne and Swisher (1995). As in Fig. 6.18, note the lowered resolution in the middle Eocene.

decreasing biochronological uncertainty towards the present. The two factors limiting oceanic biochronology are precision (apparent diachrony; Chapter 3) in the late Neogene and resolution (insufficient well-calibrated sections) in the Miocene. Although the Neogene Period spans major global environmental changes that might be expected to cause variations in biochronological resolution and precision through time, this survey by Spencer-Cervato *et al.* is too influenced by the highly skewed sampling to enlighten biochronology over longer scales.

Why are good index fossils, good index fossils?

It is hardly news that some fossil groups are better than others in biostratigraphy – in meeting the criteria both of precision and of reliability.

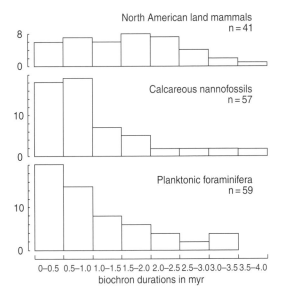

Figure 6.18 Histograms of biochron durations, using calibrations in Berggren *et al.* (1995a); and Woodburne and Swisher (1995). Unlike the previous figure, this compilation emphasizes the difference between oceanic microfossil and NALMA biochrons.

What is a good index fossil? One by one, species by species, the good index fossil marks a well-tested and narrow interval of time. It is a species whose presence tells us that we are in the rocks of a thin slice of the lower Albian or the middle Miocene stratigraphic record. The species spread rapidly and far, soon after its birth, and/or it disappeared right across this spatial range when its time came. More recently, the phrase refers to an horizon of first or last appearance of a species whose total range may actually be very long. Why is this index species 'good'? The usual reply is to do with dispersal, for it is in the rapidly dispersing groups that we find the 'best' species – the floaters, such as the skeletonized planktonic protists and the graptolites; the skeletonized swimmers, such as the ammonites and belemnites, and the conodont animals; and the blowers, such as the cuticle-bearing spores and pollen grains. It helps that the species can be found and identified without pain; and the requirements of subsurface and submarine geology make microfossils attractive, because their sample residues often of enormous numbers of specimens can be scanned with comparative ease.

In terms of taxonomy and evolution we speak of 'successful' groups, for it is they that are relatively abundant and well-dispersed. But they must be rich in evolutionary rates too – in the speciations and extinctions that give the pool of

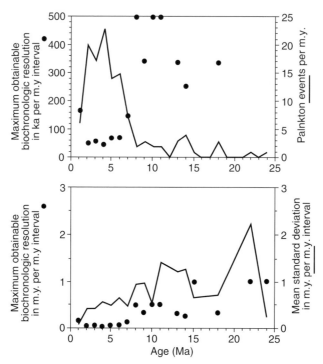

Figure 6.19 Neogene oceanic plankton events (Spencer-Cervato *et al.*, 1994, Fig. 3 and 5): using only events calibrated magnetostratigraphically in at least four sections. (Upper) Solid line, events per million years. Dots, frequency distribution of events gives average theoretical time resolution in thousands of years. (Lower) Dots, maximum obtainable biochronological resolution as in upper diagram. Line, mean standard deviation of all plankton event age calibrations per million years. The SD is the offset from the correlation line in age/depth plots for the respective sections. With permission.

events from which to choose the datums and zonal definitions. Is there a paradox here, between success, as measured in diversity, and evolutionary volatility indicating poor survival?

In one generalization about this Boucot (1983) (stated: ' … rates of speciation (speciation is used here to to include anagenesis as well as cladogenesis) tend to be highest among the species of the rarer genera, which also tend to be more endemic and stenotopic, whereas they tend to be lower among the species of the more common genera, which also tend to be more cosmopolitan and eurytopic.' However, Boucot cited ammonites as tending to be an exception to this generalization, as too the inoceramids and many oysters, among the bivalves. One of the more comprehensive treatments of the palaeobiology of biostratigraphy is Kauffman's (1969, 1977, *inter alia*). What is a 'new biostratigraphy', in the 1969 title? Kauffman summarized the main shortcomings of the biostratigraphy of the time, thus: 'The discipline was rapidly reaching a plateau in concepts and methodology and consequently in the rate of progress in zonal refinement, at a time when biogeohistorical questions were demanding more precision and accuracy. The Code was confused in its treatment of the zone: for example, the prime value of biostratigraphy is in time-correlation, yet at the same time zones are stated to exist without relationship to time – and so biofacies are stranded, excluded in one breath but accommodated in another.' Again, the firm anchoring of the zone in a body of rock is impossibly constraining for the needs of the day. As to the 'best' index fossils, Kauffman lists a 'highly prejudiced list of champions' as the usually cited groups good for correlation, and warns that we risk over-reliance on the graptolites, cephalopods, and so on while neglecting others. Finally, he contrasted in grossly simplified terms the two ways to do palaeontology. One way is to describe '*whole faunas*' which treats species in isolation from their inclusive taxa (so that there is 'a strong dominance of biological units derived from non-evolutionary systematics') and is more vulnerable to lapses into typology and loss of insights and of information. In contrast, the study of *evolving lineages* uses population systematics to avoid all such problems, and Kauffman made no bones about the reasons for the most refined and successful zonations of the Cretaceous of the North American western interior – its biological units are derived mainly from the evolutionary studies of evolving lineages. The new biostratigraphy was multiple, holistic and comprehensive, and not tied to rocks but defined only by their position in lateral and vertical space and contact relationships. And the biology of the organisms is important, particularly their habitat and strategy in relationship to physical stress (defined as the degree of non-predictability of environmental perturbations for organisms living in any particular area), for this largely controls rates of evolution. A quote *in extenso* from Kauffman (1977) encapsulates this stretch of the palaeobiology of biostratigraphy:

> In selecting organisms for construction of a biostratigraphic system
> where rapid evolutionary rates leading to biostratigraphic refinement
> are of greatest importance, one can predict which groups would
> initially be excluded from consideration because of their slow
> evolutionary rates: (1) normal inhabitants of freshwater, brackish
> water, intertidal, and very shallow subtidal environments, all eurytopic
> organisms adapted to a broad range of environmental fluctuations;
> (2) deep-water and deep-infaunal taxa protected from environmental
> perturbations by the buffering effects of sediments or a thick water
> column; and (3) environmental generalists, such as detritus feeders

utilizing a broad food base and having high tolerance for chemically poor environments. All these taxa might later be incorporated into complex assemblage zones once their origin and extinction points are known.

Rapid evolutionary rates could be predicted, however, in shallow burrowing, epi-benthic, or pelagic organisms of the marine shelf zone or in the upper water layers of ocean or epicontinental sea environments, as well as for taxa with specialized feeding or other behavioral traits, all of which would be strongly affected by unpredictable environmental perturbations through time. These easily identified taxa could be selectively chosen for the initial construction of a refined biostratigraphic system, shortcutting the trial and error methodology.

Finally, it can be theorized, for the adaptive gradient between stenotopic and eurytopic organisms living in time-stressed areas, that certain unique suites of organisms will be expected to respond to each major phase or level of intensity in a stress gradient. Given a long-term environmental decline (lowering regional temperature, global marine regression, changing water chemistry, etc.), with progressively increasing intensity of stress, certain kinds of organisms (stenotypes) will undergo rapid evolutionary bursts during initial stages of environmental decline, others at a second stage, and so on, until the final evolutionary burst among the most adaptive eurytopes as stress factors become severe. If this theory is valid, then it should be possible to predict and preferentially select fossil taxa for building a detailed biostratigraphic system, based on rapid evolutionary turnover, for every stage of a major environmental fluctuation, and thus to construct a system of zonation and correlation in the most efficient manner.

Kauffman tested his theory in the Cretaceous of the western interior. The summary in general terms is displayed here in Figure 6.20 which is based in abundant and stratigraphically tightly constrained data on trophically assigned molluscs. Transgression is accompanied by ameliorating marine environments, maritime climates and reduced seasonality and raised temperatures, lowered physical stress, expanding ecospace, niche opportunities and resources, decreasing biological competition, and relatively low rates of evolution. As regression sets in after the transgressive peak, there is a succession of peaks of evolution from the most stenotopic, or specialists, to the most eurytopic (~generalist) organisms, in successional response to two types of increased stress: a relatively abrupt increase in surface temperatures and epicontinental

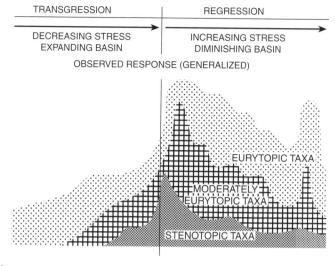

Figure 6.20 Kauffman's (1977) model of evolutionary response to environmental impact, using molluscan lineages from the Cretaceous of the Western Interior. Vertical, speciation rates; horizontal, a transgressive–regression cycle, time running to right. There are staggered responses according to ecological characteristics from stenotopic to eurytopic (broadly adapted generalists to narrowly tolerant specialists). Speciation increases during transgression as opportunities increase. The early-colonizing opportunists show the first success (in increasing diversity) but are over-taken by eurytopes as stresses build up during regression. (Peak in speciation near peak regression is in nearshore clastics.) In its pronounced asymmetry from trans-gression to regression in contrast to the simple symmetry in the Israelsky lithostrati-graphic cycle (Fig. 5.10), this model anticipates third-order sequence stratigraphy. With permission.

salinity; and an initial shock of epicontinental regression with lowered tem-peratures and salinity. The biostratigraphic model based on this analysed mar-ine cyclothem is shown in Figure 6.21, demonstrating how optimum refinement is relayed from opportunists to highly stenotopic taxa, to moderates and then to eurytopic groups. It would seem that Kauffman's model has no room for the idea that competition rises during specialization and niche partitioning under phys-ically stable and predictable conditions which are found around peak transgres-sions. Instead, the model invokes a direct impact of environmental stresses on the respective trophic groups.

Kauffman's scenario dealt with succession in a marine cyclothem, a trans-gression–regression pattern antecedent to the third-order sequence (Chapter 5). We can approach the same problem of the palaeobiology of biostratigraphy from another direction, beginning in the most basic fact of evolutionary life – that the fossil record reveals, not stately progression, but a series of pulsations

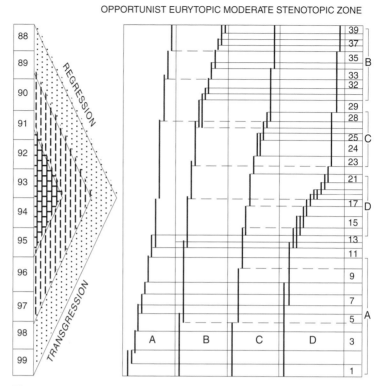

Figure 6.21 The generalizations of Figure 6.22 are here translated into a biostratigraphic model for a simple marine cyclothem (Kauffman, 1977). Optimum biostratigraphic refinement is relayed from opportunists to highly stenotopic taxa, to moderates and then to eurytopic groups. Species ranges give no less than 39 zones for one cycle, using all fossils. Ecological groups A–D give ecostratigraphic zonal groups A–D. With permission.

produced by radiation, extinction and convergence by relay (Simpson, 1964). McGowran (1968a) pointed out that the planktonic foraminifera have just this pattern, exemplified by the multiple reappearance of angular and keeled tests: *Rotalipora* and *Praeglobotruncana* in the middle Cretaceous, *Globotruncana* in the late Cretaceous, *Morozovella* in the Paleocene–Eocene, *Globorotalia* in the Neogene (to ignore some details, even then; nomenclature has proliferated subsequently). Indeed, the history of bioclassification and nomenclature is very largely the history of disentangling convergences and clarifying the anatomy of separate evolutionary radiations – carving nature ever closer to the joints. Contemplate Vrba's sketch of a burst of speciation (Fig. 6.22) constructed to illustrate her 'effect hypothesis' – the by-production of a directional trend in a trait by speciation rather than by some form of orthogenesis or orthoselection.

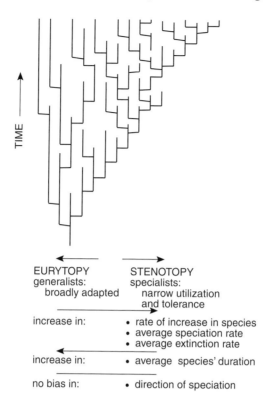

Figure 6.22 Clades and turnover in the effect hypothesis: Vrba (1980, 1985) put the ecological duality of eurytopy:stenotopy into a macroevolutionary context in this sketch of an evolutionary radiation, showing how various species' properties can suggest direction even though there be no bias in direction of speciation.

There are many more events – speciations and extinctions – of potential biostratigraphic utility in the latter part of the burst, as stenotopy becomes attractive. (Parenthetically, the non-direction in speciational frequency as shown here was tested by Pearson (1998c) who found evidence of increased survivorship among the descendant relative to the ancestral species, implying an adaptive drive.)

Thus, within one 'successful' taxonomic group we should expect to find that biostratigraphic resolution will vary through time, increasing as 'success' increases on the criterion of increased diversity, for not only speciation but also extinction increases along with diversity. Can we examine the data on zonal durations in the context of this configuration? There was a broad correlation between taxic radiations in the late Cretaceous and Palaeogene planktonic foraminifera and an environmental indicator in the form of an oxygen-isotopic gradient (Fig. 6.23) perhaps signalling more available environments. It may be

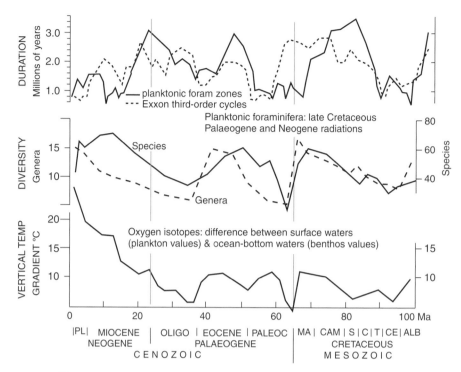

Figure 6.23 Comparisons of indicators in late Cretaceous and Cenozoic biostratigraphy using 1970s and 1980s data. (Top) Durations of planktonic foraminiferal zone and subzones (Berggren *et al.*, 1985c) and third-order sequences (Haq *et al.*, 1987). Both curves were smoothed by a three-point moving average. Peaks indicate decreased biochronological resolution and eustatic frequency and troughs the reverse. (Middle) Diversity patterns in species and genera show three planktonic foraminiferal radiations: late Cretaceous, Palaeogene, Neogene. (Bottom) The vertical temperature gradient (surface-bottom) is also three-part but displaying a radical departure in the Neogene from the previous 'cycles.' Middle and Bottom from a diagram by R. G. Douglas and S. M. Savin in JOIDES (1981).

that biochronological resolution is higher in the early parts of the radiations. The Neogene clearly is different, for the long-term trend from greenhouse to icehouse dominates any cyclicity.

Norris (1991a, b, 1992) investigated longevity among planktonic foraminifera, building on the anatomy of the radiations, reconstructed ecologies including depths of habitat (Douglas and Savin, 1978; see review by Corfield and Cartlidge, 1991), and seeming correlations with global environmental trends and events (Frerichs, 1971; Hart, 1980; Vincent and Berger, 1981; Caron and Homewood, 1983; Stanley *et al.*, 1988; Wei and Kennett, 1983, 1988; Leckie, 1989; Leckie *et al.*, 2001). The radiations display iterative emergence of a small

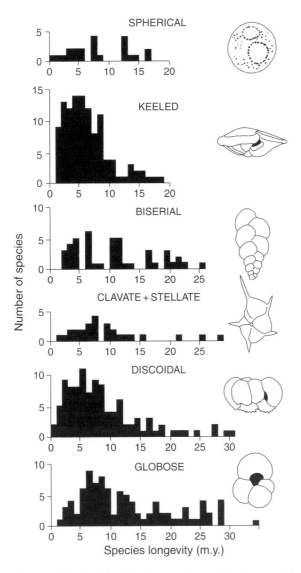

Figure 6.24 Norris (1992, Fig. 2, with permission) compiled histograms of species longevity for planktonic foraminifera from all radiations, assigned to six morphogroups. All morphogroups contain short-ranging species but the spherical and keeled groups rarely contain long-ranging species.

range of shell forms – some ten morpho-forms lumped into the three, namely the globose, discoidal, and keeled in Figure 6.24. Norris showed that all groups contained short-ranging taxa but that some – the keeled and spherical species – were almost always short-ranging. He contrasted the keeled species group with the discoidal group for each of the Cretaceous, Palaeogene and Neogene

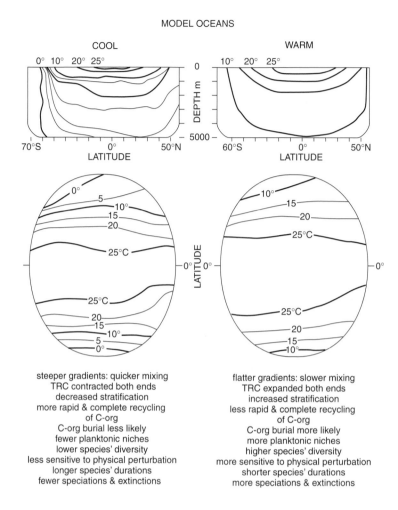

MODEL OCEANS

COOL   WARM

steeper gradients: quicker mixing
TRC contracted both ends
decreased stratification
more rapid & complete recycling
of C-org
C-org burial less likely
fewer planktonic niches
lower species' diversity
less sensitive to physical perturbation
longer species' durations
fewer speciations & extinctions

flatter gradients: slower mixing
TRC expanded both ends
increased stratification
less rapid & complete recycling
of C-org
C-org burial more likely
more planktonic niches
higher species' diversity
more sensitive to physical perturbation
shorter species' durations
more speciations & extinctions

Figure 6.25 Model oceans, in an heuristic pioneering attempt to contrast greenhouse with icehouse oceans (Lipps, 1970). The added generalizations on environments and biotas have been discussed for several decades, but are still useful and broadly meaningful. TRC, see Figure 6.28.

radiations, the mean species longevities being respectively (myr): 5.5 vs 11.0; 5.5 vs 8.8; and 7.2 vs 14.1.

Lipps (1970) contributed an early, heuristic view of model oceans in what would later be called greenhouse and icehouse worlds (Fig. 6.25). I add a few simple generalizations.

The first-order, 100-myr change in the world from Cretaceous to present is punctuated by changes brief enough to be called 'events', of which several are shown in Figure 6.26. I take the summary by Hart (1990) as starting point. The

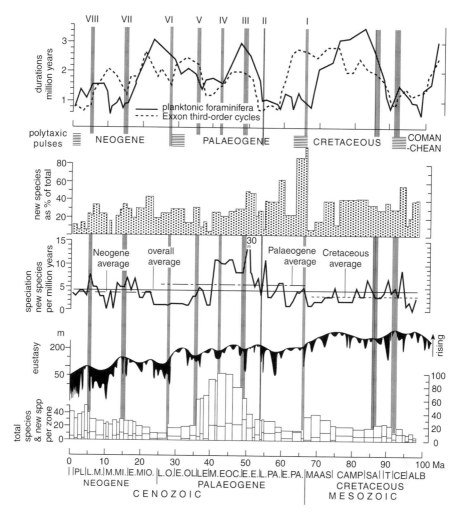

Figure 6.26 Durations of biochrons and third-order sequences vis-à-vis indicators, Late Cretaceous to Neogene. Durations, as in Figure 6.25. Polytaxic pulses and Grabau series, from Figure 6.1. New species as percentage of total, new species per million years, and total species and its subset new species per zone are all from a review by Hart (1990). 'Eustasy', from Haq *et al.* (1987). Vertical divisions running through the diagram are events I–VIII (as in Fig. 6.12) plus two major anoxic events in the early Late Cretaceous.

histogram of total species shows a very strong peaking in the Eocene, which results from a sharp increase in provincialism following the cooling and a burst of speciation outside the tropics which exceeds that effect in either the late Cretaceous or the Neogene (and which is, as a subjective opinion, taxonomically more split, thus also exaggerating speciation per myr in the early Eocene). By plotting new species as a percentage of the total, Hart provided another useful view of the radiations – it subdues the Eocene distortion but emphasizes the rebound after the mass extinction. The curves of planktonic foraminiferal zones and third-order sequences are extended back to the late Albian and plotted against Hart's curves. Simple inspection suggests that periods of lowered resolution alternate with periods of heightened resolution (troughs). The latter are accompanied by increased frequencies of third-order sequences (though weakly in the Miocene), inviting speculation that increased frequency of perturbation in sea level as a level III phenomenon is an important environmental influence – more so than is a single sharp change. The Cenozoic periods of heightened resolution are associated with, in succession, a strong warming, the reversal after the Eocene cooling, the Monterey effect in the Miocene, and the Pliocene reversal. In the Cretaceous there were the oceanic anoxic events which affected pelagic communities via an expanded oxygen minimum layer. Perhaps we can say that when planktonic foraminiferal communities increased in diversity in more stratified water, their rates of change increased, and so too did the resolution afforded by their fossilized remains. There is only one serious item of special pleading here – explaining away the high diversities and speciation rates of the early-middle Eocene, as noted above.

Fischer and Arthur (1977; Fischer, 1981; see Fig. 6.1) distinguished polytaxic pulses, characterized by transgression, climatic equability and, in the biosphere, heightened diversity and extended food chains capped by superpredators, essentially coinciding with Grabau's (1940) transgressive pulses; and oligotaxic times of crisis and the blooming of opportunists. Provoked largely by this global vision, I sketched a speculative pair of divergent pathways from subcrustal drive to modes of terrestrial duricrust formation, via the now-popular 'greenhouse'/'icehouse' dichotomy (Fig. 6.27). Within these flowcharts is the notion that the stratigraphic record and the fossil record are biased in their richness towards polytaxic times.

The trophic resource continuum (TRC)

Perhaps we have here a shadowy response in the biostratigraphic succession to the alternation of more stratified waters with more mixed waters through geological time. That alternation is superimposed on the transformation of the Cretaceous world to the Neogene. There are sufficient matches of environmental impact, speciation, and increased resolution to encourage the notion that good index fossils get better as their original communities become more stenotopic, which happens in warmer but more metastable times. This brings us to the *trophic resource continuum* – we include nutrient, along with water temperature and water depth, in relating biotic change to environmental

		TECTONISM	SEA LEVEL	CLIMATE		BIOTA	RECORD		DURICRUSTIN CENOZOIC
MODE ONE	rapid seafloor spreading	*amplified:* uplift, subsidence, igneous activity, CO₂ exhalation	marine trans- gression	marine climate, global warming, more humid, more equable	end result: GREENHOUSE	*increased:* habitat diversity, biotic diversity, evolutionary overturn	increase chance of sedimentary accumul- ation and fossilization (marine and nonmarine)	laterite and deep weathering	Middle to Early Miocene Late to late Middle Eocene Early Eocene to Late Paleocene
MODE TWO	slow seafloor spreading	*subdued:* uplift, subsidence, igneous activity, CO₂ exhalation	marine regression	continental climate, global cooling, more arid, less equable	end result: ICEHOUSE	*decreased:* habitat diversity, biotic diversity, evolutionary overturn	decrease chance of sedimentary accumul- ation and fossilization (marine and nonmarine)	silcrete	Late Miocene Oligocene early Middle Eocene Early Paleocene

Figure 6.27 Different crustal modes, leading speculatively to contrasting states in the exogenic system (McGowran, 1986a).

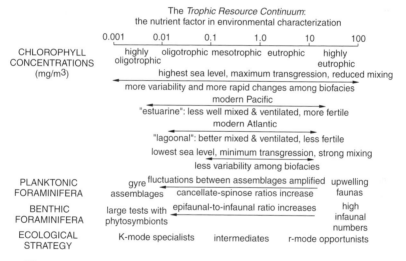

The *Trophic Resource Continuum*:
the nutrient factor in environmental characterization

Figure 6.28 The trophic resource continuum (TRC), scaled to chlorophyll concentration (Hallock, 1987; Hallock *et al.*, 1991; Boersma *et al.*, 1998). Cretaceous and Palaeogene oceans tended to stretch at both ends whereas Neogene oceans tended to shrink except, most notably, at the Miocene climatic optimum; the 'lagoonal' Atlantic is better mixed and ventilated but less fertile than the 'estuarine' Pacific (Berger, 1970). The interplay between the *Globoturborotalita woodi* group and the *Globigerina bulloides* group in the Miocene (see Fig. 6.33) reflects not only watermass temperature but a TRC effect as well. (McGowran and Li, 1996.)

change. Oceanic waters can be arranged as a spectrum from *oligotrophic* to *eutrophic* (Hallock, 1987; Hallock *et al.*, 1991). 'Oligotrophic' and 'eutrophic' are qualitative terms for regimes characterized by low and high availability of

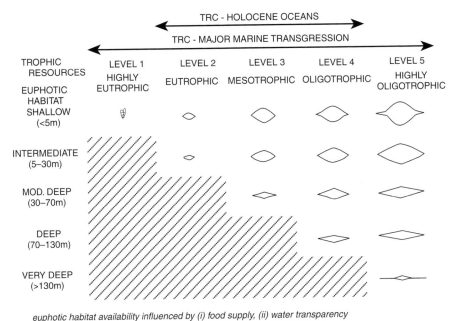

euphotic habitat availability influenced by (i) food supply, (ii) water transparency

Figure 6.29 The trophic resource continuum controlling large benthic foraminiferal distribution in the euphotic habitat by controlling food supply and water transparency (Hallock, 1987, Fig. 5). This model predicts that stratification-related diversity (and size) will increase during expanded-TRC and polytaxic times (warming and transgression) and contract during oligotaxic times (regression and cooling), with permission.

biolimiting nutrients, respectively, and their associated levels of primary productivity (reviewed by Brasier, 1995).

The central concept of the TRC is that it does not simply shift in response to a climatic and oceanic change, such as from cooling and mixing at one extreme to warming and stratification at the other: instead, it expands at both ends and contracts when the oceanic-environmental trends reverse (Fig. 6.28). If this expansion is so, then there should be evidence of it during times of transgression and amelioration. One kind of evidence might be found in stronger swings in signals of the expanded TRC, and this is shown in the next chapter. As another prediction: there should be increased stratification in both benthic and planktonic habitats. Hallock (1987) suggested that trophic resources might control both food supply and water transparency (Fig. 6.29). To the right in this sketch there will be both increased diversity and increased susceptibility to environmental perturbation, hence, more bioevents per slice of time and more biostratigraphic refinement. Thus, TRC arguments support the trends outlined above.

Taxic stratigraphic ranges and quantified biostratigraphy

Recurring themes in this enquiry include the phyletic emergence and extinction of taxa, their biogeography and dispersal, and facies barriers and environmental exclusions. A prime objective of biostratigraphy is accuracy and refinement in ordination and correlation. Since ordination involves the 'true' stratigraphic range of a taxon, the latter never quite departs our consciousness. We have considered several lines-of-development in biostratigraphy as an outgoing discipline. One is the emphasis on tested bioevents calibrated geomagnetically and radiometrically (Chapter 3). Another is the promise of bioevents constraining physical-stratigraphic sequences and meaningful hiatuses (Chapter 5). Yet a third is the search for natural packaging of the biostratigraphic record – biofacies meets geological time (Chapter 6). In all of these lines quantification has been developed and urged as a way to progress. (However, I should point out that developing the IMBS extensively used graphic comparison of sections against a timescale (Chapter 3) as promoted from the late 1960s in the early legs of the Deep Sea Drilling Project. These age-depth plots are solidly numerical, if not statistical.)

Exploration micropalaeontology, with its high numbers and relative ease of sampling, is well suited to quantitative methods (Gradstein and Agterberg, 1985). Advocates of quantitative stratigraphic methods have focused particularly on one problem in fossil distribution – with so many taxa seemingly available in a given assemblage, how come there are so few zone-defining events? In all constructions of biocorrelation lines between sampled stratigraphic sections, it does not take many operations of matching first or last appearances (bioevents) to produce crossovers – a tangled fence and a hopeless mess in which one cannot distinguish diachrony from synchrony (e.g. Worsley and Jorgens, 1977; Sadler and Cooper, 2003). Causes of the tangled fence range from biological patterns of diachrony and fluctuation, through taphonomic factors and insufficient sampling, to human error in systematics and identification. The upshot is that species' ranges can be highly conflicting from section to section – the raw data are of little use in their raw state. We cull severely and discard most taxa as facies fossils. Numerical reactions to this perceived neglect of data are broadly twofold, reflecting two 'opposing philosophies' (Edwards, 1982b; Cooper *et al.*, 2001): *deterministic* methods which seek the total stratigraphic ranges of taxa, and *probabilistic* methods which seek the most probable range (Fig. 6.30, Table 6.4). Among the probabilistic methods, ranking & scaling (Agterberg and Gradstein, 1999) attempts to identify the average or most probable tops, bases and ranges of taxa (Fig. 6.30, Table 6.4). All methods seek to untangle the fences, retaining instead of culling the lines.

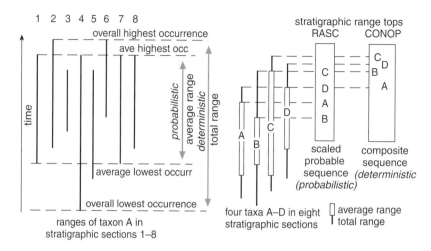

Figure 6.30 Left, two ways of seeking the meaning of a taxon's stratigraphic range from several records: lowest and highest occurrences. Right, given total and average ranges for four taxa in eight sections, the two methods RASC and CONOP produce different ordinations of range tops. Adapted from Cooper *et al.* (2001, Fig. 3 and 4) with permission.

Quantitative biostratigraphy and the search for finer resolution in global correlation have been reviewed recently by Sadler (2004). The best-known quantified method is deterministic – graphic correlation (Shaw, 1964; F. X. Miller, 1977; Mann and Lane, 1995) which has been used in microfossiliferous oceanic sections for some decades, if sporadically (*inter alia*, Prell *et al.*, 1986; Dowsett, 1988; MacLeod, 1991; MacLeod and Sadler, 1995), and in neritic sections (Moss and McGowran, 2003). Graphic correlation builds a composite section of events (bioevents, chemoevents, sequence-stratigraphic surfaces) in repeated testing by new sections, always searching to close the gap – the 'missing section' – between the 'known' and the 'real' top/base of the single taxon and hence the 'true' ordering and spacing of the events (Chapter 3; Fig. 3.9).

Graphic correlation is problematic in its labour-intensive iterations section-by-section against the composite, whereas the automated graphic correlation technique known as constrained optimization treats all sections and events simultaneously and automatically fits a correlation line (Kemple *et al.*, 1995; Cooper *et al.*, 2001). CONOP employs unitary associations (UA), a deterministic mathematical model for constructing concurrent range zones (Guex, 1991). Guex's unitary associations are equivalent to Alroy's (1992) conjunctions. The UA method constructs – *ordinates* – a discrete succession of concurrences of species- pairs in a UA range chart. This matrix filters the conflicting species' associations by deciding statistically which species' range is 'wrong'.

Table 6.4 *Comparison of the three quantitative stratigraphic techniques used by Cooper et al. (2001, with permission) – graphic correlation, constrained optimization, and ranking and scaling*

GRAPHCOR	CONOP	RASC
Deterministic method, graphic correlation	Deterministic method, constrained optimization	Probabilistic method, ranking and scaling
Uses event ordering and spacing	Uses event ordering and spacing	Uses event ordering only
Large datasets require much labour	Can process large datasets readily	Can process large datasets very quickly
Requires selection of an initial 'standard' section, then section-by-section comparison with the composite in repeated rounds	Treats all sections and events simultaneously	Treats all sections and events simultaneously
LOC fitting in section-by-section plots, partly automated	Fully automated	Fully automated
Attempts to find maximum stratigraphic range of taxa among the sections	Attempts to find maximum stratigraphic range of taxa among the sections	Attempts to find average stratigraphic range of taxa among the sections
Builds a composite by interpolation of missing events in successive section-by-section plots via the LOC	Uses simulated annealing to find either the 'best', or a very good, multidimensional LOC and composite sequence	Uses scores of order relationships to determine single most probable sequence of events
Relative spacing of events in the composite is derived from original stratigraphic spacing	Relative spacing of events in the composite is derived from original stratigraphic spacing	Relative spacing of events is derived from pairwise crossover frequency
Does not correlate sections automatically	Correlates sections automatically	Automatic correlation of sections by sister program (CASC) using RASC output

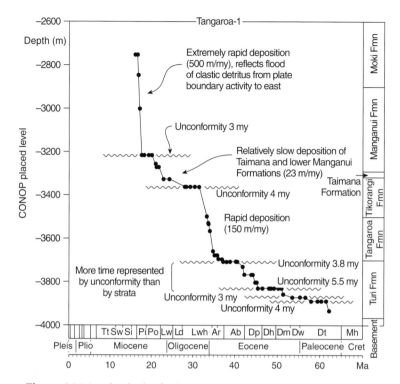

Figure 6.31 Age-depth plot for Tangaroa-1, Taranaki Basin. Biostratigraphic range tops ordinated by CONOP were graphed against the time-calibrated composite for New Zealand (NZ stage symbols: see Fig. 7.22). Terraces imply unconformities or condensed sections (Fig. 5.12) and steep slopes, rapid rates of accumulation. From Cooper *et al.* (2000, Fig. 6) with permission.

R. A. Cooper led an intensive study of the intensively-sampled Taranaki Basin in New Zealand (Cooper *et al.*, 2000, 2001; Sadler and Cooper, 2003). The samples from these offshore exploration wells were rotary cuttings, so that stratigraphic tops only were used. Four hundred and eighty seven taxa (foraminifera, calcareous nannofossils, dinoflagellates, spores and pollens) were culled to 87 range-top events, of which only 16 were found not to have 'relatively good biostratigraphic reliability'. The three techniques outlined in Table 6.4 were compared in the Taranaki Basin and shown to be not alternative but complementary. This work developed CONOP in its fusion of graphic correlation and unitary association and added an outgroup missing from the three methods – the time-calibrated composite. In the time-calibrated composite, events in the CONOP composite were graphed against their ages as 'known' elsewhere in New Zealand. The regression line could also project the boundaries of the New Zealand stages into the Taranaki Basin. An age-depth plot for the Tangaroa-1 well used CONOP-placed levels against the New Zealand chronology (Fig. 6.31).

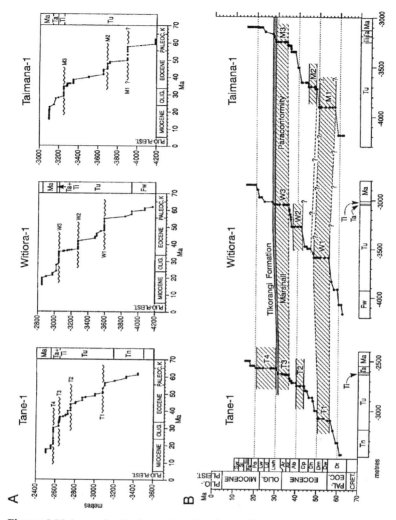

Figure 6.32 A, age-depth plots using the time-calibrated composite as in Fig. 6.31 (where lithostratigraphic abbreviations are spelt out) for three wells in the Taranaki Basin. Terraces signalling unconformities are labelled independently in each section. B, the *x*- and *y*-axes are reversed to show correlation of unconformities. From Cooper *et al.* 2000, Fig. 7) with permission.

Three other plots revealed parallels across the basin (Fig. 6.32) which seem to demonstrate the potential of this method in this kind of sampling situation. Cooper *et al.* (2000) claimed that biostratigraphic resolution in the Taranaki Basin was increased by an order of magnitude by CONOP, and previously unseen unconformities were revealed.

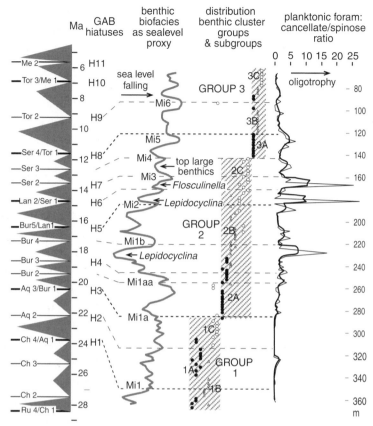

Figure 6.33 A third-order framework for Miocene developments. The chart of third-order sequences to the left is 'global' (Hardenbol *et al.*, 1998); the two curves centre and right are from the Lakes Entrance section in East Gippsland in SE Australia. The third-order glacials Mi1 to Mi6 are adapted from a composite deep-oceanic curve of $\delta^{18}O$ *Cibicidoides* (Wright and Miller, 1993). Cancellate/spinose ratio (sample-by-sample and three-point moving average): the ratio of *Globoturborotalita woodi* plus *G. connecta* to *Globigerina bulloides* plus *G. falconensis* are the dominant members of the broader and probably monophyletic cancellate-spinose and spinose-non-cancellate groups of plankonic foraminifera, respectively (Fig. 5.30). The palaeodepth curve (Li and McGowran, 1997) is based on proportions of inner, middle and outer neritic biofacies permitting estimates of where sequence boundaries fall and of correlations with Mi glacials (Fig. 5.30). Although more rigour in correlation is desirable, the palaeodepth curve can be matched with the global curves plausibly, using biostratigraphic events for correlation. These correlations are corroborated by adding hiatuses H1–H11 from the Great Australian Bight (GAB) (Li *et al.*, 2004). The three larger-foraminiferal horizons characterized by *Flosculinella* and *Lepidocyclina* correlate excellently with the strong spikes in the cancellate/spinose ratio, indicating parallel shifts on the TRC. All the third-order glaciations to Mi6 can

Just as deterministic methods produce concurrent range zones, so do probabilistic methods produce assemblages or assemblage zones. Probabilistic stratigraphy (Hay, 1972; Worsley and Jorgens, 1977; Hay and Southam, 1978) produces the optimum stratigraphic sequence, the most likely order of bioevents. Using the same data – events frequent enough to occur in the most likely sequence – it also produces the optimum cluster based on interfossil distances in time. In a major study preceding the development of RASC, Gradstein and Agterberg (1982) compared optimum clustering solutions with conventional planktonic foraminiferal biostratigraphy on the northwestern Atlantic margin, finding relatively and comparably conservative resolution in each case. The clustering was strongly and cleanly successional in time.

Miocene profile from East Gippsland

Cluster analysis has probably been more popular on the biofacies side of biostratigraphy than in the problems of event-ordination and correlation. In southern Australia we have used it routinely for Holocene neritic biofacies (inter alia, Li et al., 1996) and in the Oligocene and Miocene carbonate facies (Moss and McGowran, 2003). The Miocene foraminiferal succession in East Gippsland in southern Australia has been discussed already (Chapter 5). Here (Fig. 6.33), we present a scenario which seems to bear upon the Boucot model of recurring community groups (Fig. 6.4). The Lake Entrance section has a proxy sealevel curve based on ratios of inner:outer neritic-benthic foraminifera – this is a recurring record produced by biofacies tracking. The inferred shoalings are consistent with sequence boundaries and third-order glaciations Mi1–Mi6, as shown. Added to this environmental scenario of eustatic and climatic fluctuation are two indicators of warm/oligotrophy vis-à-vis cool/eutrophy – the planktonic cancellate/spinose ratio and the large tropical-type benthics in the pelagial and the neritic,

Figure 6.33 (cont.)

be discerned in southern Australia, implying substantial third-order environmental shifts which surely impacted on vegetation and rates of weathering at this third-order (10^6 years) scale. The same biofacies data are clustered into groups and subgroups (Li and McGowran, 2000, modified from McGowran and Li, 1996). The three clustered groups ('second-order') are almost perfectly successional. Within the main groups ('third-order') there is very little coherence among samples from similar environments, such as glacials, or upwellings or warm intervals (see McGowran and Li, 1996). Instead of communities recurring when times are right, there is a very strong sense here of ongoing faunal rehash and reassembly, with permission.

respectively. This section spans two major carbon excursions (the Oligo-Miocene and Monterey; Fig. 8.7). Clustering > 200 benthic species and their abundances produced a remarkably clean stratigraphic pattern of three successional groups (McGowran and Li, 1996; Li and McGowran, 2000). The subgroups within the respective groups are somewhat more mixed stratigraphically, but still they are largely successional. On the Boucot model one might expect recurrence to compete more strongly with succession – for 'time's cycle' to assert itself more strongly against 'time's arrow' in the clustering; for Eltonian patterns to be visible along with Gleasonian. Not so! The clustering seems to favour propinquity in time (therefore in space) over similarity of habitat. The very clean breaks between Groups 1, 2 and 3 are (by correlation) at glaciation Mi1a (at the Aquitanian–Burdigalian boundary) and at glaciation Mi4. Clean though they are, the breaks seemingly have no striking correlations with chronofaunal patterns discussed in this chapter. This clustering break lags by one glacial (Mi4) the most influential one that we have labelled Chill III (= Mi3).

7

Biostratigraphy and chronostratigraphic classification

Summary

Timescales could be as diverse as major taxic groups (e.g. an ammonite scale and a foraminiferal scale), realms (marine and terrestrial scales), or provinces (Paratethyan and Australasian provincial scales) themselves are diverse. Clearly we need a central reference controlling such potential exuberance. The key distinction is between the framework category, the decisions and rules and stratigraphic codes that make orderly progress possible by increasing communication and reducing confusion, and the phenomenon category, in which all the actual scientific disciplines comprising biogeohistory can cross-fertilize and flourish. Thus we can always improve the date of an event, we can always clarify the succession of speciations and extinctions (especially between major taxa) and such matters cannot be resolved by democratic processes. Therefore, we need agreed reference sections in the rocks (boundary stratotypes) for boundaries in the standard timescale (e.g. the Cretaceous–Palaeogene, Eocene–Oligocene, Palaeogene–Neogene boundaries) which are decided by vote but only on the basis of all available rigorous science. Meanwhile, there remains a role for regional scales which have to tie together neritic and terrestrial facies, or which are essentially biochronological. Such scales are subject to codes but the actual active use of each scheme in its unique context is the only test of its value.

Introduction: why do we need a geological timescale?

The core and soul of geology are captured in a terse phrase: *rock relationships and Earth history*. Stratigraphy is the study of successions of rocks and their interpretation as sequences of events in Earth history. 'In stratigraphy, correlation is the heart of the matter. This requires a rigorous standard, which is provided by chronostratigraphy' (Holland, 1978). There has to be an disentangling of complex geological relationships into a succession of events through time, and this succession has to be correlated with parallel successions in other rocks in other places and other continents or oceans. Such endeavours are absolutely impossible without standards, reference points, and an agreed nomenclature, all of which comprises a geological timescale or geochronologic scale. The early scales, based in lithology (eighteenth century) then palaeontology (nineteenth century), were relative timescales; the addition of meaningful numerical (radiometric) estimates of geological age (twentieth century) produced the unfortunately named 'absolute' timescale. The modern *geochronologic* scale, then, is a joining of two different kinds of scale, *chronostratic* (or chronostratigraphic) and *chronometric* (Harland *et al.*, 1990). Harland *et al.* continued: 'The chronostratic scale is a convention to be agreed rather than discovered, while its calibration in years is a matter for discovery or estimation rather than agreement. Whereas the chronostratic scale once agreed should stand unchanged, its evaluation will be subject to repeated revision. For this reason, no geologic scale can be final . . .' The place of the rule of law – codifying – in the historical science of stratigraphy was expressed by Harland (1973, 1975) most usefully in his distinction of a *phenomenon category* from a *framework category*. The phenomenon category includes the many ways of characterizing, describing, classifying, and correlating strata; above all, it is concerned with how we might interpret them. It is in the phenomenon category that the scientific questions are asked and research programmes are mounted. This is historical science. It is in the framework category on the other hand that the results of the historical science are expressed and communicated – it is in this category that rock materials and time frameworks need to be organized. Based though they are in discoverable patterns in the stratigraphic record (including chronometric estimation in years), the stratigraphic timescales *themselves* are not discovered as some sort of scientific truth but, having arisen historically, are decided by agreement in convention, by political and even polite processes during a prolonged consideration of numerous scientific lines of evidence. Hence the notion of the 'golden spike' which marks the base of the entity in the *chronostratigraphic* scale – the scale that has developed since early in the nineteenth century – and of the standard second for the *geochronometric* scale.

As Harland and others have pointed out, the respective operations, *classification* and *correlation*, use largely the same observations and so there is an inevitable overlap between the phenomenon and framework categories and an inevitable source of confusion and controversy. Harland made the 'British' point that only the two major framework classes, rock and time, need distinctive international regulation, so liberating all the others to respond to the normal development of science. Correlations are always provisional and liable to criticism and improvement. They are theories of pattern in space and time.

Similarly, Ager (1984): 'There are rocks, which remain, and there is time, which has passed and can never be recovered. All the rest is semantic confusion. We need time terms and rock terms and nothing more.' Or here is Allan (1966), even more pungently: 'Remember how a Code comes into being; there are committee meetings, lengthy discussions, endless letters, and I seem to remember the publication of minority reports. I suppose the Code finally takes form as a result of majority decisions and on the assumption that wisdom resides in numbers. But, I remind you, science does not operate on democratic principles; the really fruitful scientists are the nonconformists, the revolutionaries, those who refuse to accept the majority point of view. To such people, and they are the great ones, Codes are anathema.'

One conclusion from an earlier chapter was that formally defined zonations should be disciplined and that their proliferation should be sternly resisted. My preference has been for an absolute minimum of formally defined zonations – perhaps only one set per major group of planktonic microfossils. To impose definitions and requirements on biostratigraphy is to risk imposing crude constraints on our perceptions of the fossil record. Here, we must ask whether the same or a similar response is appropriate for multiple, parallel chronostratigraphic systems. Is 'the' geological timescale sufficient for all purposes for biogeohistory, present and in the immediate future, or do we need regional scales to cater for the complex global mosaic of environments and biotas? This question is in the framework category and therefore the test of actual pragmatic usage is more pertinent than is any 'truth'. I summarize these pragmatic questions in Figure 7.1. The core of a modern late Phanerozoic geochronology will consist of biochronology (meaning, realistically, the biozonations and/or datums based in the skeletonized microplankton), magnetochronology, and radiochronology, and it is this triumvirate in the IMBS which opportunistically cobbles together the correlations required in global biogeohistory – a cobbling now sharpened and strengthened by cyclostratigraphy. More limited in their geographic range of influence – not in their importance or relevance or information content – are such groups as the molluscs, benthic foraminifera, dinoflagellates, spores and pollens, large and small mammals. One practical question is: is the regional focus labelled 'regional geochronology' needed at all, or are we more inclined to ignore it and relate the local biostratigraphies directly to the triumvirate?

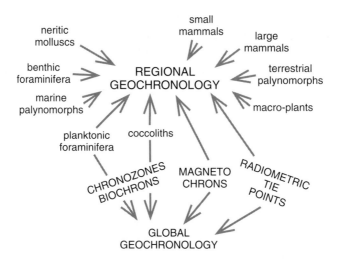

Figure 7.1 The core of late Phanerozoic geochronology, incorporating its regional component. The fossil groups have all contributed biostratigraphically and several still do. Only some of them contribute significantly to the 'global' system in which biochronology works in triumvirate with magnetochronology and radiochronology in the IMBS. Cyclostratigraphy now pervades all of this diagram.

Forests have been pulped to accommodate the writings on the nature of time–stratigraphy and the systems on which it is nourished, especially biostratigraphy. The dominating question has been the mutual relationship of zones, chrons, chemostratigraphic events and radiochronological events to 'the' geological timescale. A second question of more than academic interest is the survival of multiple and regional timescales – are we saddled forever with provincial and biotaxon-based scales, or is the time nigh upon us when all the evidence will feed into a single, stable, universally accepted numerical timescale? The literature on the development of the philosophical underpinnings of stratigraphy was listed and summarized by the two editions of the *International Stratigraphic Guide* (Hedberg, 1976; Salvador, 1994), Schoch (1989), and Harland *et al.* (1990). Here, I examine some crucial slices of the geochronologic scale, their development and their problems. There are good reasons for concentrating on the Cenozoic Erathem (beyond my own predilections). One is that the Cenozoic has the most diverse and complex stratigraphic record of the three Phanerozoic eras, and that is due partly to the increased preservation and partly to the increased provinciality and facies variation compared to the Mesozoic (Glaessner, 1943). A second reason is that the rapid recent advances in biochronology, magnetochronology and radiochronology culminating jointly in the IMBS have focused most intently on problems in the Cenozoic

(Chapter 3). Aubry *et al.* (1999) expressed the contrast elegantly if provocatively: 'Insofar as the hierarchy of time terms is made to rest on such an inappropriate base [palaeontological only], or where series and system boundaries are separately defined and correlated as paleontological entities ..., pre-Cenozoic chronostratigraphy remains in what might be called a primitive and unformed state, incapable of knowing itself.'

The nested-hierarchical geological timescale

'Hierarchies in natural science are ranked and nested structures such that units at each rank include parts that are units at lower ranks' (Valentine and May, 1996). There are four hierarchies in biogeohistory, two biological, two stratigraphic. The *genealogical hierarchy* (genome, deme, species, monophyletic taxon) is stable, it supplies the players, and it records the differential results, namely the outcome of the game of life as phylogeny, as the growth of the genealogical bush. But the game of life is actually played out in the *economic hierarchy* – the economic aspects of organisms, avatars, local and regional ecosystems (Eldredge, 1989). These biological hierarchies were considered in Chapters 4 and 6. The geological timescale is built from the *chronostratigraphic hierarchy*, the subject of this chapter. Recently there has been a resurgence of interest in sedimentary cycles and a *sequence-chronostratigraphic hierarchy* has been constructed: first-order megacycles, second-order supercycles, and third-order and higher-order cycles (Chapter 5). The chronostratigraphic hierarchy has its roots in the Wernerian succession of Primary, Secondary, Tertiary, and Quaternary which was supplanted in due course by the fossil – based succession of Palaeozoic, Mesozoic, and Cenozoic. There was an interchangeability among ranks and between rocks and time in such words as 'zone', 'stage', 'formation', 'series', 'epoch' which some saw as being flexible and responsive, others as being loose and confusing. Also, the enormous advances in regional and stratigraphic geology in the nineteenth century (not least as white men's empires spread) brought forth many competing stratigraphic schemes and some order was essential if communication and progress were not to suffocate. Stratigraphy moved 'from an essentially Lyellian descriptive framework to a bipartite [time and time-rock], hierarchical chronostratigraphic subdivision of the rock record with the formulation at the 2nd International Geological Congress in Bologna (1878) of a 5-fold stratigraphic subdivision with appropriate corresponding chronologic terms' (Aubry *et al.*, 1999). Schenck and Muller (1941) introduced the tripartite differentiation of time, time-rock and rock units. The *Guide* (Hedberg, 1972, 1976; Salvador, 1994) established the holy trinity of lithostratigraphy, biostratigraphy and chronostratigraphy. Table 7.1 summarizes this succession of timescale hierarchies.

Table 7.1 Nested stratigraphic hierarchies (Aubry et al., 1999), each level in its time-rock (chronostratigraphic) and time (geochronologic) manifestation, from the International Geological Congress in Bologna 1878 to the stratigraphic Guide

2nd IGC Bologna 1878		8th IGC Paris 1900		Schenck and Muller 1941		Guide (2nd edn., Salvador, ed., 1994)		rank
stratigraphic	chronologic	stratigraphic	chronologic	time-rock	time	chronostratigraphic	geochronologic	
						Eonathem	Eon	1st order
Group	Era	-	Era	-	Era	Erathem	Era	2nd order
System	Period	System	Period	System	Period	System	Period	3rd order
Series	Epoch	Series	Epoch	Series	Epoch	Series	Epoch	4th order
Stage	Age	Stage	Age	Stage	Age	Stage	Age	5th order
Assise	Phase?	Zone	Phase	Zone	-	Substage	Subage	6th order

What is a stage?

Exegesis of d'Orbigny

Meanwhile, there is a recurring debate over the nature of the stage which boils down to several skirmishing concepts: stage as a natural sedimentary cycle; stage as a biostratigraphic zone; stage as a facies bundle; and stage as the fundamental unit in the chronostratigraphic hierarchy. This state of affairs began with the astonishingly fecund Alcide d'Orbigny (Vénec-Peyré, 2004): in Chapter 1 I quoted Arkell's (1933) opinion of d'Orbigny's theory of global stages bounded by stratigraphic discordances and extinctions. We are indebted to Monty (1968) for an excellent explication of d'Orbigny's notions of stage and zone, correcting what he (Monty) saw as a general misunderstanding of elaborate and intricate concepts. Monty teased out the numerous contradictions and ambiguities in the writings of d'Orbigny, whom he regarded with abundant justification as one of the first modern stratigraphers, much ahead of, say, the eminent Belgian Omalius d'Halloy as a sytematician of stratigraphy, but at the same time one of the last catastrophists, much behind d'Halloy as a theoretician of faunal successions. Monty deconstructed d'Orbigny's concept of the stage into no less than five concepts:

i.   The stage as a *'palaeo-today'*, meaning a state of rest in ancient nature, a scenario of landscapes and seascapes with their faunas and floras. The palaeo-today was analogous to an historical epoch such as the Middle Ages in human Europe. Monty showed that this was d'Orbigny in 'fervent actualist' mode: fossils are indicators of biogeography and bioenvironment and our insights into ancient scenarios are guided by our insights into modern biotas.

ii. The stage as a *natural division of Earth history*, having universal meaning and thought to be isochronous everywhere. Here, fossils become tools of correlation and stages are bounded by abrupt termination. 'The combination of an actualistic philosophy underlying the stage *concept* with a catastrophist theory underlying the method of *separation* of stages constitutes the meat of d'Orbigny's geology' (Monty, 1968; emphasis added).

iii. The stage as an *accumulation of rocks*: here it is the actual belt of rocks preserving ancient resting states and preserving the earth history spanning the long period of relative stability between gaps, discordances, geological disturbances. These unconformities *defined* the stages; fossil contents *characterized* the stages.

iv. The stage as a *chronological unit*: this is the time-equivalent of the belt of strata.

v.   *The stage as (part of) a biostratigraphic unit*: latent in d'Orbigny's work were three biostratigraphic notions: one later named the epibole or abundance zone; a second not dissimilar to the subsequently named biozone and characterized (recognized?) by taxa appearing at the base or disappearing at the top; in a third, the stage seemingly synonymous with the zone, a 'discrepancy and negligence' on d'Orbigny's part since so much else in his writings showed that they were not the same thing.

Apropos of d'Orbigny's 'catastrophism', Blow (1979) pointed out that as early as 1839 d'Orbigny had found 228 species of fossil foraminifera in the Miocene of the Vienna Basin (Papp and Schmid, 1985) and recognized that ~27 species were still living in modern seas. Major faunal overturns between stages were not (in this case, anyway) comprehensive extinctions. (However, this figure of ~10% is not dissimilar to the estimates for hangover between the units of coordinated stasis (Chapter 5)). The 'zone' was developed by Oppel as an interval of strata characterized by few fossil species (Chapter 1). But whence the stage after Oppel hived off the zone? Let us approach that question through a famous text by a compatriot of d'Orbigny's a century later (Gignoux, 1955).

Stage as a natural cycle?

For Gignoux, 'science' begins when we put together and synthesize two sets of relationships, such as palaeontology and palaeobiology with lithology and petrology. When we come to synchronize such sets from different parts of the world to show patterns in time and space, and such patterns yield harmonious and coherent pictures, then we are working in the 'proper sphere of stratigraphy'. (Gignoux would have recognized the 'consilience of inductions' in Chapter 8.) The first synthesizing efforts in stratigraphy were to recognize similar successions in different localities and different regions: hence *formations* or *stages*. The second advance was to see that each lithologic stage had its distinctive suite of fossils, attributed initially to successive creations (d'Orbigny), then to evolution 'in a single continuous current of life'. Thus arose the *lithological and palaeontological* concepts of the stage, which Gignoux counterpointed with the *palaeogeographical* concept of the stage arising out of faunal or floral provinciality: whilst the same fauna in two sediments probably indicates that they are coeval, the reverse is not always true, for contemporaneous sediments of similar environments can carry quite different faunas due to climatic effects (e.g. Arctic province) or to geography and isolation (e.g. Australian province). Invoking Walther's law – that facies vary analogously in the horizontal and vertical dimensions – Gignoux tied together sedimentary successions and palaeogeography, citing as examples the Wealden stage, the Old Red sandstones and the New

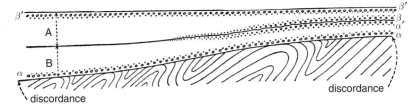

Figure 7.2 Two stages corresponding to two sedimentary cycles (Gignoux, 1955, Fig. 4, with permission). This explanation is Gignoux's own: A and B: Stages corresponding to two cycles. α and β: Littoral or continental formations at the beginning of the transgressions. α' and β': Littoral or continental formations of the regressions. To the left, beyond the region reached by the regression, sedimentation remained continuous and the stages are not palaeogeographically distinct: there we have a basin, bordered by the region of the continental platform which is to the right.

Red. Again: the 'Upper Jurassic' proceeded from being at first a white limestone excellent for building and for lime, to being a list of characteristic fossils, then to being, for us, a quite lyrically described landscape and seascape irresistably recalling d'Orbigny's 'palaeo-today'. The stage, then, is a *stratigraphic synthesis*.

But Gignoux went further. The notions of stratigraphic continuity, discontinuity and lacuna (hiatus) are integrated with the concept of marine transgression and regression to give the *sedimentary cycle* (Fig. 7.2). Not surprisingly, many of the old stages or formations, defined on their lithological facies, correspond to sedimentary cycles. And those cycles pose two questions:

(i) (To Gignoux, 'one of the greatest questions of stratigraphy'): Are the transgressions and regressions general or is their history individualized between basins? To demonstrate the former possibility by correlations 'even over the entire globe' would be to support Eduard Suess's eustatic movements; Gignoux himself leant toward the latter possibility and the influence of local or regional 'relative displacements of continental masses'. Accordingly he expected the necessary development of local or regional stages, and that 'paleontology alone' would be able to cross-correlate and integrate them into the grand synthesis.

Hence Gignoux's question (ii): are there reasons to believe that a stage conceived as a sedimentary series may at the same time be characterized by its fauna? Major changes in land-sea configuration will be disruptive and a new transgression, inaugurating a new sedimentary cycle, will bring with it a new fauna (difficulty in distinguishing the true evolutionary component from the migrational component is stratigraphically unimportant). Faunal renovation will coincide with the onset of the new cycle and the *palaeogeographic stages will often coincide with the palaeontological stages*. 'And so we shall understand how d'Orbigny, the true founder of stratigraphic paleontology, could, with good conscience and objectivity, be led to formulate his theory of the successive creations, each of which marked the beginning of a new stage'.

The stage, then, in Gignoux's words is a stratigraphic synthesis. It captures the notions of the sedimentary cycle, the packet of strata, the characteristic assemblage of fossils generated during faunal renovation as biotas respond to transgression, the coeval groups of facies distributed across broad regions and beyond the reach of any single index fossil species. Of the three 'great types of sedimentary series' – continental, epicontinental-marine and geosynclinal – Gignoux made it plain that it is the epicontinental that is crucial, for it is in this realm that there are the richest and most sensitive biotas and potential fossil assemblages, the most diverse sedimentary facies, and the most impact on palaeogeography and regional and local environments as the shallow seas advance and retreat. It is in the stratigraphic record of the epicontinental realm that we see the synthesized stage at its clearest in transgression-regression and discordant boundaries, and at its most useful for characterization and correlation (i.e. extended recognition). Gignoux (1955) and d'Orbigny (as in Monty, 1968) of a century before shared a very similar notion of the stage. They shared a biogeohistorical vision and spirit that is far more pertinent to our notions of the stage than are our opinions of d'Orbigny's rather violent scenario of successive creations and destructions. 'In Europe, where d'Orbigny's concept of a stage as a sedimentary cycle set off by 'natural breaks' has been more faithfully observed [than elsewhere], stages lead a robust existence as readily recognized, unconformity-bounded mappable units' (Aubry et al., 1999).

Stage as a biozone?

Elsewhere, strong assertions recur that the stage is no more than a biostratigraphic zone or a bundle of zones (e.g. Hancock, 1977; Miall, 1997). Rodgers (1959) stated forcefully that fossils, and fossils alone, have made it possible to determine the relative ages of rocks on different continents, and he added in a footnote: 'I still stubbornly persist in my belief that zone, stage, series, and system are all the same kind of stratigraphic unit, whatever that kind of unit may be called, and in my rejection of the currently popular distinction between time-stratigraphic and biostratigraphic categories of units. All agree that zones (in the sense in which the term was first used in stratigraphy by Oppel; nowadays some prefer to call these faunizones, assemblage-zones, or cenozones) are defined by assemblages of fossils. But stages have been defined in precisely the same way ever since the term was first introduced by d'Orbigny, by assemblages of fossils, and likewise beds can be assigned to a given series or system (except in the trivial case of beds near the type locality) only if fossil evidence of some kind can be found, either in the beds in question or in others near enough so that the local stratigraphic relations are reasonably clear.'

This view is understandable where correlation is based exclusively on fossils – as for most of the Phanerozoic during most of two centuries' labours. However, it is an overly pragmatic view which ignores the question, are chronostratigraphic units fundamentally different from biostratigraphic units? Even more, it ignores the highly opportunistic but highly integrated character of modern stratigraphy and geochronology.

Stage as the fundamental unit in the chronostratigraphic hierarchy?

This position was central to the efforts of Hedberg and the ISSC to promote the three basic ways of organizing strata – *litho*-stratigraphically, *bio*-stratigraphically and *chrono*-stratigraphically according to the respective characters or attributes lithology, fossils, and geologic age. To achieve that objective Hedberg had to do two things above all: he had to refute any synonymy or conceptual overlap between stage and zone; and he had to decouple stratigraphic classification from notions of how the Earth works and what the shape of the biogeohistorical record in the rocks actually is. As to the zone: both the assemblage zone and the range zone are limited to their facies whilst a chronostratigraphic unit is unified by and limited to representing the rocks formed during a specific interval of geologic time. (We have seen that not everyone agrees.) As to the latter strategy: Hedberg believed that the stance by d'Orbigny and Gignoux on the nature of the stage (above) still impeded our present procedures in stratigraphic classification. Although few would now openly espouse the paradigm of worldwide catastrophes in which stages were born, 'nevertheless many almost unconsciously endow the boundaries of these original time-stratigraphic units with a world-wide significance far beyond their real nature of quite arbitrary, though reasonably satisfactory and convenient, divisions of the more or less continuously developing record contained in the earth's sedimentary strata' (Hedberg, 1959). (Again, we have seen that not everyone agrees.)

Stage as a facies-bound unit?

As the basal member of the chronostratigraphic hierarchy of a designated single unique set of units, the stage of the *Guide* (1st Edn) has to include all rocks falling within the designated age-range, be they in the oceanic, neritic or terrestrial realm (for that matter, igneous or metamorphic as well), anywhere on the planet. But this is in theory: in practical application the units of higher rank will be found to be more global and those of lower rank more regional or even local although global recognition is a goal. Subsequently (*Guide*, 2nd Edn): the effective recognition of units of all ranks and their boundaries decreases with increasing distance from the type area as the resolving power of long-range time-correlation decreases. Tension beween the practical and the theoretical or the desirable underlies much of the argument between zones and stages as well as between local, regional and global units. Van Couvering (1977) devoted most of his review of the *Guide* to its application of the concepts of age and stage: in particular, to the extension from the local facies and geographic setting of the stratotype to other environments and distant locales to embrace all other coeval rocks. The hypostratotype, or auxiliary reference section, is intended to extend knowledge of the chronostratigraphic unit or boundary outwards. But van Couvering was a little concerned about how '... stages might spread across the country like mutant amoebas in a science-fiction movie, gobbling up lateral facies at a third or tenth remove by hypostratification rather than by correlation, and migrating from one palaeoenvironment to another without primary redefinition.' For example, no thoughtful stratigrapher would attempt to apply the Tortonian stage (marine, European) to continental rocks in Canada. Van Couvering identified a culprit in the mapping of the Neogene of Paratethys (see below): mapping by stage requires the a-priori assignment assignment of rocks to stages which become more and more 'omnivorous'. For van Couvering, the stage should not be extended any further than the characteristics of its stratotype can be isochronously correlated; the stage should be removed from the chronostratigraphic hierarchy; the world was not yet ready for globally extensive stages. In response, Hedberg (1977) reiterated that this *facies-bound* unit must be replaced, as in the *Guide*, by a unit controlled only by the *time span* of its type section.

Microfossils and stages

A brief review of Miocene zonations by Drooger (1966) was one of the more thoughtful papers of the earlier days in the development of planktonic microfossil biostratigraphy. Drooger pointed out the limitations inherent in exporting the planktonic foraminiferal zones from the deeper water tropical facies of the Caribbean region to less favourable regimes, so that '... one need not be surprised that the refined zoning of 10 to 18 zones of the tropical classical area cannot always be recognized elsewhere, such as Europe. As a consequence it gave rise to different zonings and often to unfortunate attempts of pigeonholing the tropical zones without direct evidence. As an example of the latter practice we may quote the zoning in Europe by Drooger (1956), who tried to recognize the *ciperoensis* and *fohsi* Zones, though he had only poor evidence for the presence of these species, or the *insueta* Zone, the nominate species of which he never observed in his material. Authors should be warned against such misleading practice which causes the laming [crippling] of zonal names, as it happened in the same way before with stage names.'

Drooger's *mea culpa* referred to the repeated and sometimes desperate attempts to relate advances in biostratigraphy to the timescale – to time terms such as 'Miocene' or 'Burdigalian' based as they were on neritic facies in restricted outcrops in extratropical Europe. Another distinguished pioneer was more scathing about the 'European standard' for the Cenozoic timescale (Reiss, 1968, p. 154):

> Although the various components of the accepted scheme were not proposed according to a 'master plan', but introduced in various places, at different times, by various authors, and on the basis of widely different criteria, the impression of a simple, universal hierarchy was implicitly or explicitly assumed. Thus, although such units as Oligocene, Miocene, Pliocene or Neogene were originally defined by the ratio between living and extinct molluscs, they were 'included' in the lithologically (!) and obsoletely defined Tertiary and 'subdivided' into stages, some of which turned out to be nothing but biostratigraphic (or even ecostratigraphic) zones, others lithostratigraphic units, and so on. Divested of their original meanings, used in a completely different sense, mostly inaccurately and often theoretically defined on the basis of sometimes uncontrollable considerations, the various units became sources of confusion and the objects of drawn-out, futile, and useless argument between stratigraphers. Furthermore, despite the persistent invocation of the 'European Standard' as the one and only, accepted, 'classical' scheme, the simple facts of nature led in various parts of the world, and even in Europe itself, to the erection of numerous time – stratigraphic units at stage, series, and even system levels.

Strong stuff. The type sections, molluscan assemblages and other criteria on which the Cenozoic timescale was compiled piecemeal in Europe are a limp frame on which to hang the rapid developments in microplanktonic biostratigraphy and the IMBS outlined in Chapter 3. And yet, that *is* the timescale, and it seemed to be necessary to make 'Pliocene' or 'Burdigalian' mean something in far-away successions rich in phylozonations and multiple zonations. A sketch by Hardenbol and Berggren (1978) capture the essence of the problem (Chapter 8). In the Palaeogene of northwest Europe the transgressions and regressions have left a record of assorted facies interrupted by hiatuses. Each type section, or stratotype, records only a small part of the elapsed time. We have here the double problem of incompleteness in the sedimentary record and of fossil assemblages too often lacking the taxa needed for correlation. The immediate environment was too cool, too shallow, or beyond the pale of normal marine salinity. Singly or in concert, those are reasons for impoverished planktonic microfaunas in too many classical sections. In the Neogene, the problems are exacerbated by the impact of Alpine tectonism on palaeogeography and the onset of climatic deterioration in the Miocene. Even rich faunas and floras were provincial in their aspect (Glaessner, 1943).

Hence the virtual absence of reference to stages and series in Chapter 2: they cluttered the story of the rise of microplanktonic biostratigraphy with a niggling sideshow. Hence, too, the low level of passion in that story – less passion was engendered in the development of the tapestry of microplanktonic successions than there was in relating those successions to the standard geological timescale.

Concurrent boundaries in a nested hierarchy?

Intrinsic to the chronostratigraphic philosophy of the *Guide* is the place of the stage: the stage and its unit-stratotype are the fundamental entity in the hierarchy. The boundary of the lowermost stage becomes automatically the boundary of each entity upwards through the nested ranks – *stage*, *series*, *system*, *erathem*. Essential though the unit-stratotype is as type or reference for the stage, definition of that stage is achieved through the boundary-stratotype (Fig. 7.3), which was given rather more emphasis in the *Guide* 2nd Edition than in the 1st. There arose the 'British' concept of the 'topless stage' (George *et al.*, 1967, 1969) whereby the base of the stage automatically becomes the top of the stage immediately underlying it: '*base defines boundary*'. The most apparent benefit of

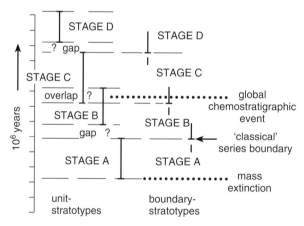

Figure 7.3 Definition of stages by unit-stratotypes and boundary-stratotypes (*Guide*, 2nd Edn, Fig. 14) with 'global horizons' added. Boundary-stratotypes avoid problems of gap or overlap that subsequent research may expose as occurring between unit-stratotypes, but research also exposes correlationally useful major events in Earth history at unforeseen levels.

the topless stage was in dealing with the reality of the highly fragmented stratal record: by defining the base, the definition was protected during any subsequent discovery of hiatus (and of finding the missing strata elsewhere). Van Couvering (1977) pointed out a further practical reason for leaving stages topless: regional stages are actively in use and accordingly must overlap in time, so that flexibility of response is needed as understanding progresses. This was part of van Couvering's distinguishing between *regional* stages and *global* series and systems. In response, Hedberg (1977) objected strongly to topless stages as being 'little more than a trick for evading the consequences of a usually avoidable original error ...', and he reiterated the 'fundamental need to designate Stage boundaries in essentially continuously deposited sequences, and only in such sequences, so that there can be no "time-gaps" at the boundaries, represented elsewhere by strata missing at the type boundary.'

The standard boundary-stratotype between two units of the Global Chronostratigraphic Scale is known as the 'Global Boundary Stratotype Section and Point (GSSP)' [note the respectful capitalization in the *Guide*, 2nd Edition]. Cowie (1986, 1990) listed requirements for a GSSP including: explicit motivation for the preference; a correlation on a global scale; completeness of exposure; adequate thickness of sediments; abundance and diversity of well-preserved fossils; favourable facies for widespread correlation; freedom from structural complication and metamorphism; amenability to magnetostratigraphy and geochronometry; and accessibility and conservation. This somewhat forbidding list is spelt out on the 2nd Edition of the *Guide*. The aim of this boundary stratigraphic procedure is to stabilize stratigraphy's framework category, 'to attain a common language of stratigraphy that will serve geologists world wide and avoid petty arguments and controversy' (Cowie, 1990).

In recent years, though, there has been some shift in emphasis away from Hedberg's bottom-up, inductive approach. For Hedberg, the stage was fundamental and *definition preceded correlation*. This was called the 'historical and conceptual approach' (Castradori, 2002). Only after boundary-stratotypification can one extend the chronostratigraphic entity geographically, using all possible means of chronocorrelation at one's disposal. That there might be a vast range in the correlational quality and usefulness of possible horizons or levels in the rock record did not ever loom very large in Hedberg's writings or in the gestation of the *Guide*, for Hedberg took great pains to avoid expressing opinion on the nature of the stratigraphic record whilst developing the *Guide* to its classification, terminology and procedure (Chapter 8).

For the International Commission on Stratigraphy (ICS), some of the emphasis is reversed (Cowie, 1986, 1990; Cowie *et al.*, 1989; Remane *et al.*, 1996; Remane, 1997). 'To define a boundary first and then evaluate its potential for long-range correlation (as has been proposed in some cases) will mostly lead to boundary definition of limited practical value.' The main criterion for a GSSP must be its capacity for correlation over long distances and wide areas. *Correlation precedes definition.* 'There is no formal priority regulation in stratigraphy. Therefore in redefining boundaries, priority can be given to the level with the best correlation potential. The redefinition will give us the opportunity to use fossil groups (such as conodonts) and methods of chronocorrelation (such as magnetostratigraphy) which were unknown or poorly developed at the time of the original definition. This does not mean that priority should be totally neglected. Practical considerations will incite us to limit changes to the necessary minimum. If, however, the interregional correlation potential of a traditional boundary does not correspond to the needs of modern stratigraphy, its position needs to be changed' (Remane *et al.*, 1996). This was called the 'hyperpragmatic approach' (Castradori, 2002).

A possible such situation is illustrated in Figure 7.3. Subsequent to the establishment of stages B and C with unit-stratotypes and their clarification by a boundary-stratotype, a chemostratigraphic event has been identified which would enable correlation between the pelagic, neritic, and terrestrial realms as well as between the north and south hemispheres. With that great, global, correlational potential, the event would make a more useful *series* boundary than is the presently stratotypified *stage* B–C boundary. Should the stage B–C boundary be reworked to accommodate this newly discovered 'time-horizon'? In that case, not only would correlation precede definition, but the series boundary would dictate the stage boundary – a double flouting of stratigraphic tradition and of Hedbergian and *Guide* principles. Or should the series (and perhaps subseries) be decoupled from the stage, which would remain forever both parochial and outside the nested hierarchy? These divergences by the ICS are reviewed at length by Aubry *et al.* (1999) and Aubry (2000), to whom we return below.

Microplankton and classical stages

The 'standard' timescale for the later Cenozoic Era includes, after a saga of correlation and clarification, the following succession of stages: late Oligocene, Chattian; early Miocene, Aquitanian and Burdigalian; middle Miocene, Langhian and Serravallian; late Miocene, Tortonian and Messinian; Pliocene, Zanclian and Piacenzian; early Pleistocene and Calabrian. Iaccarino (1985) reviewed the stratotypes of those stages and their planktonic foraminiferal assemblages, and she was able to relate the base of most stages to a biostratigraphic event (Fig. 7.4). The stages are mostly Mediterranean stages (the Aquitanian and Burdigalian were based in southwest France), widely used by

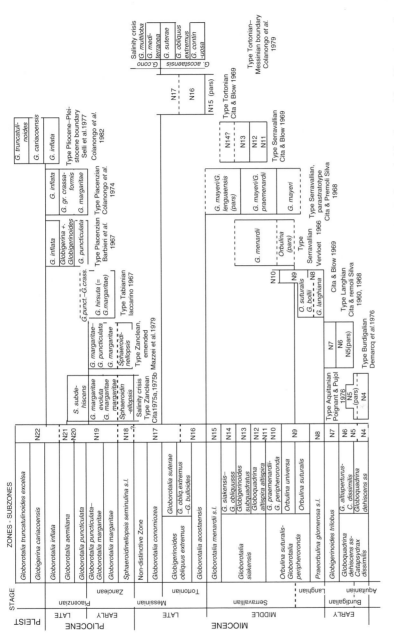

Figure 7.4 Mediterranean and western European stages of the Neogene System: planktonic foraminiferal zone identifications of stratotypes (Iaccarino, 1985, Fig. 1). With permission.

287

stratigraphers working on Mediterranean geology. That these *regional* stages are also the *standard* stages de facto, is demonstrated by the convention of their tabulations in Berggren *et al.* (1985b, c, 1995), Haq *et al.* (1987, 1988), Harland *et al.* (1990), and Hardenbol *et al.* (1998), among others. However, Aubry *et al.* (1999, 2000) tend to treat them still as of regional significance (see below).

Working within the context of the *Guide*, Iaccarino (1985) distinguished between chronostratigraphy, dealing with the *time relations of strata* and objective standard definitions of the divisions of the timescale, and the methods of *time evaluation*: bio-, magneto-, climatostratigraphy, etc. The need for unifying or integrating entities called 'stages' was due simply but compellingly to the sheer number of the separate biostratigraphic schemes in the Mediterranean Neogene. Iaccarino reviewed the correlation of Mediterranean stages to planktonic foraminiferal zonation, showing that the biostratigraphic assessment was in terms of the N-zones in the earlier part of the succession but that they are not identified in younger type sections, as geodynamically driven provinciality took hold in the Mediterranean. Iaccarino took the analysis to the next level with datums at, or very close to, the stage boundaries (Fig. 7.5). But it is clear that Iaccarino was concerned here with *necessary* regional stages (i.e. actually *used*, not merely tabulated dutifully), not with their 'classical' or global status. (Hence too the comment by Aubry *et al.* (1999) on robust European usage, above.)

As the microplanktonic fossil successions were being pieced together in the 1950s to the 1970s, an interesting sidestep occurred. Certainly, attention was paid to the stages, their stratotypes, their content (if any) of microplankton, their synonymies, and the gaps or overlaps in time of competing stratotypes. An early effort at putting the Cretaceous stages into the perspective of *Globotruncana* multiple phylozonation is shown in Figure 2.7. For the stages and epochs of the Cenozoic, Berggren (1971b) compiled a magisterial review and judgement (Fig. 7.6). However, as the planktonic zonations grew and matured, first under the aegis of petroleum geology (1930s–1960s) and then in meeting the challenges and the demands of the Deep Sea Drilling Project (1960s–1980s), it became apparent that the European stages were not indispensable to the rapid progress of marine geology and of its offspring, palaeoceanography; whilst that very progress, ironically, was due preeminently to adequate correlations and age determinations – adequate in both resolution and accuracy. A cursory glance through the modern literature will reveal many more references to a coccolith or planktonic foraminiferal zone, or to a magnetic chron, or to, say, the subseries 'middle Eocene', than to the nested stages 'Lutetian' or 'Bartonian'. We shrugged off the dead hand of the classical stages. An interesting parallel occurred in the development of lithostratigraphy in oceanic geology. Although the rule-of-law with formations, definitions, type sections,

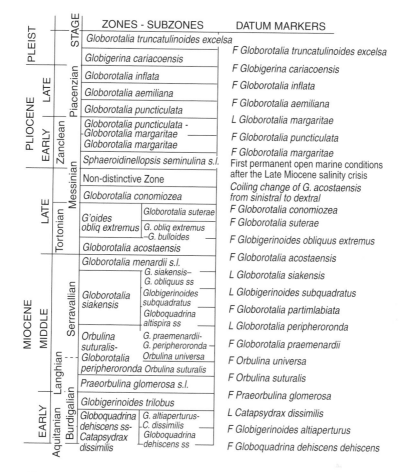

Figure 7.5 Mediterranean and western European stages of the Neogene System: planktonic foraminiferal zones and datums (Iaccarino, 1985, Fig. 2). With permission.

nomenclatural constraints and the rest of the paraphernalia is accepted in the stratigraphy of the continents, lithostratigraphy has not gone the same way in the deep oceans. There was a comprehensive lithostratigraphic nomenclature erected for the South Atlantic but the practice did not survive. Perhaps the reason was that sedimentary facies over vast areas are simpler and less variable than on the continents and their margins, lithological bodies are less distinctive or not apparent at all, and the more important problems reside in biofacies, diagenesis including dissolution and hiatus, geochemistry, and so on.

There can be little informed objection to the continued efforts to clarify the correlations of the various European stage stratotypes, to advance meaningful recommendations on the best succession of global – 'classical' – stages and on

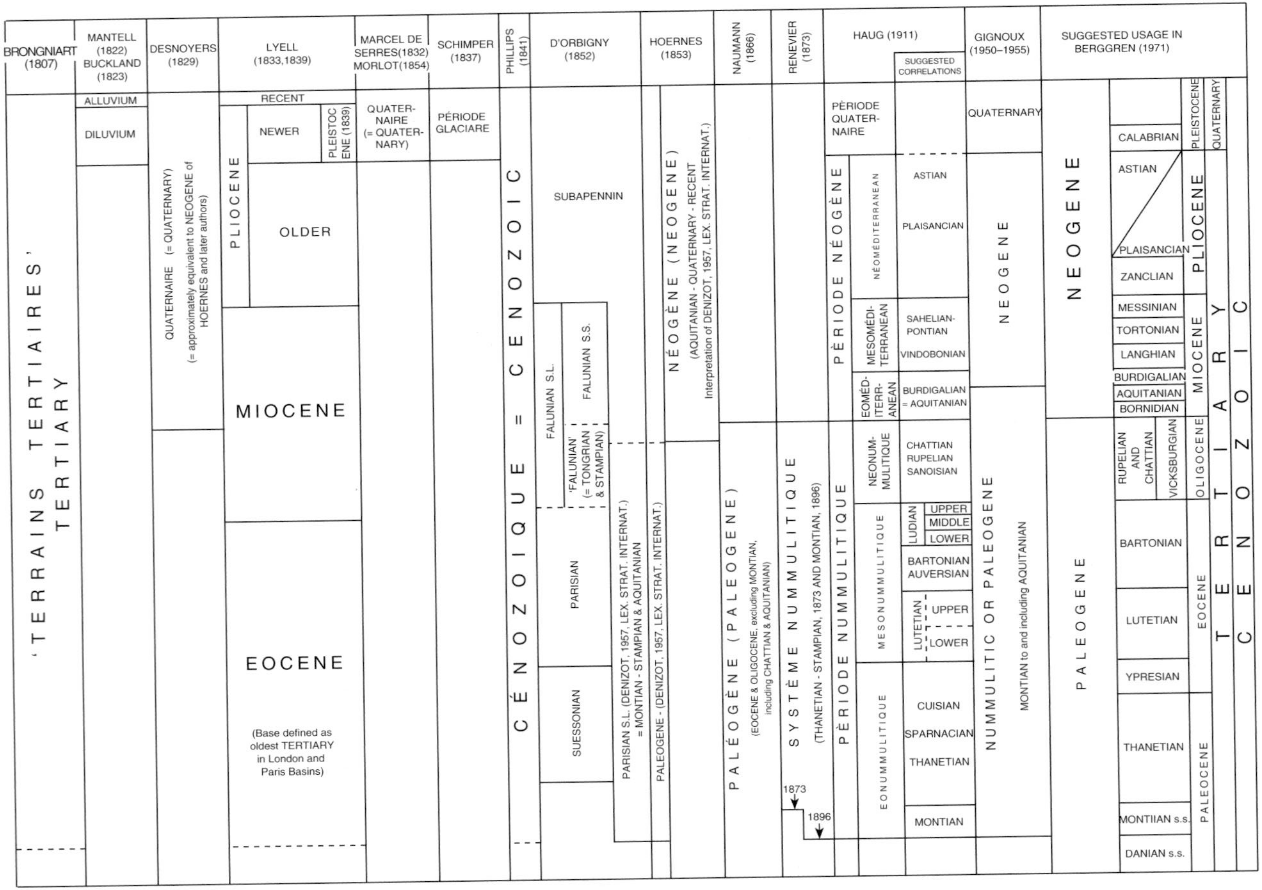

Figure 7.6 Development of Cenozoic and epochs (Berggren, 1971b, Fig. 52.1). With permission.

boundary stratotypes. All stratigraphers have experienced the confusion that arises when their colleagues trained in nonhistorical sciences attempt to take short cuts, the most common of which is to give a number, e.g. '45 myr', without citing the source and assuming that a *geochronometric* scale can stand alone without the support of a *chronostratigraphic* scale or, as Holland (1998) preferred, a *global standard stratigraphy*. That emancipation of numbers from chronostratigraphy has long been wanted and expected by the physicalists; but its time will not come in the foreseeable future. Meanwhile, one might have wondered whether we need to go further in scholarship and scholasticism in the global standard stratigraphy than establishing the necessary consensus on the series boundaries in the Cenozoic Erathem and their divisions into 'lower', 'middle' and 'upper' subseries – but the rise of sequence stratigraphy mostly in the neritic domain has changed that and stages are back (Chapter 5).

Where is the Cretaceous-Palaeogene boundary?

The relationship of the planktonic microfossil succession to the epochs and stages matters wherever it will help or hinder future research and scholarship by choosing one horizon as a major chronostratigraphic boundary instead of another. A prime example was the Cretaceous–Palaeogene boundary which is also the Mesozoic–Cenozoic boundary. That was a controversial matter ever since Desor added the Danian stage to the Cretaceous in 1846. The development of the episodic controversy was documented with his usual exuberant thoroughness by Berggren (1971b), from whom Figure 7.7 is adapted. There were two questions. The major question was: should the Danian stage be in the Cretaceous system or in the Palaeogene system? The lesser, and distracting, question was: is the Montian stage based in Belgium younger than or coeval with the Danian stage based in Denmark? In 1825 Forchhammer had regarded the strata of the (subsequently baptized) Danian Stage as Tertiary in age, and Desor correlated them with the calcaire pisolithique of the Paris, now believed to be of Dano-Montian age (Berggren, 1971b; Berggren *et al.*, 1995a). Both were 'correct', retrospectively, but where the major boundary ought to be is another matter.

The Danian stage remained at the top of the Cretaceous, where Desor had put it, for the rest of the nineteenth century and until de Grossouvre suggested that the top of the Mesozoic (Erathem) should be placed at the horizon of disappearance of the typical Mesozoic fossils – the ammonites, rudistids, belemnites, marine reptiles, dinosaurs. That level is between the Maastrichtian and the Danian stages. It was an eminently reasonable suggestion to draw the line between the age of reptiles and the age of mammals, but the problem with the

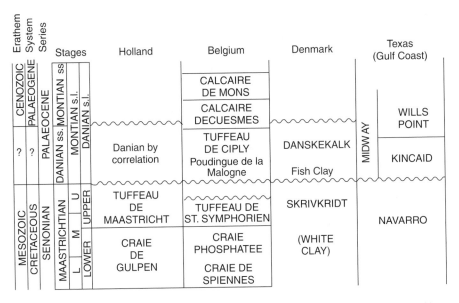

Figure 7.7 Maastrichtian–Danian–Montian Stage correlations (Berggren, 1971b).

grand gesture, the bold stroke, is that the image becomes fuzzy under magnified scrutiny. The Mesozoic organisms disappeared from the stratigraphic record at different places in different sections and regions, causing conflicting correlations. Although some workers responded positively to de Grossouvre's suggestion, others resisted the change (Berggren, 1971b).

In the 1920s Gayle Scott opined that the Danian nautiloid *Hercoglossa danica* was the same species as the Midway (US Gulf Coast) *Enclimatoceras ulrichi*, so that the Midway was Cretaceous in age – an inspired but heretical suggestion, as the adverse reaction from Julia Gardner (1931) soon showed, even though she too took the type Danian as being in the Cretaceous. The Danian was Cretaceous; the Midway was Tertiary; therefore, the two cannot be correlated! The biggest faunal break in the succession in the Gulf Coast was between the Navarro and the Midway – much bigger than between the Maastrichtian White Chalk and the overlying (and also highly calcareous) Danian in Denmark, as Berggren (1971b) pointed out, so that Scott's correlation had far-reaching implications. The position of the Danian vis-à-vis the erathem boundary was reopened when it became clear that the great planktonic foraminiferal break – the extinction of *Globotruncana* and many others – occurred between the Maastrichtian and the Danian whereas the microfauna of the latter evolves smoothly into the later Paleocene; see, as a particularly clear example, Glaessner's range chart (Fig. 2.3) showing the great contrast between the Maastrichtian and the Danian faunas at his zone VIII-IX boundary. By the

1960s the disjunction in the calcareous nannofossil succession was clearly at the same horizon and even more spectacular as a record of mass extinction (Bramlette and Martini, 1964).

And yet Brotzen (1959) and other palaeontologists, with many years' experience on the late Cretaceous and early Tertiary of Scandinavia, could argue that there was no faunal evidence to support allocation of the Danian to either era and that there was accordingly no reason to shift it from its conventional ('classical') position at the top of the Cretaceous. (This is an understandable stance, for the faunas reflect the fact that post-Danian neritic carbonates are very rare in NW Europe, so that there is little above the Danian to compare it with.) Rama Rao (1968) took the same line, urging that we must take account of the total fossil evidence. One of the last sustained attempts to hold the classical line was Eames's analysis of the faunas of the stratotypes (Eames and Savage, 1975). 'These leave us with no option but to retain the Danian in the Cretaceous, and to regard Danian faunas as being of latest Mesozoic age.' Although Eames argued a case for best practice in making a decision in the framework category, that case was weakened by an appeal to lithofacies similarities over fossil contrasts and by a failure to acknowledge the contemporaneity of the Tuffeau de Ciply in Belgium and the Danskekalk in Denmark (Fig. 7.7) (Berggren *et al.*, 1985a).

But the issue was not primarily the assessment of differences in total aspect between the Maastrichtian, Danian and Montian stratotypes and nearby sections and the resolution of the presence of dinosaurs or ammonites in strata said to be of Danian age. It was more than the question of where the strongest differences and similarities lay between the stage stratotypes and among the molluscs, brachiopods, bryozoans, echinoids. The issue was one of ease and accuracy of worldwide correlation (McGowran, 1968b). In the framework category though it was, 'The problem involves the basic issue of the distinctness, on a world-wide scale, of boundaries between major units of geologic time' (Hornibrook, 1958). This case was argued strongly by Berggren (1964, 1971b): that the boundary should be at the horizon between the global extinction of marine microplankton and nekton at the end of the Maastrichtian and the beginnings of the new radiation seen in sediments of Danian age. The evidence of the calcareous microfossils supported the contention by de Grossouvre at the turn of the nineteenth century, that the extinction of the ammonites marks the close of the Mesozoic. Sheer convenience and the needs of global correlation and age determination won out over convention and tradition. The end of the Maastrichtian stage is the end of the Cretaceous series and the end of the Mesozoic Erathem – and the extinction horizon not only of the ammonites but also of the rudistids, inoceramids, belemnites, marine reptile groups and dinosaurs.

The question of where to put the higher-order chronostratigraphic bound-aries (Senonian–Paleocene Epochs, Cretaceous–Palaeogene Series, Mesozoic–Cenozoic Erathems) as to the stage boundaries probably was settled (in the minds of most stratigraphers) several years before there emerged the big issue of the 1980s – the renewed consideration of events at the boundary and of the hypothesis of extraterrestrial catastrophe as a mechanism of mass extinction at the selfsame horizon (Christensen and Birkelund, 1979; Silver and Schultz, 1982). Scrutiny of the boundary at centimetre-by-centimetre scale in a deep-water section near Gubbio in the Apennines had revealed how sharp the micro-faunal break is (Luterbacher and Premoli Silva, 1964). The diverse fauna of large specimens of *Globotruncana* etc. is succeeded by an assemblage of very small, low-diversity forms, the '*Globigerina*' *eugubina* zone. Luterbacher and Premoli Silva found a clay precisely at the faunal interface, and it was this clay that yielded the iridium spike (Alvarez *et al.*, 1980). That discovery triggered the current phase of boundary science – of the search for palaeontological evidence of mass extinction in various environments and of chemical and sedimentary evidence of its causes, all in the essential matrix of rigorous, 'high-resolution' chronocorrelation. Suffice it here to say that the boundary has held up very well at the horizon with the iridium anomaly in geomagnetic chron C29r and dated numerically (^{40}Ar–^{39}Ar) with some precision and consistency (Berggren *et al.*, 1995a, Tables 2–6).

The Cretaceous–Palaeogene boundary, the IMBS and cyclostratigraphic calibration

The 'settled' boundary was placed at the very sharp disappearance of the most prominent Maastrichtian genera and species in the communities of calcareous microplankton, such as the coccolith genus *Micula* and the plank-tonic foraminiferal genus *Globotruncana* (Fig. 7.8). Even so, the re-invigoration of theories of catastrophic bioextinction demanded answers to several urgent questions. Could the evidence of sudden marine extinction be an illusion due to truncation of the fossil succession by hiatuses (McLean, 1981)? Did marine communities actually suffer extinction at the same time as the terrestrial dinosaurs? How rapidly did it all happen? This development is a very good example of how pattern-questions that bear very heavily on theories of process, such as the terrestrial versus extraterrestrial theories of asteroidal impact, are raised by the emergence or resurrection of those theories. Repeated demands on correlation and age-determination force us back to the rocks, time and again in the past and the foreseeable future – for precision, accuracy and robustness of correlation are a moveable feast.

In this instance, magnetobiostratigraphic correlations of oceanic cores and correlations with marine sections on land seemed to show that the sudden

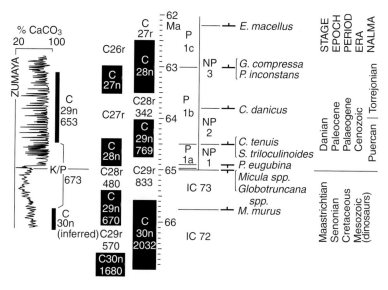

Figure 7.8 Correlations at the Cretaceous–Palaeogene boundary. Centre, numerical scale in 10^5 and 10^6 years (Ma). Right of numerical scale: Biozones: planktonic foraminifera (P) and calcareous nannoplankton (NP), defining events and estimated ages, from Berggren *et al.* (1995a); Cretaceous IC zones from Bralower *et al.* (1995). Chronostratigraphy: on nesting and bottom-up criteria, the age, epoch, period and era boundaries are concurrent. NALMA, North American Land Mammal Ages: this boundary too is concurrent. Left of numerical scale: two magnetostratigraphic scales, Berggren *et al.* (1985b) left and Berggren *et al.* (1995a) right. Numbers in the chrons in the latter are estimated durations in 10^5 years from Berggren *et al.* (1995a, Tables 2 and 3): compare with earlier estimates of both duration and age of magnetochrons. Far left, geomagnetically calibrated carbonate cycles and precessional chronology at Zumaya in Spain (Ten Kate and Springer, 1993) used by Herbert *et al.* (1995, Fig. 7 and Table 1) to estimate: C29N, 32 ± 2 cycles spanning 653 ± 41 kyr; C29R (Paleocene), 13 ± 1 cycles, 296 ± 41 kyr; C29R (Maastrichtian), 18.5 ± 1 cycles, 377 ± 20 kyr (total 673 kyr).

disappearance of marine microplankton was not sharpened by unrecognized unconformities (MacLeod and Keller, 1991a,b) and that the event occurred within an interval of reversed magnetism identified as Chron C29r in the pelagic limestone sections exposed in the Apennines (Alvarez *et al.*, 1977). Also, the mineralogical and geochemical evidence for bolide impact seemed to be clustered at that horizon, as did stable-isotopic evidence for suppression of marine productivity, as inferred from a catastrophic impact on the biosphere (Zachos *et al.*, 1989). However, that same inference demands that the effects should have been felt in terrestrial communities simultaneously with disaster in the neritic and pelagial realms whereas early work implied a younger boundary in the terrestrial realm, so that dinosaurs seemed to go extinct perhaps 1.5 million years after all the truncated marine lineages – implying that catastrophic extinction at the end of the Cretaceous was unlikely. It was suggested (Lindsay *et al.*, 1978) that the boundary defined by the highest stratigraphic occurrence of dinosaur remains was in Chron C28n or C29n. A summary by Berggren *et al.* (1985a) of the copious ensuing writing concluded that the problem could not be resolved by the presently available magnetostratigraphic data. The conflict was resolved when the normal interval identified as C29n was recognized as a normal polarity overprint of that part of the section, and the biostratigraphic boundary could be reassigned to Chron C29r, consistent with the marine boundary (Butler and Lindsay, 1985; Hailwood, 1989).

Chron C29r had a duration of the order of 10^5 years, too long to bear upon theories of global catastrophe. Also, this part of the geomagnetic record relied on the seafloor pattern in the South Atlantic where there were significant changes in spreading rates, not well constrained radiochronologically (Cande and Kent, 1992). Thus we have uncertainties within the web of biostratigraphy, magnetostratigraphy, numerical ages, and durations; and added to that is the question of upward reworking of microfossils, which becomes more significant as increased resolution is attempted. There have been two strategies for refining its chronology and correlation: cyclostratigraphy and graphic correlation.

Cyclostratigraphy employed the carbonate–marl couplets of the oceans, including Tethyan outcrop (Herbert and D'Hondt, 1990; Park *et al.*, 1993; Ten Kate and Springer, 1993; Herbert *et al.*, 1995; Herbert, 1999). The cycles are orbital and Herbert *et al.* believed that precessional cyclicity (~20.8 kyr) in pelagic sediments could place the K–P boundary and associated radiometric dates in C29r to within 10–20 kyr – down from 10^5 to 10^4 years' resolution. By stacking and correlation, cycles could be counted from below the top of Chron C30n in the Maastrichtian, across the biostratigraphically identified K–P boundary, into the Danian and across the C29r–29n boundary. Herbert *et al.* could estimate the duration of Chron C29r as 673 kyr, with 18.5 precessional couplets below the boundary and 13 above it. The microfossil record of mass extinction is accompanied by a pronounced drop in accumulation rates of biogenic and pelagic sediment; these events are confined to within one precessional period. Thus, three estimates of the duration of Chron C29r are shown in Figure 7.8 – 480, 833 and 673kyr – showing that independent approaches are converging well within the same order of magnitude, allowing mutual testing. Herbert (1999) suggested that the cyclochronological calibration of the geomagnetochrons necessitates some revision of the latter – modelling by Cande and Kent (1995) might have overestimated the South Atlantic spreading rates used.

The good agreement between the Spanish and South Atlantic sections suggested that the precessional clock was reliable, but it also suggested that the deepwater sections studied were not significantly incomplete. This conclusion contrasts starkly with a central outcome of graphic correlation of the K–P boundary interval (MacLeod and Keller, 1991a, b, 1993; MacLeod and Sadler, 1995): that oceanic sections are hiatus-ridden. The template for MacLeod's study is a composited succession of 64 datums spanning 7000 kyr, 64.3–65 Ma. It is difficult to relate this high-resolution succession to the high-resolution succession of datums in the IMBS because two of the latter, top *Parvularugoglobigerina eugubina* and base *Subbotina triloculinoides*, calibrated respectively against Chrons C29r and C29n (Fig. 7.8), are reversed by graphic methods.

Where is the Paleocene–Eocene boundary?

This series boundary has presented problems since the nineteenth century. There are two long-standing definitions arising out of the usual difficulties in western Europe of unconformity-bounded stages and shallow-marine and terrestrial facies (London–Hampshire, Paris, Belgian and North Sea Basins); and a third possibility promising to cut through the problems of long-distance correlation (Fig. 7.9). Still other problems have been geomagnetic – the interval in question is within the 2.5-myr-long Chron C24r; and radiochronologic – new dates have forced a change in estimated age of one boundary from 57 or 57.8 Ma in the 1980s to 54.8 Ma in the 1990s. Aubry *et al.* (1998, 1999, 2000; Aubry, 2000; Aubry and Berggren, 2000) could use this boundary to illustrate chronostratigraphic problems and procedures because they have been well aired by a Working Group; this is the last of the Palaeogene boundaries to be anointed by a GSSP (Aubry *et al.*, 1996; Berggren and Aubry, 1996).

One series boundary is at the base of the Ypresian Stage, either the Ieper Clay in the Belgian Basin or the London Clay in the London–Hampshire Basin. The base of the London Clay falls at either the base of the Walton Member or the base of the underlying Harwich Formation, the latter recording one of the key tephro-horizons of NW Europe, the – 17 Ash and its equivalents at ~54.5 Ma. This level is significant in marking a major marine transgression responding to tectonic relaxation after widespread eruption and tectonism in the North Atlantic region (Knox, 1996). Since base defined boundary, the underlying Thanetian Stage was extended up from the type Thanet Sands to include several units, such as the Woolwich and Reading Beds, scattered and isolated as small synthems. However, 'Ten years ago, the base of the Ypresian Stage ... was seen as an unconformable, undatable surface that was hopelessly uncorrelatable outside NW Europe' (Aubry *et al.*, 1999). But, against unlikely odds as Aubry *et al.* went on to show, the base of the London Clay could indeed be dated, first by tephra-biostratigraphic correlation, then by the initial occurrence of the calcareous nannofossil species *Tribrachiatus digitalis*, which existed for only ~0.2 myr (Aubry *et al.*, 1996).

Another preference for the Paleocene–Eocene boundary was at the base of the Sparnacian stage, based on the Lignites du Soissonnais in the Paris Basin. Also in the Sparnacian was the mammalian fauna of the Conglomérat de Meudon, a fauna displaying a 'major break' from the fauna of the underlying (Thanetian) Conglomérat de Cernay. Thus a Thanetian–Sparnacian boundary was particularly attractive to vertebrate palaeontologists as the series boundary, an attractiveness redoubled in North America because the basal Wasatchian Land Mammal Age (zone Wa0) records the greatest of all overland dispersals between Europe and North America and the highest level of generic-level similarity of the Cenozoic era (e.g. Savage and Russell, 1983; Woodburne and Swisher, 1995). (However, zone Wa0 is just a tiny bit younger than the base of the Sparnacian.)

Selecting yet a third horizon would acknowledge major environmental and biospheric changes at the global level (Rea *et al.*, 1990; Kennett and Stott, 1991; Stott, 1992; Thomas, 1992; Koch *et al.*, 1992; Berggren *et al.*, 1998, with references; Corfield and Norris, 1998), changes which have become clearer from the stratigraphic study of boundary events in space and time (Berggren and Aubry, 1996; Aubry *et al.*, 1998). Such changes include pronounced global warming and weakening of amospheric circulation – the Late Paleocene thermal maximum (LPTM). There were several extinctions marking the end of the *Stensioeina beccariiformis* benthic foraminiferal assemblage at abyssal and bathyal depths; it was replaced by the *Nuttallides truempyi* assemblage (Tjalsma and Lohmann, 1983). This is the (deep-sea) benthic extinction event (BEE). In the neritic realm of western Tethys, the base of the Ilerdian Stage is marked by a turnover in large foraminifera (LFT) (Hottinger, 1998; Orue-Etxebarria *et al.*, 2001). There was a rapid decrease (~10^3 years?) of ~4% in δ^{13}C in deep-marine carbonates and, seemingly coevally, in soil carbonates and mammalian teeth near the base of the Wasatchian NALMA. This is the carbon-isotopic excursion (CIE). Cyclostratigraphic analysis by Norris and Röhl (1999) showed that the entire, transient, δ^{13}C anomaly occurred over a span of ~150±20 kyr, and it seems clear now that the environmental impact of its cause – a catastrophic release of methane? – was felt keenly and simultaneously in the terrestrial, neritic, and pelagic realms.

The three horizons are dated respectively at ~54.37 Ma, ~55.8 Ma, and 55.5 Ma, and are shown in Figure 7.9 as the 'classical', 'French', 'vertebrate' and 'series-dominated' options respectively, along with biostratigraphic and other events (from top Thanet Sands to base *Tribrachiatus digitalis*) of potential use in long-distance correlations. Aubry *et al.* presented several options for relating

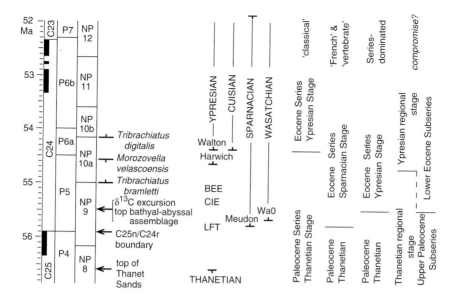

Figure 7.9 Correlations and summary of the Paleocene–Eocene boundary, compiled after Berggren *et al.* (1995a) and Aubry *et al.* (1996). From left, integrated geochronology: ages in Ma, geomagnetic chrons, foraminiferal (P) and calcareous nannofossil (NP) zones; three biostratigraphic, one carbon-isotopic and one geomagnetic events are shown. Middle: regional (but also 'classical') stages: shown are correlations of the bases of three stages Ypresian (two choices), Cuisian and Sparnacian, the top of the Thanetian, and the base of the NALMA Wasatchian Stage. Right, location of boundaries of: 'classical' series and stages; the 'French' and 'vertebrate-palaeontological' series and stages; the top-down or series-dominated boundary at a global and marine + continental environmental event; and a compromise in which stages and subseries are not in nested hierarchy. BEE, deep-benthic extinction event. CIE, carbon-isotopic excursion. LFT, larger-benthic foraminiferal turnover (Orue-Etxebarria *et al.*, 2001).

stages to series. Since the Ypresian long since has been the standard lower Eocene stage, and according to the precept that 'base defines stage' it should be in a nested chronostratigraphic hierarchy, the base of the Ypresian and the Thanetian–Ypresian stage boundary should also be the Paleocene–Eocene series boundary. A GSSP decision to place that boundary at base *Tribrachiatus digitalis* would approximate and therefore 'respect' the classical relationships. Placing the subseries boundary at the level of the carbon-isotope excursion would facilitate recognition of the boundary in the neritic, pelagic and terrestrial realms. In this case the regional stages would have to be redefined to fit in with (nested in) the series. A compromise would be to abandon the *stages* to regional or parochial employment and to use the *subseries* as the working unit in global standard stratigraphy. (See the end of this chapter and Fig. 7.26.)

Where is the Eocene–Oligocene boundary?

Whereas the position of the Cretaceous–Palaeogene boundary was settled before the rejuvenation of arguments over mass extinctions – their *discoverable* reality, synchroneity and precision, and terrestrial and/or extraterrestrial causes – those questions surged into the next-youngest of the extinction horizons, the extinctions of the Late Eocene, before the Eocene–Oligocene boundary was addressed using modern, multidisciplinary, integrated lines of enquiry (Pomerol and Premoli Silva, 1986; Premoli Silva *et al.*, 1988). For it did not take long for a parallel situation to develop – an association of evidence for impact events with major changes in the biosphere, implying cause and effect (references in Pomerol and Premoli Silva, 1986). The classical position of the Eocene–Oligocene boundary coincided with (or was close to) the following changes in the biosphere and in the global environment (see the summaries by Cavelier *et al.* (1981); Berggren *et al.* (1985b); Prothero and Berggren (1992); and Prothero (1994a, b)).

Terrestrial

The beginning of the Oligocene succession of mammal faunas in the Old World has long been associated with the *Grande Coupure*, a rapid episode of faunal overturn and reorganization named by H. G. Stehlin in 1909. Animals representing many mammalian families made their first appearance in the region of western Europe for the first time, and there has been much discussion of the actual stratigraphic level, the speed and the diachrony of the *Grande Coupure*, and the importance of climatic and sea-level change and the breakdown of biogeographic barriers at that time (discussion: Woodburne and Swisher, 1995). Major palaeobotanical changes occurred then, particularly in well documented floras in North America where broad-leaved evergreen forests were replaced by temperate deciduous forests (Wolfe, 1978).

Marine

Surveying the monographed record of marine animals in general, using a coarse-meshed chronology which sieved archival data only to the nearest stage, Sepkoski (1986) showed an extinction peak in the Late Eocene. Neritic molluscan assemblages in the Gulf Coast displayed major changes (Hansen, 1988). In the plankton, the rosette-shaped discoasters and other calcareous nannoplankton disappeared, as did the planktonic foraminiferal clades of *Globigerapsis*, *Hantkenina* and the *Turborotalia cerroazulensis* group. Among the large benthic foraminifera (see the East Indian stages, below) the lineages of *Discocyclina* and *Pellatispira* disappeared and the dominant group of the genus *Nummulites* was replaced by another (Adams *et al.*, 1986; Barbin, 1988).

Environmental

Close to the boundary is 'Chill II' (McGowran *et al.*, 1997; Lear *et al.*, 2000), the premier climatic event of the Cenozoic heralding the development of the oceanic psychrosphere and the onset of well-established third-order glacial cycles (*inter alia*, Kennett and Shackleton, 1976; Zachos *et al.*, 1996; Miller *et al.*, 1998). A plausible fit can be made between sea-level fall and climatic fluctuation (see Chapter 5). And now part of the environmental scenario for those times are the 'impact events' cited above.

Those physical and biological changes collectively are part of the *Terminal Eocene Event*, *sensu lato*, an appellation encompassing a rather thick slice of time, perhaps eight million years. The 'Khirthar restoration' (McGowran *et al.*, 1997) begins at 43–41 Ma with a strong clustering of tectonic, transgressional and biospheric events or changes in mode (McGowran, 1990, 1991b). Later, at the Middle–Late Eocene boundary at about 37 Ma (and distinctly preceding Chill II at about 34 Ma) there is an array of palaeobiological events in all realms (terrestrial, neritic, pelagic), sufficient to convince Prothero (1994b) that: 'The most fundamental biotic division of the Cenozoic is not between the Tertiary and Quaternary, or Paleogene and Neogene, but between the middle and late Eocene.' At Chill II, in contrast, 'Despite the publicity given to the "Terminal Eocene Event" [i.e., *sensu stricto*] in the past, it is an embarrassingly small extinction', now highlighted by a most pronounced lack of response among well-studied North American land mammal clades (Prothero and Heaton, 1996; Prothero, 1999).

Our question is twofold. The first is in the phenomenon or discoverable category: how good is the geochronology upon which all of this is based? Second, in the framework or bureaucratic category: where should the series boundary between the Eocene and the Oligocene be placed? It became generally accepted that the Upper Eocene was typified by the rich molluscan faunas of Barton Cliffs in Hampshire (described by Solander as long ago as 1766); hence, the Bartonian stage, with its Tethyan–Mediterranean counterpart the Priabonian stage based in northern Italy, both established in the nineteenth century (Berggren, 1971b). However, the Bartonian has calcareous microplankton assemblages clearly older than the Priabonian, and Hardenbol and Berggren (1978) showed that correlation would be best served, and confusion minimized, if the Middle Eocene consisted of the Lutetian and Bartonian and the Upper Eocene, the Priabonian – upon which attention became focused. In northwestern Europe there are problems arising from the sedimentary facies – evaporitic and brackish facies in the Paris Basin, shallow-marine to brackish in the Hampshire Basin, and mostly unfossiliferous sands etc. in the Belgian

Basin – which make correlations uncertain. Another difficulty is that Lyell and the other pioneers did not emphasize a precise separation of the Eocene, Oligocene and Miocene series as we need to today (Berggren and Prothero, 1992). Since the Oligocene stages including the classical lower Oligocene, the Rupelian stage in Belgium, were based in northern Europe (Fig. 8.1), we still have the question: how do we arrive at a meaningful (and above all else, a *useful*) Priabonian–Rupelian boundary which ought to be the Eocene–Oligocene boundary?

At the Eocene Colloquium of 1968, the Priabonian stage *sensu lato* was considered to encompass planktonic foraminiferal zones P15 to P17, so that an interval spanning zones P16 to P18 will contain the boundary stratotype with parallel succession and zonation in other microfossil groups and a magnetic record that can be correlated to the GPTS. International Geological Correlation Program #174, 'Geological events at the Eocene–Oligocene boundary' concluded with the book *Terminal Eocene Events* (Pomerol and Premoli Silva, 1986). In their review of the proceedings summarizing significant biological and physical events (Fig. 7.10), the Editors recommended, 'as apparently the most reliable biostratigraphic event upon which to define the Eocene–Oligocene boundary', an horizon marked by a cluster of events in the planktonic foraminiferal succession – the highest occurrences of *Hantkenina* and *Cribrohantkenina*, of *Globigerinatheka* and of *Pseudohastigerina danvillensis*. It was well correlated with the geomagnetic record (just above the younger normal event in upper Chron C13r) and well dated radiometrically (33.7 ± 0.4 Ma). However, it was up to the International Subcommission on Paleogene Stratigraphy to decide on a GSSP – for the decision moves into the framework category when the necessary science of characterization and correlation has been done. The next initiative was a comprehensive study of the Massignano section in the Apennines (Premoli Silva et al., 1988) including arguments for locating the boundary at the 19 m level (Fig. 7.11) which is the last occurrence of the genus *Hantkenina*, falling within Chron C13r, and about 4 m above a biotite-rich level which yielded a seemingly reliable K-Ar age of 34.3 ± 0.3 Ma. This well documented submission was accepted; the age of the boundary was estimated as 33.7 Ma (Berggren et al., 1995a).

This stratotypified Eocene–Oligocene series boundary should, by stratigraphic convention based in the principle of nested hierarchy, be the same as the Priabonian–Rupelian stage boundary, but there was no serious reference to the boundaries of the Priabonian or Rupelian. However, Brinkhuis (1994; Brinkhuis and Visscher, 1995) raised some important questions about the type Priabonian. Using a stratigraphic succession of dinoflagellates, Brinkhuis demonstrated that in the series-boundary stratotype at Massignano and in other sections the last *Hantkenina* falls between base *Achomosphaera*

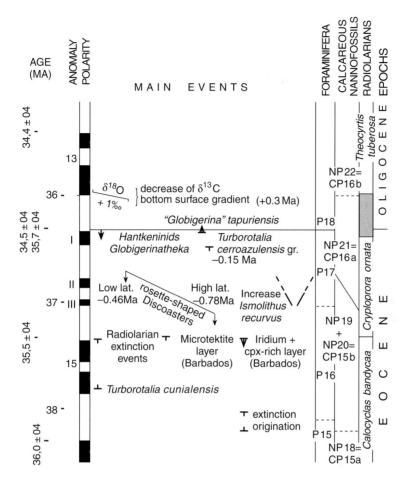

Figure 7.10 Summary of stratigraphic events at the Eocene–Oligocene boundary (Pomerol and Premoli Silva, 1986); sources, mostly papers in this volume. Estimated dates of various relevant physical and biological events are given as older or younger than the advocated boundary. With permission.

alcicornu and base *Glaphyrocysta semitecta*, which interval itself is by firm biostratigraphic correlation within the Priabonian type section. The traditional top of the Priabonian stage at Priabona and nearby is at the Bryozoan Limestone and just below top *Areosphaeridium diktyoplocus*; a revised top zone is slightly higher but firmly correlated with Chron C13n. Meanwhile, dinoflagellates in the lowest units of the type Rupelian in Belgium (Berg and Ruisbroek sands) suggested that base stratotype Rupelian in Belgium is close to top stratotype Priabonian in Italy.

Brinkhuis and Visscher differed from the proposers of the series GSSP at Massignano on two matters. The first was that the latter's top-down approach

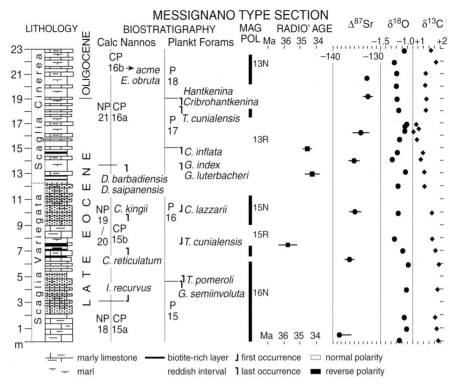

Figure 7.11 Summary of Eocene–Oligocene boundary GSSP at Massignano (in Premoli Silva *et al.*, 1988). Sources of this chart are papers in that volume. The advocated boundary here as in Figure 7.10 is at the level at which the planktonic foraminifera *Hantkenina* and *Cribrohantkenina* disappear. This same horizon is marked in Figure 7.12 as the subseries boundary between the Upper Eocene and the Lower Oligocene.

intended by GSSP designation to settle the Eocene–Oligocene series boundary first, and only then to sort out the Priabonian–Rupelian stage boundary problem – *series preceded stage*. Brinkhuis on the other hand cleaved to the bottom-up precept that boundaries of lower rank serve to define boundaries of higher rank – *stage preceded series*. This suggested that the GSSP decision creates a new Priabonian–Rupelian boundary problem 'at the very time the old uncertainties and controversies on the mutual delimitation of these stages are being resolved'. The second difference is in one's attitude to the meaning and utility of this GSSP. As it stands, the GSSP is about half a million years older than the cluster of major events – the oceanic and global chilling, the onset of ice marked by the third-order glaciation Oi1 = Ori − 1 (Fig. 3.29) and the downcutting sequence boundary 4.3–4.4 = Pr4–Ru1 (Fig. 3.29), the *Grande Coupure*, and so on, all of which centre on Chron C13n and several of which Brinkhuis could correlate with the upper boundary of the

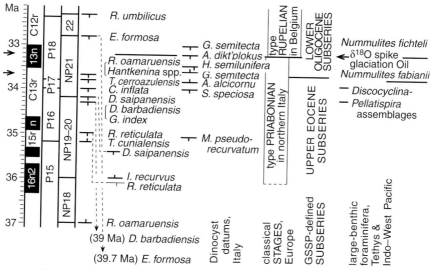

Figure 7.12 Eocene–Oligocene boundary a decade later. Geochronology after Berggren *et al.* (1995a), as are estimated ages of selected first and last appearance datums. Note the strongly allochronous tops in the nannofossils *R. reticulata*, *D. barbadiensis*, *D. saipanensis* and *E. formosa*. Dinocyst event correlations (Brinkhuis, 1994; Brinkhuis and Visscher, 1995) demonstrate that (i) the bioevent at the GSSP-based series boundary, topmost *Hantkenina* spp., is well within the type Priabonian stage, and (ii) the type Priabonian is succeeded contiguously by type Rupelian, at or very close to oceanic δ^{18}O-defined glaciation Oi1 and the downcutting sequence boundary 4.3–4.4 = Pr4–Ru1. Thus the stage boundary not only is 0.5 myr younger than the GSSP-based series boundary but it is at a biogeohistorically more meaningful level.

type Priabonian. (Another is the more-than-useful LAD of the large benthic foraminiferal family Discocyclinidae.) Brinkhuis and Visscher concluded that this tight clustering of global events makes the upper boundary of the type Priabonian a considerably better horizon to put the series boundary. Ironically, a Priabonian–Rupelian stage boundary is more easily recognizable (e.g. in southern Australian neritic; McGowran *et al.*, 1992) than is the GSSP level based upon a warm-pelagic single-lineage extinction.

Where is the Palaeogene–Neogene boundary?

The name 'Tertiary' survives from the old and largely obsolete succession, Primary, Secondary, Tertiary, Quaternary (Harland *et al.*, 1990). The Tertiary was divided by Lyell in 1833 into the Newer Pliocene, Older Pliocene, Miocene, and Eocene; the Oligocene was added in 1854 by Beyrich and the Paleocene in 1874 by Schimper (Berggren, 1971b) (Fig. 7.6). Hornes introduced

Table 7.2 *Divisions of the Cenozoic Era (Harland* et al., *1989). The Neogene Period and System should absorb the Quaternary, which should be abandoned along with the Tertiary: these neptunian hangovers are more misleading than useful.*

Era Erathem	(Sub-era) (Sub-erathem)	Period System	Epoch Series
Cenozoic Era	(Quaternary Sub-era)	(Anthropogene or Pleistocene Period)	Holocene Epoch Pleistocene Epoch Pliocene Epoch
	(Tertiary Sub-era)	Neogene Period	Miocene Epoch Oligocene Epoch
		Palaeogene Period	Eocene Epoch Paleocene Epoch

the Neogene by combining the Pliocene and Miocene in 1853 (emphasizing the Miocene–Eocene contrast in the molluscan faunas, before the Oligocene was erected), also the Palaeogene, at first synonymous with Eocene, later including the Paleocene to Oligocene. To Harland *et al.* (1989) the Tertiary was useful and unambiguous – even though it spans more than 97% of Cenozoic time – and they presented a classification retaining both the Quaternary–Tertiary and Neogene–Palaeogene pairs (Table 7.2). To Berggren *et al.* (1995a) the Tertiary is inappropriate and they would retain only the Palaeogene and Neogene, including the Quaternary in the latter. I concur entirely on these points. If this much more balanced, meaningful and useful Neogene–Palaeogene division of the Cenozoic is to flourish, then the boundary GSSP is needed, and Steininger (1994, 1997) presented the results of two decades' work to establish it.

The Cretaceous–Palaeogene and Eocene–Oligocene boundaries are at or close to major changes in the exogenic system. The Palaeogene–Neogene boundary does not fall at or near comparable clusters of events and the Oligocene–Miocene transition seems to record more development and less disruption. The lack of dramatic faunal, palaeoceanographic and climatic changes has made it difficult to establish definitive biostratigraphic criteria for global correlation at this important boundary, which has made the boundary one of the most difficult and controversial in the Cenozoic and accounted for ongoing polemics (Berggren *et al.*, 1985b).

The Neogene System has been well served by scholarly collections in recent years (Montanari *et al.*, 1997; van Couvering, 1997).

Steininger and the proponents of the Working Group (1994) outlined developments after the group was formed in 1976. First, they agreed in 1979 that the

Palaeogene–Neogene boundary coincides with the Oligocene–Miocene boundary which must be situated between the Chattian and Aquitanian Stages. Second, there should be a single continuously outcropping section spanning the *critical time span concept* and the *boundary interval concept*. The Chattian and Aquitanian Stages are located in northern and southwest Europe respectively and mutual correlation is very difficult. Third, GSSP candidates in several parts of the world were evaluated, sections in the Mediterranean region and Paratethys were studied in some detail, and, in 1988, the Working Group settled on two sections in Italy. Detailed study led to the excluding of one, because it did not cover the entire critical time span, and finally (1992) the Lemme–Carrioso section in northern Italy was chosen as the official candidate for the GSSP. The proposal included reports on several microfossil groups as well as lithostratigraphy and magnetostratigraphy (Fig. 7.13) and chemostratigraphy.

Among the calcareous microplankton, critical events in the vicinity of the GSSP are top *Sphenolithus delphix* and base *Paragoloborotalia kugleri*. As an anagenetic pseudospeciation, the latter event could be problematical (Pearson, 1998b). Subsequent work showed some inconsistency in the distribution of *S. delphix* and some doubt about the identification of the normal polarity interval as Chron C6Cn2n (Fornaciari and Rio, 1996; Raffi, 1999; Shackleton *et al.*, 2000). However, Shackleton *et al.* concluded that the Lemme–Carrioso section was useable and could be correlated with deep-sea sections by nannofossil species. Their cyclostratigraphic analysis used eccentricity cycles, suggesting that the age at $\sim 22.9 \pm 0.1$ Ma is some 0.9 myr younger than the accepted 23.8 Ma age (Cande and Kent, 1995; Berggren *et al.*, 1995a).

The $\delta^{18}O$ record and the warm/cool signals in planktonic foraminifera and dinocysts all suggested a cooling trend bottoming at about the level of the GSSP, which may be the oceanic signal of glaciation Mi1. Cyclostratigraphy and high-frequency oxygen-isotopic studies have now pinned down a transient ice age (Mi1) at the boundary (Zachos *et al.*, 2001a), the dataset consisting of a rare orbital anomaly (low amplitude variance in obliquity conflating with a minimum in eccentricity) tightly correlated with two superb oxygen curves displaying the sudden brief cooling. Mi1 falls at one of the periods of low eccentricity marking the 400-kyr cycle – but no more extreme than the neighbouring minima in eccentricity; what was crucial in encouraging ice growth was the low amplitude variability in obliquity at about the same time. Also of interest is the use of epiboles to constrain correlations between two oceans (DSDP Site 522 and ODP Site 929) and the GSSP in the Mediterranean region. Three successional peaks in relative abundances of the sphenolithid nannofossils *Sphenolithus ciperoensis*, *S. delphix* and *S. disbelemnos* hold true as interocean epiboles. Such epiboles work as tools of correlation

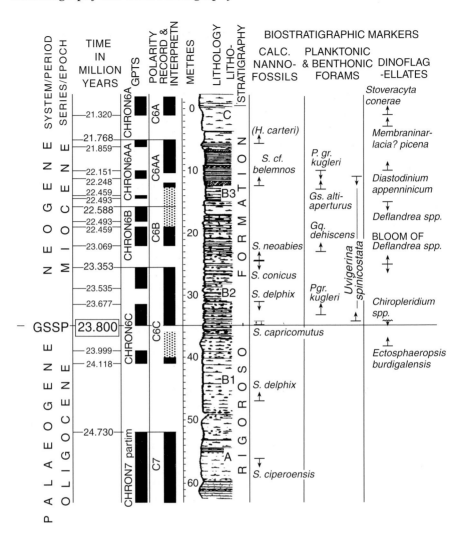

Figure 7.13 GSSP Palaeogene–Neogene – chronostratigraphy, geochronology, magnetostratigraphy, lithostratigraphy and biostratigraphy of the Lemme–Carrioso section (adapted and simplified after Steininger *et al.*, 1997, Fig. 7).

because they follow, respond to, physical changes in watermasses at high frequencies at similarly interoceanic scales. For example, Billups *et al.* (2002, Fig. 7) demonstrate an excellent pattern match in the Oligo-Miocene between equatorial Atlantic and Southern Ocean sites in both oxygen and carbon patterns.

This boundary stratotype was settled without reference to the Aquitanian Stage – 'certainly a mistake' in Castradori's opinion, but not actually causing any great inconvenience or conflict of concepts.

Palaeogene–Neogene boundary: ecostratigraphic characterization of a sequence and series boundary

The arguments developed by Loutit *et al.* (1988) address the problem: how can we use physical stratigraphy to correlate between the terrestrial-neritic-pelagic realms – between the often mutually exclusive biofacies domains of microfossil groups? A parallel question is: how can we correlate across the planet? using the same kinds of argument based on the recognition of sequence-stratigraphic surfaces, in this case surmounting biogeographic constraints. The specifications for the Palaeogene–Neogene Series boundary stratotype included a sequence boundary and glaciation Mi1. The Janjukian second-order cycle in southern Australia spanning this boundary has played a part in developing sequence-stratigraphic concepts (Haq *et al.*, 1988; Reeckmann, 1994). If third-order physical surfaces can (i) be recognized ecostratigraphically and (ii) be reasonably constrained by biocorrelations, then we should have a useful test of a sequence boundary at a series boundary.

The succession comprises several well-marked formations bounded by unconformities – Anglesea, Angahook, Jan Juc, and Puebla – and a cluster analysis showed that these sedimentary packages are also strongly characterized and cleanly sequential biofacies packages or assemblages in two drilled sections about 4 km apart (Fig. 7.14). Deconstructing the biofacies into inferred inner-, middle- and outer-neritic assemblages yields a proxy for a sea-level curve, also consistent between the sections and displaying falls at the litho- and clustered-biofacies boundaries. (As an aside, this third-order pattern was compared with a fourth-order pattern in a selected 1.3-m section; the fractal effect in biofacies is obvious.) Constrained by the identification of nannofloral zones NP23–NN2 and the presence of the earliest Miocene marker *Globoquadrina dehiscens* in the basal Puebla, it is clear that the Jan Juc/Puebla contact at least approximates the Palaeogene–Neogene boundary. If, as implied in Fig. 7.14, the chronostratigraphic boundary is a glacio-eustatically controlled sequence boundary, then the biofacies fluctuation and implied shallowing surely consolidate that correlation. Likewise, the biofacies fluctuations are strongly consistent with the Late Oligocene scenario of sequences and glaciations. The warm-oligotrophic association of *Pararotalia mackayi* and *Amphistegina lessonii* in the upper Jan Juc represents an ecostratigraphic horizon across southern Australasia in the Late Oligocene. In conclusion: (i) neritic, southern-extratropical biofacies reflect and confirm third-order sequences; and (ii) ecostratigraphic fluctuations, constrained by first and last events, can be powerful sequence-biostratigraphic and chronostratigraphic tools.

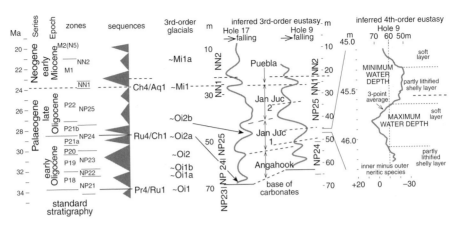

Figure 7.14 Oligocene–Miocene boundary in the Torquay Basin, southeastern Australia. Curves under 'Inferred eustasy at the third order' are ratios of outer-neritic to inner-neritic benthic foraminifera (categories are partly actualistic, plus other criteria), such ratios thereby enticing the notion of a local proxy for a eustatic curve (as also in Fig. 6.33). Shallowings under biostratigraphic constraint are consistent in age with sequence boundaries, themselves consistent with ice ages Oi1 to Mi1a inferred from pelagic isotopes. The same pattern is seen at higher resolution (right). The intervals labelled Angahook, Jan Juc 1, Jan Juc 2, and Puebla are benthic foraminiferal faunal assemblages based on Q-mode cluster analysis. They are strongly and cleanly successional, not oscillating or cyclical, implying that palaeocommunities have something strongly in common with stratal packages. (Li et al., 1999, 2003).

Where is the Miocene–Pliocene boundary?

The Miocene–Pliocene boundary, the base of the Pliocene Series, was traditionally placed at the base of the marine strata overlying the evaporative, alluvial and lacustrine strata in the Mediterranean Basin. This extremely sharp lithological and faunal break between the brackish-water Arenazzola Formation of the Messinian Stage and the deep-sea Trubi Marls of the Zanclean Stage was acknowledged from Lyell in 1833 onwards (reviewed by Berggren, 1971a; see Fig. 7.6) and was formally proposed as the Miocene–Pliocene boundary by Cita (1975). Extending across more than 3800 km of ocean basin, the disconformity marking the (apparently very rapid) refilling after dessication of the basin may not mark a significant hiatus and is effectively isochronous throughout: '... there cannot be a physical stratigraphical level anywhere in the world that has been more widely and consistently identified with a series boundary, than the essentially isochronous basal Zanclean contact has been used to signify the beginning of the

Pliocene' – more than meeting the strong and repeated recommendation of the Guide for continuity and stability in stratigraphic nomenclature (Aubry et al., 1999).

However, this splendid series boundary comes at a cost. The succession of calcareous-planktonic zones and datums in the Mediterranean have a prominent gap in the latest Miocene – Iaccarino's 'non-distinctive zone' (Figs. 7.4, 7.5) – so that biostratigraphic characterization within the region is difficult and correlation with the wider world more so. This is a two-way problem between the Mediterranean region and the planet at large: in one direction global chronostratigraphy is impeded; in the other; the roles of glacio-eustasy and tectonic changes or tectono-eustasy in isolating, dessicating and flooding the basin are obscured. Bioevents in the vicinity of the Miocene–Pliocene boundary in the global ocean are not useful in the Mediterranean. Events are missing or allochronous for biogeographic or environmental reasons of exclusion or delayed entry by species (Rio et al., 1991; Berggren et al., 1995b; Kouwenhoven et al., 1999; inter alia). Thus, both a lack of bioevents and a too-contrasting change in lithofacies from the Messinian to Zanclean stratotypes in the Mediterranean urged a GSSP outside that basin (Rio et al., 1991). There were two approaches to solving that problem (Berggren et al., 1995a). In one, Hilgen and Langereis (1993, 1994) positioned the base of the Zanclean in Sicily cyclostratigraphically and founded it in a marine section outside the Mediterranean. They proposed that the base of the Pliocene (base of Trubi Marls unconformably on Messinian sediments) was five precession cycles below the Thvera Subchron (C3n.4n) at 5.33 Ma, orbitally tuned. In a second approach, Benson and Hodell (1994; Benson, 1995) argued that the Messinian and Zanclean Stages are regionally limited in character and usefulness and that the series boundary should be decoupled from the stage boundary. They proposed that the series boundary be based on a GSSP on the Atlantic coast of Morocco (i.e. outside but close to the Atlantic–Mediterranean portal) using the base of the Gilbert magnetozone (C3r–C3An) (which is Messinian). This action would place the Miocene–Pliocene Series boundary ~0.4 myr below the Messinian–Zanclian Stage boundary.

The Miocene–Pliocene GSSP has been voted as at the Messinian–Zanclean boundary in Sicily (Eraclea Minoa section in Rossello composite section). (See van Couvering et al., 2000.) ' ... The shortcomings of having such a peculiar stratotype-section, with non-marine sedimentation below the boundary, were compensated in practice by the recent advancement of new stratigraphic techniques, thus rendering the stylistic flaw more digestible' (Castradori, 2002).

Where is the Pliocene–Pleistocene boundary?

This has been the locus of the biggest controversy of all (van Couvering, 1997; Aubry *et al.*, 1999; Castradori, 2002). Defined and ratified in the 1980s at a boundary stratotype in the Vrica section in Calabria, Italy at ∼1.8 Ma, it suffered further vicissitudes. One was the fall and resurgence of the classical Calabrian Stage and backward-fitting of its boundary to the Series boundary. The other was a proposal to lower the Series boundary to the base of the Upper Pliocene at ∼2.6 Ma, on the 'hyper-pragmatic' principle (Castradori, 2002) that this is stronger and more widely applicable for correlation in different environmental domains, being as it is in a major climatic transition. The proposal was not successful, being disruptive and historically unjustified.

Regional timescales

The Neogene of central Paratethys

A problem of long standing has been the chronostratigraphy of the sediments of the great seaway extending from the region of the Alps across central and southeastern Europe and beyond the Aral Sea. That Miocene seaway was known to the nineteenth-century workers and its faunas featured strongly in Eduard Suess's *Das Antlitz der Erde*. In the Miocene, Suess distinguished a 'first Mediterranean Stage', broadly Aquitanian–Burdigalian, and a 'second Mediterranean stage', Vindobonian. Those divisions did not survive; but nor did the recognition of the classical Mediterranean or western European stages in central and eastern Europe.

That region was designated as a biogeographic concept under the appellation *Paratethys* by V. Laskarev in 1924 (Steininger *et al.*, 1976; Baldi, 1980). Paratethys and the Mediterranean were successors to the perished Tethys. Their evolving palaeogeographies have been synthesized elegantly by Rögl (1998, 1999). As the complexities of Paratethys became clearer, Senes (1960) proposed a regional division into Eastern, Central and Western Paratethys. Eastern and Central Paratethys have parallel successions of regional stages. The Oligocene was a time of biogeographic transition. The earlier and middle Oligocene faunas were more comparable with the faunas of the Mediterranean and northwestern European bioprovinces than was the case in the late Oligocene, but the biotas already showed signs of isolation. In the plate tectonic terms of later discourse, collisions took place along the length of Tethys, between Africa, Arabia, and India to the south and Europe–Asia to the north. Microplates formed, elongate fold belts rose as mountain chains, and the basins that received the shed sediments took on two important characteristics: their stratigraphic record was highly complex in its facies patterns (a typical molasse characteristic); and provincialism increased as biotic communication lessened. In Steininger's resonant phrase (Steininger and Papp, 1979), geodynamic evolution gave rise in the Paratethys to an endemic faunal revolution. Thus we have the twin themes of variation and isolation, and both were enhanced through Neogene time as the stratigraphic record developed, mostly as a regional response to global climatic deterioration and fall of sea level. Under those circumstances, efforts to force the record into the 'classical' framework went too far. Molluscan assemblages recognized a century ago were correlated on the basis of mutual similarities which, all too often, reflected similar facies more than similar ages. There was 'no possibility of effectively using the classic European stages … in the central Paratethys' (Steininger, 1977). Today, neither of the GSSPs established (in northern Italy) for the Eocene–Oligocene and Palaeogene–Neogene boundaries is directly applicable in central Paratethys (Rögl, 1998).

A new concept of regional chronostratigraphic stages was adopted in the late 1960s (Papp *et al.*, 1968). Based in the first instance on the old molluscan assemblage succession, the regional stages developed conceptually into a still strictly biostratigraphic concept christened the '*integrated assemblage zone*' (Steininger, 1977). The integrated assemblage zone is a biostratigraphically characterized interval defined by the first appearance, first concurrent appearance, total range zone, and/or partial range zone of taxa derived *in situ* by speciation in local evolutionary lineages, or of taxa introduced by immigration. The taxa are not expected to occur together in a single rock unit, because they are drawn from various coeval environments (Steininger, 1977). All available lithostratigraphic and biostratigraphic data – going back to 1840 – were collected and collated in a very large and very impressive cooperative project on the stratigraphy of Paratethys (Steininger *et al.*, 1985).

Figure 7.15 illustrates the idea of the integrated assemblage zone, using three of the regional Miocene stages which span major transgressions and regressions. Note the variety of taxa; note too the few planktonic data available for correlation with the wider world. Rögl (1985, 1996) assessed the planktonic foraminiferal succession in central Paratethys and its correlation with the calcareous nannofossil record.

As Steininger (1977) told it, the Paratethys working group established the regional chronostratigraphic stage system using the following criteria: (i) The time interval of a stage should correspond to one or more integrated assemblage zones unified by their *overall faunal character*. (ii) The lower boundaries of a stage should be defined by *isochronous levels of high confidence* marked by the concurrent first appearance of many different organisms. (The upper boundary is the lower

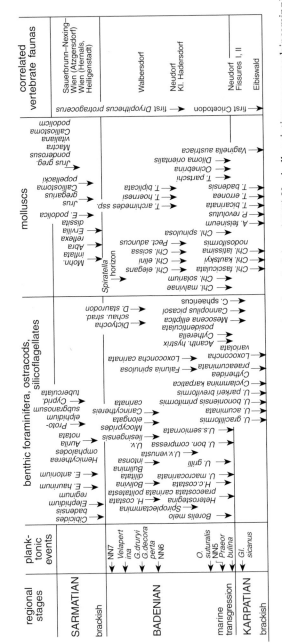

Figure 7.15 Integrated assemblage zones, mid Miocene, central Paratethys, from Steininger (1977). Vertically pointing arrows mark incoming horizons of species of several microfossil groups as well as molluscs. Note the clustering of events at or near regional stage boundaries and at certain levels within the Badenian Stage. inviting comparison with the patterns later called coordinated stasis. However, the integrated assemblage zone is of its nature a composite, and the act of compiling would tend to produce clustering.

boundary of the next stage.) (iii) Each stage is represented by various formations in the various basins of deposition in central Paratethys. Steininger concluded, ebulliently: 'These formations can be rather precisely correlated using the integrated-assemblage-zone biostratigraphic concept.' The boundaries and various facies have stratotypes.

The faunal characterization of the stages illustrated in Fig. 7.15 demonstrates that the criteria are initial appearances, not extinctions or merely disappearances. They are also composite – not just because you will not find autochthonous remains of e.g. terrestrial mammals with marine molluscs as a routine expectation, but also because the molluscs themselves, and the microfossils, will not occur together in their full diversity all the time. That is the point – the integrated assemblage zones are intended to encompass all the coeval fossiliferous horizons known. Thus they are oppelzones, *sensu latissimo*, and surely are as guilty as are the New Zealand stages to Carter's charge (below) of confusing definition with recognition and correlation.

How have the integrated assemblage zones stood the test of two decades' usage? There are two regional stage successions in Paratethys – an eastern and a central-western. There is a system of European large-land-mammal ages (=continental stages, depending on the author) and a nested system of zones, the MN micromammal zones, as discussed in the next section. There are two ways of comparing the continental and the marine biochronological timescales (Berggren *et al.*, 1985b). One is the classical, opportunistic, direct correlation wherever the biostratigraphic relationships permit the establishment of crossties. That is the basis for the holistic philosophy of the integrated assemblage zone. It is also a basis for relating the regional scale to the Mediterranean standard scales – for example, using mammal faunas intercalated with the marine assemblages to identify the mammal stage or zone, and taking that by correlation to the marine succession on the other side of the terrestrial tract. The second strategy is the more recent one of finding the geomagnetic polarity reversals of the oceanic succession in mammaliferous strata. Palaeomagnetic tie points provide the most precise correlations available (Berggren *et al.*, 1985b). Paratethys stratigraphers have dallied with a third way: the direct use of radiometric determinations to effect correlations. It has not been conspicuously successful because the power of radiochronology resided in calibrating scales in the total biogeohistorical matrix, not in simply comparing numbers, as Berggren and his colleagues have shown in a series of papers on geochronology (as mentioned in Chapter 3, the advent of ^{40}Ar–^{39}Ar dating changed that situation). The calcareous microplanktonic record of Paratethys has not been expressed as local zones, but as correlations between the regional stages and the standard planktonic zonations (Rögl, 1985).

Thus, the stages of Paratethys are functioning divisions. They act as a focus for whatever is relevant to the biogeohistory of the region, such as the major palaeogeographic syntheses. As an excellent example, consider the synthesis by Rögl and Steininger (1983). Their transgressive–regressive cycles are shown in Figure 7.16. The regional stages are being *used* – a more significant point than all the esoteric discussion on the nature and validity of a separate, regional chrono-stratigraphy. However, the advent of sequence stratigraphy reopens the question of the need for a regional chrono-focus, and Figure 7.17 illustrates this point for the Pannonian Basin in Central Paratethys. Integrated stratigraphy, including sequence and cyclostratigraphy, has come to Paratethys (Norzhauser and Piller, 2004).

The East Indies letter classification

In the great days of empires administered out of western European, the work of geological surveys and mining ventures and, later, of petroleum exploration revealed vast tracts of Cenozoic sediments in what was known Eurocentrically as the 'Far East'. Problems of correlation emerged very early in the Dutch East Indies and were repeated in neighbouring areas – problems residing in the realities of very thick sedimentary suites ranging, often rapidly, from bathyal and neritic carbonates to molasse-type detrital suites, and of extremely difficult access and observation in rainforested, karsted terrains presenting very little good outcrop. Not least, the policy of company confidentiality kept much hard-won data locked up in files and microfossil collections left on the mountain or dumped in the harbour (these are not metaphors), even after the competitive urgency of exploration had dissipated. Problems in the bigger picture were dominated by the scarcity of similarities between the tropical marine assemblages and the classical European assemblages.

Four strategies of correlation and age determination were tried (Glaessner, 1943; also van der Vlerk, 1959). In one, K. Martin, the molluscan monographer over five decades, used the Lyellian percentage method, which gave a general indication of Early, Middle, and Upper Neogene. The method was open to the criticisms that had been made of it in Europe; also, there were the additional problems of insufficient stratigraphic field evidence on the succession and relationships of mollusc-bearing strata. Another approach, by H. Douvillé in 1905, compared assemblages of larger foraminifera with assemblages in western Europe. It encountered the confusion among classical stages and stratotypes and their assemblages, it suffered from a lack of species in common and of knowledge of the intervening 'Middle East' faunas, and it did not prosper. The third strategy was the celebrated letter classification and the fourth was the collaboration between molluscan, vertebrate, and foraminiferal specialists, and

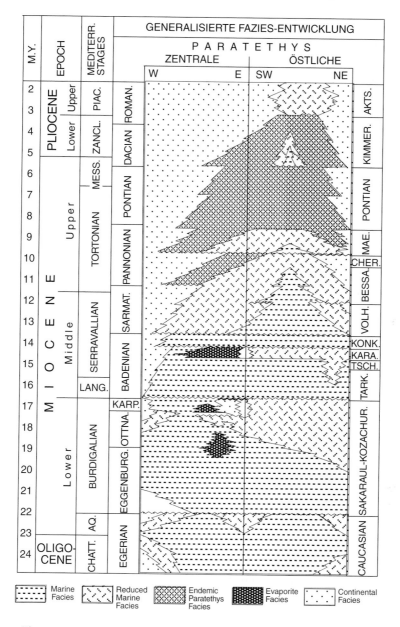

Figure 7.16 Stratigraphic patterns in central and eastern Paratethys (Rögl and Steininger, 1983). Scaled against geological time, this diagram emphasizes not only the perception of unconformity-bounded stratal packages (synthems) but the great geographic extent of hiatuses succeeded by rapid transgression: note specially base Eggenburgian, base Badenian, intra-Badenian, base Sarmatian, and intra-Pannonian. This major synthesis was the antecedent of sequence stratigraphy in Paratethys (Fig. 7.17).

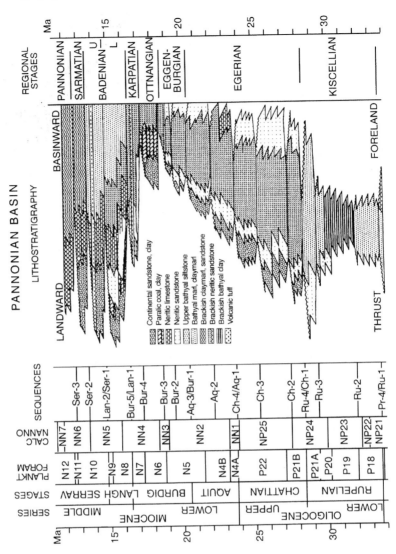

Figure 7.17 Lithostratigraphy and chronostratigraphy of the Palaeogene Hungarian Basin and the Early Neogene Pannonian Basin in Hungary. Left, geochronology (Berggren *et al.*, 1995) and sequences (Hardenbol *et al.*, 1998). Centre, lithofacies organized as synthems – note the balance between classical diachronous facies patterns and unconformity-bounded packages (Chapter 8). Right, the regional stages of central Parathethys. Simplified after Vakarcs *et al.* (1998) with permission.

field geologists, who together 'placed the study of the Tertiary in the East Indies and indeed throughout the Indo-Pacific Region on a firm base' (Glaessner, 1943). Van der Vlerk (1959) stated that Martin's subdivision was intuitive and Douvillé's even more so, although 'it was Douvillé who drew attention to another factor. He strongly stressed the moments of extinction and of first appearance of genera, as well as their assemblages'.

However, van der Vlerk (1959) recounted how in the early 1920s he recommended that 'for so clearly autochthonous a region' both Martin's and Douvillé's correlations with Europe be abandoned. In their place, but taking Douvillé's work as the starting point, he introduced a classification based entirely on assemblages of genera of foraminifera. That became the basis of the letter classification (van der Vlerk and Umbgrove, 1927). It worked, as is demonstrated by its prominent place in the regional geologies of that most difficult terrain (van Bemmelen, 1949; Visser and Hermes, 1962; Australasian Petroleum Company, 1961). It began with six divisions, based on different combinations of the (large foraminiferal) nummulitid, orbitoidal and alveolinid genera, and named Ta … upwards to Tf, with two more added later, to Th, and with subdivisions ('zones') especially of the Neogene divisions Te and Tf (Leupold and van der Vlerk, 1931). The 'most probable comparison' between Europe and the East Indies was of Ta, Tb with the Eocene; Tc, Td, Oligocene; Te, Tf(part), Lower Miocene; Tf(part), Tg, Upper Miocene; Th, Pliocene.

The letter classification was based at least in part on samples that had not been collected in stratigraphic order and partly on uncertain stratigraphic data. The stages were not typified by strata. Testing and criticism by several authors (Tan Sin Hok, 1939c; Glaessner, 1943, 1953) pivoted on three issues: (i) whether the assemblages – their taxonomic identifications and the inferred, highly composite ranges of the taxa – were correct, and the great need for detailed field sections; (ii) whether the letter classification was of any regional use outside the East Indies (Indonesia); (iii) whether robust correlations with western Europe could be achieved. The answers were a lot less clear than is that set of questions, but they were discouraging, for by the late 1940s van der Vlerk pronounced his creation 'more or less failed', proceeding, nonetheless, to give it a new lease of life by simplifying the Neogene Te and Tf and using 'only those fossils whose identity is not in dispute' (van der Vlerk, 1955; see Fig. 7.20, herein). The range of answers to the problems of chronstratigraphic division, correlation and age determination for the Neogene for the period 1939–1969 is shown in Figure 7.18. For petroleum exploration and geological mapping in the Far East it was deemed necessary to use local stages. Responses to the correlations with Europe varied from van der Vlerk (1955) who concluded that, for most letter stages, they were quite impossible, and Tan Sin Hok, who sidestepped by using Neogene subdivisions, through Glaessner, who was gloomy about the defects in both sets of stages and advocated using lower, middle and upper Miocene (not a solution, so long as those divisions are simply agglomerations of stages, as he later (1966) acknowledged), to the Clarke and Blow

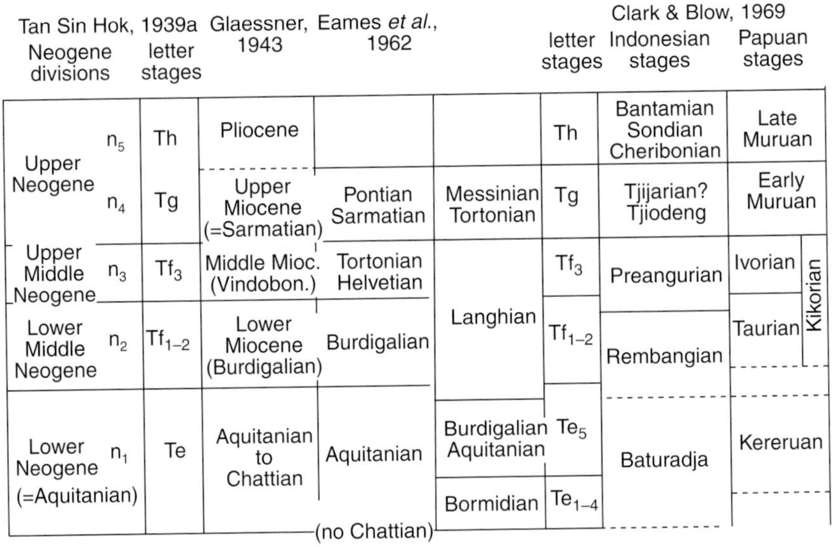

Tan Sin Hok, 1939a Neogene divisions	letter stages	Glaessner, 1943	Eames et al., 1962			Clark & Blow, 1969 letter stages	Indonesian stages	Papuan stages
Upper Neogene — n_5	Th	Pliocene				Th	Bantamian Sondian Cheribonian	Late Muruan
Upper Neogene — n_4	Tg	Upper Miocene (=Sarmatian)	Pontian Sarmatian	Messinian Tortonian		Tg	Tjijarian? Tjiodeng	Early Muruan
Upper Middle Neogene — n_3	Tf_3	Middle Mioc. (Vindobon.)	Tortonian Helvetian	Langhian	Tf_3		Preangurian	Ivorian — Kikorian
Lower Middle Neogene — n_2	Tf_{1-2}	Lower Miocene (Burdigalian)	Burdigalian		Tf_{1-2}		Rembangian	Taurian
Lower Neogene (=Aquitanian) — n_1	Te	Aquitanian to Chattian	Aquitanian	Burdigalian Aquitanian	Te_5		Baturadja	Kereruan
			(no Chattian)	Bormidian	Te_{1-4}			

Figure 7.18 Classification and correlation of the Neogene of the Far East, 1939–1969. The letter 'stages', although assemblage zones, are used chronostratigraphically in the SW Pacific region; the Indonesian and Papuan stages were used more in the past than the present.

correlations which had the advantage of some tiepoints with the subsequently developed planktonic N-zones.

Adams (especially 1970, 1984; but many papers during the past thirty years) rescued the East Indian letter classification of the Tertiary, put it on a sound modern footing and extended it to cover the Indo-West Pacific region between latitudes 40°N and 40°S. His comparison of the European stages and the letter stages is enlightening (Table 7.3). Adams put the two reasons why the letter classification worked in practice: (i) the sequences of assemblages were broadly correct even though only two taxa retained their original ranges from 1927 to 1970 (Adams, 1984); (ii) the stages were defined quite unambiguously in terms of larger foraminifera. On the other hand, subdivision has always been a problem – the units are seen to be large (temporally extensive), and yet lacking successional detail, 'Although it has always been easy to determine faunas including *Biplanispira* and/or *Pellatispira* as Tb, it is as difficult today as it was forty years ago to say whereabouts in Tb such faunas are situated' (Adams, 1970). One reason is that there are very few descriptions of thick sections, or local composites, spanning stage boundaries (Adams *et al.*, 1986). These problems are still with us, but there is progress (Boudagher-Fadel and Banner, 1999).

The long-acknowledged problem of fitting the letter classification to the classical timescale is essentially a problem of biogeographic provinces. As the

Table 7.3 *Comparison of European stages and East Indian letter stages (Adams, 1970).*

European Stages	East Indian Letter Stages
1. Based on a section or sections in a specified locality or area.	Based on assemblages of larger foraminifera thought to be characteristic of particular divisions of Tertiary time.
2. Type sections were, or could be, designated.	No type sections exist.
3. Total flora and fauna available for investigation.	Nothing available.
4. Further collecting from stratotypes usually available.	Further collecting impossible.
5. Ranges of stratotypes theoretically determinable by reference to stratotypes.	Ranges not determinable in this way.

Tethys closed and was succeeded stratotectonically by the Mediterranean and Paratethys, so did the connections between the Indo-Pacific and the west tend to close off (with sporadic and important reconnections: Rögl, 1998, 1999). Accordingly, the problem of species-in-common, always present, was exacerbated. To make successful correlations requires three things (Adams, 1970): (i) a planktonic zonation at low latitudes with which larger foraminiferal assemblages can be correlated; (ii) an acceptable European chronostratigraphy; (iii) an adequate knowledge of the ranges of the defining foraminifera of the letter stages. By invoking the planktonic foraminifera, we are adding facies exclusions to biogeographic – provincial – exclusions. As Adams (1970) emphasized, not only are the habitats of the larger and the planktonic foraminifera mutually exclusive, but the former are a polyphyletic group of forms that occupy a range of habitats themselves, from open shelf, through fore-reef shoals and reef, to back-reef and lagoonal (see, *inter alia*, Eames *et al.*, 1962; and especially Hottinger, 1983). There are also carbonate turbidites that interleave with planktonic facies. Clarke and Blow (1969) presented correlations of the larger foraminiferal associations with the planktonic zonation, and the later assessment by Adams (1984) is repeated here (Fig. 7.19).

   There remain the questions: what are the divisions of the letter stages, and are they now part of yesterday's stratigraphy and biogeohistory? As pointed out by Glaessner (1943) the divisions are not stages in the sense of being referred explicitly to type sections, but were only related in a very general way to the stratigraphic sequence; and later, Leupold and van der Vlerk generalized the stages and zones so that they finally included stratigraphic units in which

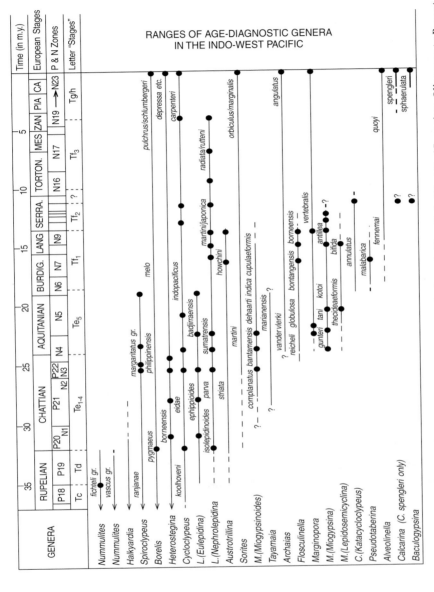

Figure 7.19 Range chart and correlations, larger neritic benthic foraminifera, Indo-West Pacific region, Oligocene to Recent (Adams, 1984). Solid line, proven range; broken line, uncertain range; black spot, planktonic control on these inner- mid-neritic facies and assemblages. Adams's caption continues: 'A few commonly occurring Neogene species, including the first and last in each genus, are shown in their relative stratigraphic positions. They do not necessarily represent single lineages nor are they always controlled by plankton. Some species intergrade; others have overlapping ranges.'

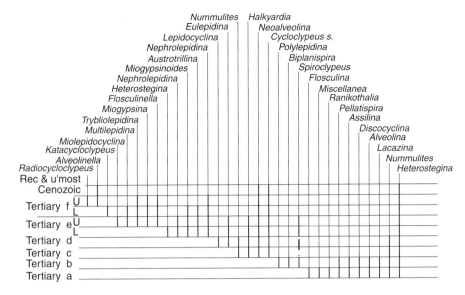

Figure 7.20 Van der Vlerk's last version of the letter 'stage' classification (van der Vlerk, 1955, with permission). These divisions are biozones, not stages, because they are associations of taxa, mostly genera and subgenera (species discrimination and identification has been more tentative than in most micropalaeontology; but see caption to Figure 3.21). Indeed, the units are akin to the unitary associations and conjunctions discussed in Chapter 6.

the distinctive units had not been found. The letter stages were described as based entirely on assemblages of genera (van der Vlerk, 1959) and diagnosed as 'wholly biostratigraphical', i.e. as zones (Clarke and Blow, 1969), and as assemblage zones (Adams, 1970, 1984). The ranges presented in van der Vlerk's last revision (Fig. 7.20) are all terminated at the boundaries – a symptom of assemblage zones, as is the lack of well-studied or even available boundary sections; Adams' (1965) study of the Melinau Limestone in Sarawak was a noteworthy exception. In concentrating on the matter of datums, Adams (1984) was less than enthusiastic about the first and last appearances, or defining events, as being more or less isochronous, and he noted also the instances of alternative events (e.g. last *Eulepidina* and last *Spiroclypeus* both distinguishing Te from Tf). As noted above, the desirability of good biostratigraphic boundaries in the succession of these faunas goes back to Douvillé but they are still rare. However, the letter stages would seem to be very similar to the North Land Mammal Ages in being subdivisions of time (not strata) thus fulfilling the definition of the *biochron* (van Couvering and Berggren, 1979; Berggren and van Couvering, 1978). And do they prosper? Adams (1970) stated as one aim of his major overhaul of a creaking system was to contribute to its accurate

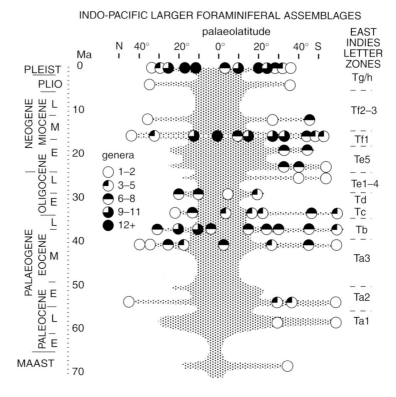

Figure 7.21 Extratropical excursions by neritic, essentially tropical larger foraminifera (modified after McGowran *et al.*, 2000, Fig. 14), mostly using sources in Chaproniere (1975, 1981, 1984), McGowran (1979), Hornibrook *et al.* (1989) and Adams *et al.* (1990). There were always larger photosymbiotic forms on the shelves, platforms and atolls of the tropical Indo-West Pacific region (Adams, 1970–1990) and dispersal into southern Australasia is a well-corroborated notion. There is, however, always the question as to how much an episodic fossil record records episodic events such as immigrations versus how much it displays merely a sporadic fossil record. I incline strongly towards the former of those two views. With permission.

correlation with the planktonic and classical-chronostratigraphic successions, 'thus, paradoxically, contributing to its eventual abandonment in favour of a single system of universal applicability'. Three decades later, we can say that the letter classification is a stable and useful system in its own right, without resiling thereby from the need for further research into its structure and its integration with other Cenozoic scales (Boudagher-Fadel, 2002). It has a close similarity with the North American Land Mammal Ages, differing significantly in having been in the hands of far fewer workers down the decades.

A latitude-time envelope of the Indo-Pacific Letter Classification is sketched in Figure 7.21. Its aim is to draw attention to the response of the essentially tropical larger foraminifera to fluctuations in climate and in sea level. It is a theory of pattern and is eminently falsifiable (McGowran, 1986a, b). The elements of the pattern are the rapid, shortlived extratropical excursions and the tentative correlations of the letter stages – shown as less-than-satisfactory defining events – with the planktonic foraminiferal P- and N-zones. There were always larger photosymbiotic forms on the shelves, platforms and atolls of the tropical Indo-West Pacific region (Adams, 1970, 1984, 1992) and dispersal into southern Australasia is a well-corroborated notion. There is, however, always the question as to how much an episodic fossil record records episodic events such as immigrations versus how much it displays merely a sporadic fossil record. I incline strongly towards the former of those two views (McGowran *et al.*, 2000); Hornibrook *et al.* (1989) and others, towards the latter.

The Austral Cenozoic of New Zealand and southern Australia

Southern Australia and New Zealand together comprise the onshore areas of the southern, extratropical, Indo-Pacific region but they have hosted separate chronostratigraphies (Fig. 7.22). Whereas the stratigraphic record in southern Australia is a thin, poorly outcropping cap on a relatively stable passive continental margin, New Zealand Cenozoic geology has a stronger tradition of recognizing the need for a timescale that will cater for the needs (especially including the mapping needs) of a region that is remote (from western Europe), tectonically active, and displaying complex patterns of sedimentary facies. 'The ideal of correlation using international stages has never been rejected by New Zealand workers but most have had strong reservations about attempting to apply a stage classification based on a different biogeographic and palaeoceanographic region in the northern hemisphere some 12,000 miles distant' (Hornibrook, 1976). Datums arose in their modern sense most overtly in New Zealand, which was also the source of the blast by Allan (at beginning of this chapter) against too much stratigraphic bureaucracy.

As outlined by Scott (1960), Hornibrook (1965), and Carter (1974), the New Zealand regional scale began with the introduction of eight Cenozoic stages by J. A. Thomson in 1916. Thomson was in no doubt about his aims: 'There are two objects to be aimed at in framing a classification of the younger rocks of New Zealand and it is important to distinguish them. The first is to set up a standard reference by which rocks from different parts of the country may be correlated with one another; the second is to correlate by various divisions of the classification thus established with their equivalents in the classification in the other parts of the world, and particularly in the accepted time scale based on the rocks of Europe.' By 1933 the number of stages had tripled, there was overlap, and definitions left something to be desired. But it was clear that Thomson and

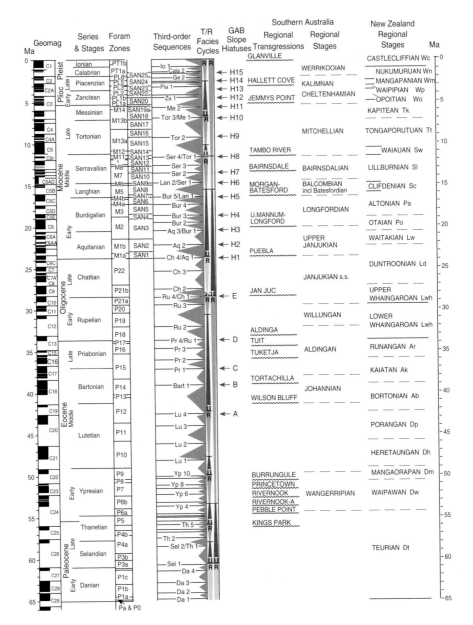

Figure 7.22 A Cenozoic timescale for southern Australasia. Regional transgressions and stages in southern Australia and their correlation with integrated geochronology, from McGowran *et al.* (1997). New Zealand stages, from Hornibrook *et al.* (1989) and Morgans *et al.* (1996).

subsequently Allan (1933) were establishing time units: 'Of course, both these writers, and their contemporaries who so readily adopted this local stage classification, used the classical techniques of correlation (particularly bio-stratigraphic zonations) in order to recognize the stages at different localities within New Zealand, but they avoided the trap that was to snare so many later workers – that of confusing their *means of correlation* with the logically separate problems involved in *definition* of a standard time scale' (Carter, 1974; emphasis added). The 'trap' was to include in the definition of the stages material from sections away from the types. The correlations of mudstones and clay-stones, not rich in macrofossils, were improved greatly by H. J. Finlay's studies of microfossils in the 1930s. The stages were still defined as intervals of time, but there was a shift in emphasis away from the type locality and toward the palaeontological characteristics which were based also on assemblages included in the stages by correlation. The modern New Zealand stage classification dates from the papers by Finlay and Marwick (1940, 1947). They are no longer being *defined*; instead, their faunal characteristics are being *described*. To Carter (1974) they were similar if not identical to d'Orbigny's stages and Oppel's zones. However, Carter also concluded that there is no clear answer to the question, are those units ages–stages or zones; and to assert one or the other 'represents unwarranted simplification of an important and complex question'. The shift from stratigraphic *definition* to biostratigraphic *characterization* was ascribed to the influence of the paper by Schenck and Muller (1941) in which the zone was placed in the time-stratigraphic hierarchy.

Whatever the stages are, and whether they are needed, are matters for pragmatic test. To Hornibrook (1965) rigid definitions based on type sections will lead to absurdity as gaps and overlaps in the succession become apparent. Type sections are useful and necessary anchors to the stages, which are them-selves useful and necessary unifiers in New Zealand stratigraphy. Isochronous buffer-zones are more realistic as boundaries and allow more necessary flex-ibility than do isochronous surfaces. To which Carter (1974) rejoined: true, but the necessary corollary is that these units are not stages, but zones.

In Hornibrook's last compilation and discussion (in Hornibrook *et al.*, 1989) the stages are alive and well: they are being used as more than an adornment to correlation charts. But are they stages? Every one of the twenty-four Cenozoic stages has a stratotype, all shown in two superbly drafted compilations in a chapter headed 'Foraminiferal basis for New Zealand Cenozoic Stages' which emphasized the faunas of the stratotypes. The practice in New Zealand, Hornibrook reminded us, 'has been to include all useful macro- and microfos-sils within the working definitions of stages, emphasizing one or more specified taxa (key species of Finlay and Marwick) to define their boundaries. Most stages are multiple biostratigraphic units characterized by taxa with restricted ranges (range-zones …) and also those with longer ranges, partly overlapping the ranges of other taxa (concurrent-range zones: they approach Oppel-zones defined by the International Stratigraphic Guide …' There is still the boundary problem: 'Correlation with the body of the stages is usually able to be made most confidently. It is the placement and correlation of the boundaries between contiguous stages that requires the most careful biostratigraphic resolution.' Hornibrook acknowledged the importance of stratotypes (including boundary stratotypes) as 'a tangible stratigraphic anchor for the definition of the standard working correlation units …' But inflexible practices that do not permit selected boundary stratotypes to be moved if and when they are found to be uncorrelatable would only result in impracticable stratigraphic units. Hornibrook was sanguine about the mixing of categories: 'It is often persuasively argued that there is no real difference between bio- and chronostratigraphy in practice and that stratotypes are not a valid means of defining stages since they are invariably selected on the basis of an already established biostratigraphy.'

Carter predicted in 1974 that in New Zealand, 'local chronstratigraphies will serve no useful purpose in the future, if indeed they do so now, but they could be usefully reformulated in biostratigraphic terms, using the oppelzone as their basic category'. It would seem that the stages are actually being used, that they are biostratigraphic in their essence, but that the overt reformulation has not happened. There is a nod toward the stratotypes demanded by the *Guide* and the units are still called stages. That New Zealanders have not been not merely parochial in all this is easily demonstrated by a long modern history of grappling with the problems of international correlation, from Hornibrook (1958) onwards.

A somewhat different approach was taken by Jenkins (1985). Where Iaccarino (1985) and Rögl (1985) related the regional planktonic foraminiferal successions to the regional stages (in the Mediterranean and central Paratethys respectively), Jenkins in the same volume dismissed the New Zealand, regional stages altogether, as being the outcome of a 'parochial attitude' which must defer to the adoption of a microplanktonic biostratigraphy.

Since the above was written, a superb compilation of the New Zealand timescale has appeared (Cooper, 2004). Regional Cenozoic stage boundaries are defined mostly on fossils; a proposal to switch geomagnetic polarity events (Canter and Naish, 1998) has not progressed below the late Neogene.

Chronostratigraphy in southern Australia

Local stages were erected in this region for the same reasons as in New Zealand: in part as a reaction to the confusion ensuing from the difficulties in comparing faunas with the classical succession; in part to facilitate local progress whilst offshore correlations remained contentious or impossible. The stages were biological in their essence, based on distinctive molluscan assemblages whose overall character was the basis for correlation (Darragh, 1985). How successful this regional chronostratigraphy was, can be gleaned from the following:

> The long history of the exploration of the Tertiary strata in Australia is marked by persistent controversy about the sequences as well as the age of strata. In the heat of this controversy the need for careful stratigraphic observation, full description of stratigraphic details, and collecting from clearly described and measured beds was often overlooked. Discussion generally centred not on sequences of sedimentary rocks developed in varying thicknesses and facies, and characterized by assemblages of fossils differing according to age and ecologic conditions, but on 'Stages' which generally were nothing but convenient labels for collections of fossils from certain well-known collecting grounds. Some of these were limited in area, thickness, and lithological composition (e.g. 'Balcombian'), while others designated outcrops several miles long exposing strata hundreds of feet thick, representing almost the entire Tertiary system and including major unconformities (e.g. 'Aldingan'). The late F. A. Singleton reviewed the history of these discussions and brought some order into chaos by his monograph published in 1941 (Glaessner, 1951, p. 273).

Although foraminifera were being described since the nineteenth century, and Crespin (1943) made some progress in shifting the definition and character-ization of the stages towards foraminifera and lithostratigraphy, it was this paper by Glaessner that provided the real basis for subsequent progress in correlation and age determination. There was progress, but without the benefit of regional stages, which never recovered their health after that blast from Glaessner (*inter alia*, Glaessner and Wade, 1958; Wade, 1964; Ludbrook, 1971) or with but passing reference to stages (A. N. Carter, 1964; Ludbrook and Lindsay, 1969). O. P. Singleton (1968) became confident enough about progress in foraminiferal biostratigraphy to suggest that regional stages should be aban-doned altogether, which meant that local biostratigraphy, including the evi-dence of vertebrates, terrestrial and marine palynomorphs, echinoids, molluscs and other groups, would be tied to the planktonic foraminiferal system. McGowran *et al.* (1971) were less confident about that, believing that a local chronostratigraphic focus was needed for the diverse biostratigraphic evidence accumulating in a range of continental to open marine environments. Even so, only Lindsay (1981, 1985) among modern biostratigraphers has taken regional stages seriously and actually used them; the major review by Abele *et al.* (1976) acknowledged stages but used the Carter zones (1958a, b) as their actual working framework. Nor do the regional stages loom very large in the reviews of the stratigraphy (Chaproniere *et al.*, 1996), or the classical Tertiary marine fossils, the molluscs (Ludbrook, 1973; Darragh, 1985). It is of more than passing interest, though, that vertebrate palaeontologists found a regional chronostratigraphy useful for framing terrestrial fossil assemblages (Woodburne *et al.*, 1985).

The mammals of North America and Europe

Biostratigraphic and chronostratigraphic systems based on terrestrial mammals in non-marine strata have developed along somewhat different lines than have those based in the marine record of fossils and strata. Most of the nineteenth-century advances in stratigraphy and Earth history were made in the latter realm and so it is on marine facies and enclosing fossils that the founding documents of this discipline are based and the classical, standard timescale is built. Non-marine environments and their lithostratigraphic units are less extensive in their lateral continuity and nonmarine strata inherently are more vulnerable to subsequent erosion. As well, vagile quadrupeds are not well known for their ability to walk on water. All of these factors have necessitated the recognition of regional terrestrial biotic successions even more than in the marine realm. Yet another differentiating factor is more or less cultural, in that the vertebrates traditionally are invoked to tell the story of the evolution of life on Earth, whereas to the invertebrates and protists have fallen the more applied tasks of applied palaeontology (i.e. ages and environments): hence the dismal exchanges of some decades ago on 'vertebrate palaeontologist: biologist or geologist?'

Although the study of fossil mammals in their stratigraphic context goes back to Cuvier's work in the Paris Basin (e.g. Buffetaut, 1987), a comprehensive, biostratigraphically significant succession of divisions was erected in North America (Tedford, 1970). That achievement was a major outcome of the heroic age of palaeontology in the American west which, in the collective public mind, has been associated more with Marsh and Cope, their controversies and their dinosaurs. The mammalian faunal succession developed from the pioneer work by Hayden but is associated especially with the names of Osborn and Matthew (Tedford, 1970). Osborn and Matthew saw their main purpose to be 'faunistic rather than geologic' (in Lindsay and Tedford, 1989). Beginning with the 'golden age' of palaeontology – Lamarck, Brongniart, Cuvier, Deshayes – the work of the nineteenth century was in erecting the grand time divisions of the Cenozoic, whereas the work of the twentieth century is in precise correlation (Osborn,

1910). Osborn goes on in a passage well worth the quoting: 'It will certainly prove the best that the grandly successive series of Tertiary horizons in France should be adopted as the *chief bases of time division*, partly because of their priority of description and definition, but chiefly because in France, owing to the instability of the continent ..., there is a remarkable alternation of fresh-water deposits containing remains of mammals and of marine deposits containing fossilized shells, the shells serving as time-keepers of the evolution going on in other parts of the world. Thus in France the evolution of mammals, or the vertebrate time scale, is checked off by the invertebrate time scale. As we shall see, the Lower Cænozoic of America from the base of the Eocene to the summit of the Oligocene offers us a much more complete life story than that of France; in fact, it is an unbroken historic chapter. The same is true of our Oligocene and to a somewhat less extent of our Miocene. But the mammal-bearing series is entirely fresh-water.'

'Our first object', Osborn continued, was '... to show how far the Epochs or *Systèmes* of America and Europe can be synchronized and similar permanent limits be placed between them; our second object is to establish Stages as convenient divisions of each ... Of course the synchronizing of the stages and substages throughout will present greater difficulties and may in some instances prove impossible, owing to the absolute independence of the move-ments of the earth and of the other physical phenomena which causes these stages in the Old and New Worlds. It is obvious that the overlapping in time of these minor periods of deposition would be the rule and that exact synchronism would be largely coincidence and therefore highly improbable; all that we can reasonably hope to establish in the near future is *approximate synchronism of the stages*' (emphasis in the original).

Whilst the geologic unit was the *formation*, the biologic unit of Osborn and Matthew was the *life zone*, for which Tedford (1970) and Lindsay and Tedford (1989) demonstrated a firm biostratigraphic basis – we are considering here the actual fossil-bearing strata. Using their life zones, Osborn and Matthew achieved a sequential ordering of relatively isolated patches of continental sediments; there were few examples of superposition available at that time to facilitate the sequential ordering. As belts of strata characterized by their fossil content, the life zones were similar to Buckman's (1902, 1903) faunizones (Callomon, 1995). And since fossil associations were better known than were the stratigraphic ranges of taxa, they were assemblage zones, which is interesting insofar as they were erected primarily in palaeobiological rather than geological research pro-grammes, in contrast to most of their protistan- and invertebrate-based counter-parts. The faunal zones subsequently presented by Matthew (Tedford, 1970) differed conceptually from the life zones in being based on a single lineage – the horses.

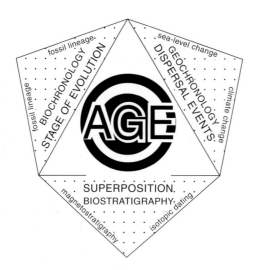

Figure 7.23 Three approaches to the age of a fossil mammal assemblage (Lindsay, 1989): superposition, stage of evolution and dispersal events. With permission.

From these great studies there emerged in due course (see Tedford, 1970) the *North American Land Mammal Ages* (NALMAs) defined in the Wood Committee Report (Wood *et al.*, 1941). Prothero (1995) emphasized a break in tradition then and a break for the worse: '... the good beginning established by Osborn and Matthew was lost, since the next generation of vertebrate paleontologists virtually ignored their pioneering work.' By this, Prothero meant that the secure grounding of the units in the rocks was abandoned, so that the Wood Committee's 'Provincial Ages' turned out to be a 'complex hybrid of local rock units and time units delineated by index taxa, characteristic taxa, and first and last occurrences of mammalian genera.' The NALMAs are not biostratigraphic but biochronological units, representing spans of time during which the characterizing fauna lived. 'This emphasis on mammal occurrences *in time rather than in rocks* distinguishes North American vertebrate chronology from other chronologic systems' (Lindsay and Tedford, 1989). [The East Indies letter classification is another example.] But the NALMAs are not ages or stages, but bichrons, as Berggren and Van Couvering (1974), among others, also pointed out. Besides the improvements, insights, and corrections that emerge inevitably as research and discovery proceed, the three problems on which NALMA studies have focused are (i) the need for some sort of stratotypification, (ii) the need to clarify boundaries, and (iii) the need to improve correlational links with other Cenozoic timescales. The problems are signified in the 'three ways of approaching the age of a fossil mammal assemblage' (Lindsay, 1989; Fig. 7.23 herein):

superposition, *stage of evolution* and *dispersal events*. The younger part of the NALMA succession is correlated with other scales in Figure 7.25.

The first of the three perceived problems, the need for stratotypes, can be illustrated with an example borrowed from Prothero (1995): the Wood Committee cited multiple, more or less coincident, criteria that led to conflicts as subsequent finds separated those criteria in time. The Late Eocene Chadronian NALMA was originally defined by two criteria – the co-occurrence of the horse *Mesohippus* and the brontotheres, on the one hand, and the limits of the Chadron Formation on the other, at a time when the top of the brontothere record was thought to coincide with the top of the Chadron Formation. But the subsequent discovery of brontotheres above the top of the Chadron Formation (in the lower Brule Formation, White River Group, Wyoming–Dakotas) now forces a choice of defining criterion: lithostratigraphic with stratotypes, or palaeontological and biochronological? When the NALMAs were established, very few had superpositional relationships to aid in their ordering – and super-position, desirable for biochronology, is downright essential for biostratigraphy (Lindsay, 1989). The situation has changed and the ages can be treated like zones; they can be stratotypified – as some have – like full-blown chronostrati-graphic divisions. While they were still largely in a state of succession without the benefit of many unambiguously superpositional configurations, the NALMAs were tested radiometrically and they passed in some triumph when their individual radiometric calibration was shown to be consistent with their inferred ordinal succession (Evernden *et al.*, 1964).

The second problem is the boundary problem and has to do with precision – precision of definition and precision in correlation. Woodburne (1977) argued strongly that the LMAs in North America, effective and successful though they were in their functioning as the non-marine timescale, nevertheless would work better – achieving more refined, more precise correlations – if their boundaries were each defined on a single taxon as a zonal boundary. Immigrational and evolutionary first appearances could be used, Woodburne's preference clearly being for the latter. The biostratigraphic triad of *definition*, *characterization* and *identification* should be kept distinct from each other (Murphy, 1977). The lower boundary is targeted – bottom-up – as in most recommendations and codes. Most of the faunal divisions and their subdivisions are defined by immigrational events; in some cases, key defining events arose from faunal changes within endemic groups (Woodburne *et al.*, 1987; Lindsay and Tedford, 1989; Woodburne and Swisher, 1995). Woodburne and Swisher distinguished ten epi-sodes of immigration/emigration of particular importance in both biogeography amd NALMA definition, plus more than thirty 'background' dispersals involving only a few taxa. Lindsay *et al.* (1987) distinguished the lowest stratigraphic datum, as observed in a local stratigraphic section, from the (inferred) first appearance datum, interpreted as the 'real' first appearance of what clearly is a biochron.

We take the discussion of events further by invoking the European Cenozoic mammalian chronology. There is an irony here – as Osborn noted long ago, the stratigraphic record in France allowed the interleaving of marine and terrestrial assemblages of fossils and encouraged the use of the same stages in both facies realms; Savage and Russell (1983) employed several marine-based stages in their Eocene framework. However, it seems, as a broad but reasonable contrast between the North American and the European situations, that the latter has a more robust, better studied taxonomic base and a weaker, less completely known stratigraphic base, whereas the reverse applies in North America (Lindsay, 1989) where high-precision ^{40}Ar–^{39}Ar dates are in widespread use (Prothero and Swisher, 1992; Woodburne and Swisher, 1995; Berggren et al., 1995a). Where both research programmes developed biochronologically and employed the three criteria, in reality contiguity and superposition have been more difficult to discern in Europe. Instead, the continental chronology that has developed in less than two decades has relied much more on stage of evolution and adjacent notions. Modern biozonation begins with Thaler (1965, 1966). Thaler's faunizones, bounded by breaks (coupures) and very similar to the life zones of Osborn and Matthew, developed into his biozones, more clearly bio-chronological and bounded by niveaux répères, or palaeontological reference levels, after Hartenberger (1969) and the nearest approach in this discipline to the marine datum plane (Lindsay and Tedford, 1989). To escape the marine thrall, as it were, there was a need for a comprehensive framework for mammal history, and it was provided to a large extent by the MN zones, based on Neogene mammal assemblages distributed through Europe and North Africa (Mein, 1975, 1979, 1989). As well as this mammal zonation, a succession of continental stratigraphic stages has been developed (Steininger et al., 1989, 1995).

Quo vadis, NALMA? Prothero (1995) discerned progress among vertebrate palaeontologists in using more rigorous biostratigraphic methods, for example, in not confusing rock units with time units, and in collecting fossils according to the actual section instead of merely recording the formation. The latter practice down the decades has made it impossible to use old collections to subdivide NALMAs according to modern insights. He saw salvation in adhering strictly to Article 54e of the North American Stratigraphic Code, which demands designa-tion of a stratotype for each new biostratigraphic unit and of reference sections for emended biostratigraphic units. In contrast, wrote Prothero, authors review-ing all the Palaeogene ages in Woodburne (Ed., 1987) were attempting to take a rigorous approach but used lineage-zones and interval-zones in the 'looser' sense of the International Guide. Such zones are still biochrons, based on the abstracted first and last ocurrences of taxa, not true biostratigraphic zones and stages, which must be based on local ranges of fossils in specific sections. These 'informal biochrological schemes' should soon, with the detailed documentation now coming to hand, be replaced by 'formal range-zone biostratigraphy', as in studies in Prothero and Emry (1995). 'When a formal biostratigraphic basis for all the North American land mammal "ages" is established, *they will become true stratigraphic stages*' (emphasis added). Thus Prothero distinguished sharply between biochronology and biochrons, on the one hand, and biostratigraphy plus chronostratigraphy, on the other, with zones and stages in mutual hierarchy and conceptually identical. Prothero's argument leads us in quite another direction from the IMBS, where stratotypification is restricted to age/stage and epoch/series (GSSP) boundaries but zones, however precise their definitions (Chapter 2) are based on rigorously collected and documented data but not on formally designated reference sections.

Walsh (1998), a dedicated nominalist, took great pains to separate rock-interval terms from time-span terms and erected a very detailed terminology (which we bypass here). He concluded that NALMAs should not be redefined as formal geochronological or 'geochronostratigraphic' units. They should be regarded as biochrons and specifically as Oppel disjunctive biochrons.

Alroy (1992, 1994, 1998a) has pushed the path of biochronology – the 'fundamental goal' of which is ordering taxonomic first and last appearance events – into new territory. He began by fastening exclusively onto one of the various kinds of taxon range zone (see the concurrent range zone in Fig. 2.9 herein) – the overlap where the first appearance event of one taxon predates the last appearance event of a second taxon (FAE < LAE, first/last or F/L statement). This is the only kind of relationship between two taxon ranges that, once observed in a local (not composite) section, cannot be falsified by subsequent discoveries (ignoring misidentification/revision, reworking, etc.). There is but one, true, global ordering of appearances – the succession of speciations and extinctions in North American land mammals awaiting discovery. Therefore F/L statements, which cannot be contradicted once made, will converge on the true pattern as the data accumulate. Again, F/L relationships can be demonstrated without recourse to stratigraphic data, because coexistence is demonstrated by co-occurrence of any two taxa in a taxonomically reliable faunal list. Using algorithms based on this F/L configuration, Alroy (2001) developed a quantitative mammalian timescale. A sequence of 6196 F/L events was generated from many faunal inventories and stratigraphic sections and it was calibrated numerically. The ordination was divided statistically into discrete time intervals which agreed with the NALMA system – Alroy was impressed that there were not many discrepancies between 'traditional, subjective correlations' and his

automated analysis. This implied that NALMA was robust, methodological disputes notwithstanding; and that 'time scales of long duration and wide geographic utility may now be viewed as explicitly testable scientific hypotheses
instead of legalistic formalisms'. Alroy severely criticized two departures in
recent decades from the biochronology of Osborn and Matthew and of the
Wood Committee. One departure was chronostratigraphic – just as fossil discoveries cannot be tied down by golden spikes, nor can 'immigrant first appearance datums' acceptably conflate baptism and naming with hypothesis testing;
these datums are a very minor part of the record and several are strongly
diachronous. The second departure was to rely on outside data, namely radiochronology and geomagnetochronology (a very different matter from integrating, as in the IMBS). Instead, this biochronology must stand on its own feet.
However, Alroy's most radical inference is the redundancy of NALMA (for all its
robustness and surprising – to Alroy – concurrence with his ordination of F/L
pairs). All of the faunal lists have their place in the ordination and have been given
their independent estimated ages (with errors), so that the 'land-mammal age
system per se has now been rendered a mere appendage of methods that use
fossils to define the flow of time; any conceivable scientific problem that might
require paleontological dating could proceed directly from the locality-specific
age estimates, and thereby avoid the age system.' Still further: this exercise on the
land mammal succession could be extended to virtually all the palaeontological,
stratigraphic, and geochronological data in the geological record – bringing it
together in the one system could make the international chronostratigraphic
system itself redundant.

Depositional sequences and regional stages

I have outlined varieties of response to the d'Orbignyan question: what
is a stage? One of the regional chronostratigraphies, in New Zealand, exemplified the tensions between the stage as a biostratigraphic unit, based in molluscan faunas, later in foraminiferal faunas then in defining events, and the stage
as a chronostratigraphic unit. Meanwhile, apropos of the stage as a cycle, Vella
(1965, 1967) perceived eight sedimentary cycles in the Palaeogene of New
Zealand and attributed their genesis to worldwide, eustatic changes in sea
level. Loutit and Kennett (1981) subsequently tested the notion of those stages
as classical sedimentary cycles bounded by unconformities (i.e. synthems) in
the new context of sequence stratigraphy. Loutit and Kennett had several aims:
to see whether the regional stages were natural sedimentary cycles; to test their
correlations with the sequence-stratigraphic scenario for the Cenozoic (which
would also corroborate that 'global' scenario); and to promote the use of

unconformities in correlation, especially in regions biogeographically somewhat isolated, such as southern Australasia. They concluded that 16 of 18
stage boundaries in the Palaeogene–Miocene appeared to correlate (within a
few hundred thousand years) with inferred global changes in sea level; that
most of those stages appeared to represent natural cycles; and that there was
indeed much potential in unconformities for global correlation, at least in those
unconformities in which there had been rapid sedimentary response to rapid
eustatic changes (thereby constraining the hiatus).

   Hornibrook *et al.* (1989) concluded: 'A new type of event stratigraphy seems to
be emerging, which, if it can be demonstrated to be related to world wide events
recognisable in sedimentary sequences, will have a profound effect on stratigraphic classification. The present reality is, however, that the New Zealand
Cretaceous and Cenozoic are divided into regional stages following the principle first clearly enunciated by Thomson in 1916 and extended initially by
Allan (1933) but principally by Finlay and Marwick who had worked out the
basic framework of New Zealand Cenozoic biostratigraphy by 1947.'

   To the advocates of sequence chronostratigraphy, depositional sequences
provide the best way to divide the rock record into time-rock units, therefore
should be prominent in the concept and definitional principles of the regional
stage, as stratigraphy turns full circle in returning to a physical division of the
rock record (e.g. Loutit *et al.*, 1988; Vail *et al.*, 1991). Early stratigraphic classification was based on lithology, unconformities and fossil assemblages. As biozones shifted from assemblage criteria to boundary and phyletic criteria, so too
were biostratigraphic defining criteria emphasized more in chronostratigraphic
units. In the stratigraphy of continental margins and especially in settings of
rapid terrigenous sedimentation with rapidly changing marginal marine environments, however, it is 'virtually impossible' to correlate using biozonal surfaces as chronozonal surfaces and the definition and extension of stage
boundaries becomes hazardous. Loutit *et al.* contrasted two authoritative treatments of the Palaeogene stratigraphy of the Gulf Coast after a century of study
including pioneering petroleum geology:

   Murray (1961) tended to define the boundary on the first appearance of
marine organisms at the transgressive surface, whereas in Toulmin's (1977)
classification the stage boundary generally is at an unconformity or surface of
subaerial exposure. This difference can be seen for the Midway–Sabine and
Sabine–Claiborne Stage boundaries in Mississippi–Alabama (Fig. 5.23). The
Midway–Sabine boundary (Upper Paleocene) is either at the base of the '*Ostrea
thirsae* Beds' (Murray) or at the base of the Gravel Creek Sand (Toulmin). The
Sabine–Claiborne boundary (near top Lower Eocene) is either at the base of the
Tallahatta Clay (Murray) or at the base of the Meridian Sand (Toulmin). It is

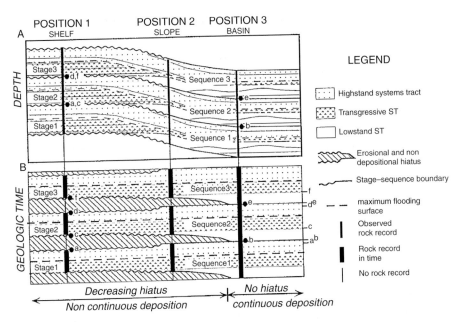

Figure 7.24 Stage stratotypes in a context of sequence stratigraphy (Abreu and Haddad, 1998, with permission). The cartoon of a marine depth–space cross-section (A) shows a less complete stratal section towards the neritic, which lacks the lowstand deposits in the basin. The time–space cross-section (B) shows (right) the ordination of critical horizons – the tops and bottoms of the neritic stratotypes and the sequence boundaries in the basin. The latter, surfaces 'b' and 'e', are younger than the tops and older than the bases of the respective stratotypes. Stage stratotypes accordingly need the support of boundary-stratotypes. (Note that this diagram does not inform us about the duration of hiatuses vis-à-vis geochronological resolution.)

apparent that these alternatives are close together and probably at the same surface across most of the schematic section. The difference is seen at downcuts, or incised valley fills, and the difference becames significant downdip into the subsurface or into settings of increased rates of accumulation. Loutit *et al.* (1988) preferred the Toulmin approach because of the attributes of depositional sequence boundaries, which: (i) separate older rocks from younger (all rocks above the sequence boundary are younger than all rocks below it); (ii) can be identified physically in a boundary-stratotype; (iii) can be correlated (traced) from the terrestrial to the neritic to the pelagic realms; (iv) can be dated by opportunistic bracketing criteria. Rather than defining the stage boundary on a biostratigraphic datum (however good), it is better (they argued) to identify the sequence boundary-unconformity and correlative conformity and date it using all available means (primarily biostratigraphic in the deep basin) and trace it landwards well beyond the reach of the key marine microfossils (as well as acting as a tie between pelagic and terrestrial fossils).

Abreu and Haddad (1998) have illustrated the sequence-stratigraphic preference (Fig. 7.24). Stages 1 to 3 might be 'classical' stages seen in outcrop, with consistent contrasts in neritic fossil content providing adequate body-stratotypes. The unconformities–hiatuses are known and useful, but the strato-typification is not complete (it turns out) until boundary stratotypes are found when more complete sections with a better pelagic fossil record can be accessed basinwards ('no hiatus' is not realistic but this cartoon is no less useful for that).

Conclusion: on fossils and time

Figure 7.25 attempts to locate two common but contrasting points of departure, namely the more 'geological' superposition and the more 'biological' stage of evolution, and two endpoints, namely regional stages and biochrons. Chronostratigraphy and biostratigraphy are compared and contrasted in Table 7.4. An incisive study six decades ago gives us a benchmark for assessing our present position on the topics of this chapter: 'Biostratigraphic correlation of Mesozoic marine deposits is based on zones which are either worldwide or

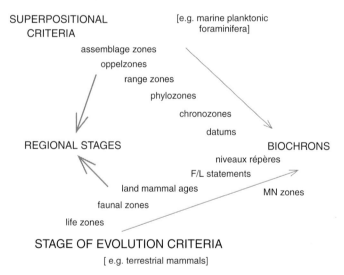

Figure 7.25 Conceptual series departing from superpositional and stage-of-evolution criteria, respectively, and leading to regional stages and biochrons respectively. This arrangement is not really to advocate some whiggish high-road of progression but to celebrate the efflorescence of concepts and units each attempting to capture some small segment of the variety in the fossil record. F/L statements (bold) are predicted by Alroy to make much of this diagram redundant in near-future research (see text).

Table 7.4 *Chronostratigraphy and biostratigraphy: comparison and contrast*

Questions	Chronostratigraphic units	Biostratigraphic units
Are they necessarily complete?	yes: they are intended as timescale: gaps are not allowed	no: they contribute to the timescale
Are they subject to codes?	yes: they are in the framework category	as little as possible: they are in the phenomenon category
Do they require type sections?	yes	no (some disagree)
Are they hierarchical?	yes and no: nested with coterminous lower boundaries, except stages	not usually; but subzones are nested in zones
Have they isochronous boundaries?	yes, by definition	yes or no, by testing and argument
How are they defined?	by golden spike at lower boundary; fixed by enlightened committee decision	by first and last occurrence events; assessed and improved as discovery proceeds; never fixed by *Diktat*
How are they characterized?	by next boundary stratotype; by other 'core' sections	other range events and presences: 'congregation'
How are they identified?	by multiple biochronology, geomagnetochronology, radiochronology, chemochronology	by biochronology: associations of species and phyletic events
What are their component parts and rock/time counterparts?	stage/age, series/epoch, system/period, erathem/era	biozones to chronozones/biochrons; or only defining events (datums)
Framework category or phenomenon category?	they *are* the framework category of this discipline	they should remain in the phenomenon category

at least useable within the wide limits of a palaeo-zoogeographic province. Correlation of Tertiary deposits is a much more difficult problem on account of climatic differentiation, topographic isolation, and close stratigraphic subdivision of deposits representing a comparatively short time interval. No worldwide scale of fossil-zones based on well-defined ranges of a set of index-species exists. A sequence of Tertiary faunal assemblages was long ago established in Europe and it is not surprising to find that workers in other continents first turned to this sequence for guidance by means of direct comparison and correlation. As long as no scale of zones is available, the next higher unit in stratigraphic classification, the stage, must be the basic unit for measuring geologic time. The recognition of the European stages in the East Indies proved so difficult that a number of workers gave up and even condemned attempts at inter-continental correlations' (Glaessner, 1943, p. 52).

The contrast between the Mesozoic and Cenozoic was correct enough and important for this reason: the great bulk of the data on the development of the timescale and its 'philosophy' came from the cephalopods, fossilized in belts of strata that are not very similar to most of the stratigraphic record, marine and continental, of the Cenozoic. There was a somewhat stronger need for multiple and provincial scales to capture the Cenozoic record. Some of the scales discussed here seem to be as useful and used as they were in the 1930s and 1940s; the southern Australian is not; and the Parathethyan scale is quite new – a telling initiative. A second point also turns on the nexus between ammonites and the birth of the science: it is the place of the zone in the scheme of things. In the extract quoted above, Glaessner was following the zone's categorizing and ranking by Schenck and Muller (1941) at the base of the time-stratigraphic hierarchy – an action that has caused problems. For Teichert (1958) as for Arkell (1933), 'the careful reader [of Oppel's] 'Juraformation' will be in no doubt that he used the term for rock units, not in a time sense'; whereas to Schindewolf (1957, 1993), as to Harland et al. (1990), it is clear that Oppel conceived zones as time divisions. That question of whether zones are the rocks that you can hit with a hammer, or the time taken for them to accumulate sporadically, is one facet of a mostly unrewarding controversy. Another facet is whether biostratigraphic (and biochronologic) units should be kept separate from chronostratic units, as in the *Guide*, or whether we should continue to acknowledge the historical reality that correlation and age determination in the Phanerozoic have been dominated, even overwhelmingly, by the use of the fossil record. In these days of highly integrated geochronologies and of new, opportunistic tools of correlation, the very fact of a unified and holistic stratigraphy demands a common time scale kept well clear of biochronologic systems. Biozones are indeed biostratigraphic, as stated by Teichert (1958), Berggren (1971b) and others, and the *Guide*.

This brief review of selected provincial scales suggests that they have been necessary. The New Zealand and the Paratethys stages, the Mediterranean planktonic foraminiferal, the North American and European mammalian, and the Indo-Pacific larger foraminiferal divisions are all alive and well. Complaints and criticisms have more to do with improvements than with abandonment. Some are biochrons (mammals, larger foraminifera) even if that status is changing. The Mediterranean stages have microplanktonic datums at their boundaries. Others are – or are based in – 'multiple biostratigraphic units' (New Zealand) or 'integrated assemblage zones' (Paratethys). Each is a focus in its respective region for all the bits and pieces that comprise that goulasch that is correlation and age determination – a goulasch whose recipe changes from region to region in response to regional needs. In this reality, the pragmatic course is to take as little action as possible in the framework category.

The rise of microplanktonic biochronology was presented in Chapter 2 as a kind of whiggish progression to the state of grace wherein reside the biochrons and datums tied to the global polarity time scale and astrochronology. It is reasonable that crispness of definition and cross-correlation demand that more attention be given to boundaries than to contents. An urge in the same direction is seen in criticisms of fuzzy boundaries in the letter stages and in the call for boundary-stratotypification and clarification of taxon ranges in the mammal ages, and also in the emphasis on the lower boundaries of the Paratethyan integrated assemblage zones. But there has been a contrary view too. Savage and Russell (1983) cast a cold eye on the datum planes/levels/events (FADs, LADs, *niveaux répères*) claimed to be recognizable in the terrestrial mammal succession: the concepts, they said, overreach the real chronological resolution inherent in chronostratigraphy and geochronology. Savage (1977) pointed out the value of the fuzzy boundaries of the Wasatchian (NALMA) stage. Adams was consistently doubtful about the isochrony of larger foraminiferal events; to him, the letter stages were stronger as assemblages than as divisions marked off by successional bioevents (which goes against my comment above that the letter stages were biochrons). Hornibrook (1965; in Hornibrook *et al.*, 1989) appreciated the fuzzy boundaries (buffer zones) of the New Zealand stages because that fuzziness gave much-needed flexibility. The strongest view was expressed by Drooger (1974) because he was addressing the 'standard' – i.e. NW Europe and classical – stages. For Drooger, the most embarrassingly weak link in the architecture of litho-, bio- and chronostratigraphy is the concept of the isochronous surface, 'which is theoretically needed to separate two adjoining intervals of the chronostratigraphic scale'. The increased acceptance of the datum concept into the 1970s was not matched by an increase in our accuracy or resolution in correlation. Accordingly, Drooger would replace the *boundary-stratotype* concept with the *isochronous-body* model for the procedures in standardizing the timescale.

Two decades later, these criticisms may have been neutralized by the emergence, first, of the IMBS with its growing lists of ^{40}Ar–^{39}Ar tiepoints and bio-datums calibrated to the GPTS, second, of cyclostratigraphic calibration of important boundaries, and third, of the powers released by astro-cyclostratigraphy (scientific, not bureaucratic!) to refine and test synchrony/diachrony. However, the application of the IMBS has revealed a deep-oceanic stratigraphic record much more fragmented by unconformities than many (not all) had thought (Aubry, 1995). This revelation impelled Aubry to defend the distinction between temporal and stratigraphic terms, thus making it essential (*contra* Harland, 1992) to sustain the dual terminology (age/stage, epoch/series, period/system, era/erathem, early/lower, late/upper, respectively) and recalling the distinction between the objectively preserved *record* and the partly subjective restored *succession* (Wheeler, 1958). The distinction is essential for a sound temporal interpretation of stratigraphic sections by as many integrated procedures as possible – which is where the (now cyclostratigraphically tuned) IMBS is unique, as Aubry asserted (Chapter 3). Even so, Aubry suggested that a chronostratigraphic framework is no longer necessary and should not intervene between the Cenozoic stratigraphic record and the operations of powerful modern tools for extracting geological time from it: 'we have come to a point in history where stratigraphy would benefit greatly from abandoning the concept of ideal sections in favour of establishing as well as possible the relationships of real sections and time.' Aubry's rejecting the ideal section does not seem to entail rejecting formal and stratotypified chronostratigraphy but, rather, rejecting the implied prime place of the ideal section as the composite section of graphic correlation, which sooner or later must assume improbably constant rates of accumulation.

In contrast to this rock/time separation is the latest proposal to end the distinction (Zalasiewicz *et al.*, 2004). They would use 'chronostratigraphy' to also include 'geochronology', allowing the latter to revert to its 'mainstream and original' meaning of numerical age dating (now including astrochronology and radiochronology). In turn, 'geochronometry' would become redundant. They do not find compelling the arguments for the dual classification which separates the evidence (in the rocks) from the inference (time).

Finally, we return to the divergence of precepts between the Hedberg–*Guide* (stage is fundamental, *definition precedes correlation*) and Cowie–Remane–ICS (long-range correlation is fundamental, *correlation precedes definition*). Seeking a way through this conflict has been the subject of numerous publications by Aubry, Berggren and colleagues, especially Aubry *et al.* (1999). Those authors reviewed

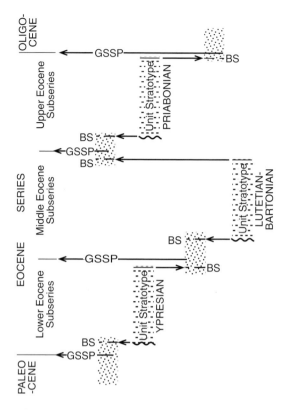

Figure 7.26 Series, stages and stratotypes – a chronostratigraphic hierarchy proposed by Aubry *et al.* (1999, Fig. 6, with permission). Series and subseries are formalized and constitute the main global chronostratigraphic divisions; their boundaries are defined by GSSPs. However, standard stages 'remain important chronostratigraphic elements, and are used in concert with series and subseries for a workable chronostratigraphic framework. 'To achieve this, the authors urged that a section yielding a GSSP should be extended enough to overlap with the two standard stages nearest to the GSSP, so that the respective top and bottom of the stages can be delineated as boundary-stratotypes (BS) in the section with the GSSP. In this example, no Eocene Series or Subseries boundary is concurrent with an Eocene Stage boundary. Aubry *et al.* saw 'a double advantage to this procedure: the standard stages would retain their chronostratigraphic significance (linked to their type area) and the whole subseries and series would be represented by their component stages plus the stratigraphic intervals between the lower and upper GSSPs and the base and top of standard ages.'

the status of the five inter-series boundaries in the Cenozoic Erathem, finding in each instance that what should be a simple nested configuration turns out to be complex and difficult (as we have seen, above). They proposed that the standard stage be relieved of its role as (in their eminently citable style) 'the obligate elemental subdivision in a rigidly hierarchical chronostratigraphic scale'. Stages must be *concrete* bodies of rock, boundary stratotypes are critically important and stages are fundamental to something less than the global ideal. Subseries, series and systems on the other hand are '*essentially virtual units* ...' (emphasis added) defined by GSSPs which are specific levels chosen for their potential in global correlation and age determination. Thus stages are not intrinsically a part of the chronostratigraphic hierarchy – boundaries of the concrete entities may or may not be coextensive with boundaries of the virtual units. At the stroke of the pen (the pen of Aubry *et al.*, 1999), we have emancipated stages to do what stages have always been intended to do – to hold together some regionally (climatic, tectonic, palaeogeographic) or environmentally (pelagic/neritic/terrestrial) bounded stratigraphic situation. Meanwhile, series and systems do *their* job – which is holding together earth and life histories at all geographic and temporal scales. Figure 7.26 was presented by Aubry *et al.* as the 'proposed chronostratigraphic hierarchy', but it is not a hierarchy in the usual sense of the word at all. Instead, it comprises two parallel systems, one virtual and hierarchical, the other concrete and stand-alone. Although Aubry *et al.* paid fulsome tribute to the advocating skills of Hollis Hedberg, the decoupling of concrete stages from the virtual chronostratigraphic hierarchy promotes views from a quarter-century ago by one of its authors (van Couvering, 1977) distinguishing between *regional* stages and *global* series and systems, to Hedberg's eloquent disapproval (1977). However, Aubry *et al.* have taken a major step forward in abandoning any quest for the universal stage, thus freeing the 'classical' stages to be no more and no less than any of the other regional or parochial systems capturing and constraining the biogeohistorical record on this polyglot planet. Stages live!

8

On biostratigraphy and biogeohistory

Summary

Biostratigraphy is a thoroughly historical science subject to several of the ongoing arguments – the power of consilience, the nature of the biospecies, and the shift away from Lyellian gradualism and easy diachrony. The answer to 'palaeontologist – biologist or geologist?' is, both, in biostratigraphy as much as in any of its sister disciplines. Biostratigraphy has three strands in the immediate future. One is to increase the density and rigour of ordination of datums among the major planktonic taxa. A second is the sequence strategy of reconciling bioevents with biofacies and networking mainstream biostratigraphy into facies lacking the index species. The third is to bridge more comprehensively the gap between 'applied' or geological biostratigraphy and palaeobiology.

The idea of Earth history

This account began with the systematic use of fossils in mapping and correlating strata. Historical geology and historical biology, geohistory and biohistory – they go together in so many contexts that I lump them together as *biogeohistory* – are remarkably young historical-scientific disciplines. Hancock (1977) briskly dismissed Smith's forerunners of the seventeenth and eighteenth centuries as of little account in stimulating any sustained research programmes in systematic stratigraphy and mapping (with all respects due to the place of the neptunians in Earth history). Even so, there had to be some preparation of the ground, some intellectual developments that stimulated the historical disciplines early in the nineteenth century. What were they?

Geology and palaeontology played no part in the scientific revolution, that phenomenon of the seventeenth century described as one of the great episodes in human experience, and which ought to rank as 'amongst the epic adventures that have helped to make the human race what it is' (Butterfield, 1949). Butterfield himself mentioned Steno, but not on geology, and discussed Hooke at some length but made no reference to his writings on fossils and strata.

Nor did rocks and fossils play any part in the Reverend William Paley's triumphal compilation of natural phenomena at the beginning of the nineteenth century, a compilation intended to demonstrate the existence of a Creator through a demonstration of the beauty, design and complexity of the world and its life. In 1802 Paley published *Natural Theology, or Evidences of the Existences and Attributes of the Deity Collected from the Appearances of Nature* (Paley, 1827), the best-known exposition of the arguments from design, ranging from animal and vegetable biology to astronomy. The structure of the world and the universe is there, along with function, the way that things work. Indeed, Paley covered much of the range of science as we know it, insofar as science consists of *structure* and *function* – how matter and things are constructed and how and for what they work. Missing from the entire exposition is any *history* – history of the earth and the solar system; history of life on Earth. (There is passing reference to fossils.) Chapter IV is headed 'Of the succession of plants and animals', but 'succession' is a variant of the *great chain of being* (Lovejoy, 1936), referring as it does to generation – from fructification in plants, to oviparous strategies in animals, to animals which bring forth their young alive. Certainly there is a succession, but it is 'from the lowest to the highest; from irrational to rational life, from brutes to the human species' and it implies nothing as to geological time or Earth history. That absence from the the best-known of all the natural theologies of time, of geological history, or of the succession of life on Earth is the most telling pointer of all to the embryonic state of Earth and life history at the turn of the eighteenth century to the nineteenth. 'There is no historicism in [Paley's] *Moral and Political Philosophy* and no geology in his *Natural History*; and the two books are good illustrations that a sense of history was as uncharacteristic of utilitarian political philosophy as a sense of evolution was of eighteenth-century natural philosophy' (Gillispie, 1951).

Thirty-five years later we have a second and almost as famous a natural theology, William Buckland's *Geology and Mineralogy, Considered with Reference to Natural Theology*, one of the Treatises funded by the Earl of Bridgewater to expound on the power, wisdom and goodness of God as manifested in the creation. Buckland clearly saw three important subjects for enquiry within those domains of natural theology entrusted to him. There is the mineral kingdom, whose composition and disposition in the Earth were wisely provided and adapted for the uses of the Vegetable and Animal Kingdoms, and especially for Man. There are the theories of origins – of the world, of systems of life from preceding systems 'by an eternal succession', of gradual transmutation of one species from another. The evidences of geology decisively oppose all such theories. Buckland's third category extended Paley's arguments, using the same kinds of investigation, '*into the Organic Remains of a former World*' (emphasis added), because there is by now a coherent Earth and life history based on rock relationships and on stratal and fossil succession, and it was put together in less than three decades.

When historians speak of *the* scientific revolution they mean the Galileo–Newton clockwork universe. They mean the assessment of data, old and new, in the context of new ideas, and the emergence of new theories. Understanding is not advanced simply or inductively by the inexorable accretion of new observations with new technology, but more by conceptual improvements. Even so, the abundant new celestial observations by Tycho Brahe were subsumed in Kepler's equations; there were the mathematical advances by Leibnitz and Newton; there were the inventions and improvements of the telescope and microscope. The central scientific problems were the problem of motion and the place of the Earth in the universe. Where science extended beyond physics and astronomy, its most noteworthy success was in another realm of structure and function – the anatomy and physiology of blood circulation.

However, there was an thread of awareness of other ways to the truth. Mayr (1982) identified Pierre Bayle (1646–1706) as being apparently the first to assert that historical certainty is not inferior to mathematical certainty, but merely different. Cited much more frequently in our context are two contemporaries, Burnet and Steno, who fare rather differently in the subsequent 'whiggish' assessment of history. In the whiggish view of history (Butterfield, 1951), one tends to do two things, both now in rather bad odour among the historians (but see Hull (1988) and Mayr (1982, 1988)). One tendency is to sort the historical figures and ideas into winners and losers as judged from our exalted present situation (thus, Steno the advocate of superposition and Burnet the biblical romancer, respectively); the other is the inductivist scenario in which – in our sciences – armchair speculation is overthrown by a busy search for facts in the field. In a timely reassessment of some of our antecedents, Gould (1987) mounted a strong case against these views or habits, employing as his central theme the discovery of *deep time* in the context of two metaphors, time's arrow and time's cycle. In *time's cycle*, events have no meaning as distinct episodes. Fundamental states are immanent in time and apparent motions are parts of repeating cycles. Actually, two notions are conflated in time's cycle: a true and unchanging permanence, or immanent structure; and recurring cycles or repeated events. These are two of the four ways of conceiving the origin and nature of the world (Mayr, 1982) – a static world of unlimited duration; and a cyclical world alternating in its state from golden to decay to rebirth. Both go back to the Greeks. In *time's arrow*, history is an irreversible series of events, which then are unrepeatable. If time's cycle is a Greek notion, time's arrow is biblical and it too conflates the ancient Hebrew notion of a string of unique events, strung out from a creation to an ending, with the later notion of inherent direction, which can either be progressive or towards decay. These then are the other two world-states – the Judaeo-Christian static world of short duration; and the gradually evolving world of Lamarck and Darwin. In this matrix Gould assessed Burnet and Steno as having impressive similarities, whatever their differences. In their seventeenth century context, they share three sets of similarities – assembling evidence from nature and from scripture; invoking similar mechanisms, the Noachian Flood, with collapse into the space vacated by the water, but with no concept of repair or renewal; and reading history as a mixture of time's arrow and time's cycle. History is a set of cycles in both theories, however different they may seem at first glance, but each turn of the cycle is very different from the preceding. These points in common are too significant to permit the simple judgement by which Steno is hailed as the advocate of stratal superposition and Burnet is an arid theorizer., and Gould argued that Steno in his 'seventeenth-century ramblings' (Hancock, 1977) anticipated advances of the eighteenth century.

The third figure in Gould's pantheon is James Hutton, and again we have a significant departure from the standard view of Hutton as 'father of geology' which received its main impetus from Geikie (1905) (see also Hallam, 1983). For Hutton established one essential strand of geological thinking, rock relationships and deep time, whilst misplacing the other, Earth history. He established more cogently than did any precursor the necessity for deep time and he asserted the necessity of cycles. Cycles plus vast spans of time are required because there must be a counterbalance to the already appreciated processes of weathering and soil formation, of erosion and deposition – without renewal, these phenomena lead in one direction toward an Earth of very low relief; hence the deduced need for uplift, the eversion of the seafloor into new land just as the old lands are worn down. All of this theory construction entirely preceded Hutton's study of unconformities and almost entirely preceded his acquaintance with granites and their intrusive rock relationships. Instead, we have here an intellectual, deductive attempt to match in earthly configurations Newton's cosmic cycles: 'Hutton's world machine is Newton's cosmos read as repeating order through time' (Gould, 1987, p. 78). (It was Hutton's bowdlerizer John Playfair who 'found the cyclic stability of the Newtonian solar system an

appealing parallel' (Rudwick, 1988).) Hutton's crucial observations of rock relationships in the field came later. But, as Gould established comprehensively,
Hutton lost history whilst discovering deep time. It does not detract from his
achievements in the importance of rock relationships and the 'rock cycle' to
point out that Hutton was not in the stream that was shortly to advance rapidly
the arts of geological mapping and erecting a geological timescale, to say
nothing of chronological correlation and the controversies of biogeohistory.
Thus Hutton, with his eighteenth-century mind, formulated an ahistorical
theory that met with a cool reception among the already historically minded
early-nineteenth-century audiences (Oldroyd, 1979).

Other developments during the eighteenth century supplied the necessary
complement to Hutton's worldview. Cartesians, mechanists – in this context,
Descartes' mathematically elegant clockwork universe – were preoccupied with
how questions – *proximate* questions. The others asked in addition *why* questions –
ultimate questions. In biology, this is the difference between the physiological
sciences or the study of proximate causes, and natural history or the study of
ultimate or evolutionary causes – a two-part division extending through thousands of years (Mayr, 1961, 1982). In the late eighteenth–early nineteenth centuries attempts to reach beyond proximate causes in biology were found also outside
biology, in such German thinkers as Herder, Kant, Goethe, but this tradition
collapsed into the sterilities of *Naturphilosophie* (Mayr, 1982), and antimechanistic
natural history retreated into 'non-problematic description' which was also intel-
lectually unrewarding.

However, this kind of research includes the documentation and classification
of organic diversity in space and time. Classification of minerals, sedimentary
strata and biological species is essential to the *comparative method*. Comparative
science flourished unobserved by the historians and philosophers of those
physical sciences based on experimentation (Mayr, 1982). The materials of the
natural sciences include stars, strata, and species (Teggart, 1925), and in these
realms diversity and variation are a central problem, not a marginal concern as
they are in the search for laws of structure and function. In the eighteenth
century Kant and Laplace could arrange different classes of stars as successive
stages of stellar development. Linnaeus, after Leibniz and ultimately after
Aristotle, could arrange organic species in the *scala naturae* or great chain of
being (Lovejoy, 1936; Mayr, 1982). Werner and the neptunians could attempt the
arrangement of sedimentary strata as a time series based on their lithology and
mineralogy. Students of humans and societies instituted comparisons between
the social conditions observed in existing 'savage' groups, on the one hand, and
conditions revealed in the earliest historical records of civilized peoples, on the
other (Teggart, 1925). All of these otherwise disparate programmes are examples of

the comparative method. They rely on the classification of presently existing phenomena to exemplify stages in a succession – a succession that may have temporal and historical significance; and there is a dynamic back-and-forth between classification and comparative science. There are later, celebrated examples of this strategy. One of Darwin's ways of historical reasoning was to arrange the observations of modern situations and structures as stages of a single historical process, as in the subsidence of coral-ringed islands to form coral atolls (Gould, 1982). In his geomorphic system of youth, maturity and old age in subaerial landforms, W. M. Davis was doing the same thing some years later.

Comparative study and the firm taxonomic grasp of diversity and variation, plus the notion of deep time, are necessary but not sufficient ingredients of the mix that spawned biogeohistory. We need a succession of events and each event must have its temporal signature – it must identify its slice of time (or its moment?) distinctively and unambiguously. And, secondly, the succession, the temporal sequence of events, must have directionality – we must know which way is toward the present – in stratigraphy and structural geology, exemplified by the so-called law of superposition. This is the essential point in the significance of fossils and the origins of what much later would be christened biostratigraphy. But, first, fossils had to be accepted for what they were (Rudwick, 1972; Gould, 2002) – neither mineralogical sports, nor inventions by the devil, nor comprehensively the victims of the Noachian deluge. In the case of modern-looking shells of molluscs or corals in strata above present sea level, they signified either raised strata or lowered sea level, arriving at that point in the debate with less vicissitudes than did such common fossils as the ammonites, which had no known living counterparts. Again, the decisive advances took place in the eighteenth century, two in particular. One was the apparent fact of extinction, finally established by Cuvier in 1796. Well before then it was well known that the ammonites and belemnites did not extend above the chalks in the succession of strata in northwest Europe – but there was always the possibility that they still lurked in distant unplumbed seas. The rapid (European) exploration of the globe reduced that possibility, as it did for the locally extinct mammoths and other locally extinct terrestrial mammals, and for the giant terrestrial and aquatic reptiles. The second development goes with extinction: it was the discovery that sedimentary strata contain fossils in distinctive assemblages (loosely termed faunas or floras): *certain fossils are associated with certain strata*. From Hooke in the late seventeenth century to Blumenbach in the late eighteenth, there were several suggestions that fossils might be used in stratigraphy in much the same way that coins and medallions were used in archeology. Robert Hooke speculated in 1705 not only that advance and retreat of the

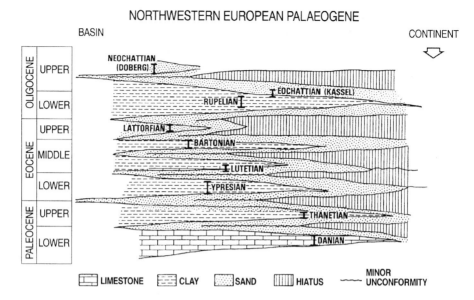

Figure 8.1 A modern sketch of the transgressive–regressive pattern of the Palaeogene in western Europe (from Hardenbol and Berggren, 1978, with permission). Stage stratotypes are located (Chapter 3). Major eustatic changes generate transgressions sweeping in from the open sea, the pattern in the region that stimulated Cuvier's theory of repeated, cautiously catastrophic extinction. Meanwhile, another grand theory of the Earth, this one historical, was brought forth by Cuvier and Brongniart. Cuvier, the first great exponent of the comparative method (Mayr, 1982, 1997) and the first great biogeohistorian was concerned less with correlating strata than with the history of life on Earth (e.g. Hancock, 1977). The theory was of extinctions triggered by the advance and retreat of the seas – the pattern shown here. Cuvier's cautious determinism was gingered up by Buckland and strengthened by Elie de Beaumont's demonstration of crustal deformations extending from deep in the geological record, well up into the Tertiaries (i.e. Cenozoic); and it reached its apotheosis in the synchronous, worldwide catastrophic extinctions of d'Orbigny's inspired, magnificent, visionary, but premature and essentially wrong theory of the 1840s (most of the adjectives are from Arkell, 1933).

sea would explain marine fossils far inland, but also that one might raise a chronology based on the fossils in strata, 'the first intimation that a history of the earth might be compiled by examining its fossil contents. This would, I suggest, have been the true beginnings of historical geology had the programme been carried through' (Oldroyd, 1996). Mayr (1982) gave an incomplete list of no less than fourteen workers in Europe to whom credit for this discovery of the association of fossils and strata has been assigned at one time or another. But it is agreed that two men more than any others integrated the study of strata and fossils into stratigraphy – Cuvier in France (Fig. 8.1) and Smith in England. They established organic extinction and consistent, repeatable fossil succession subsequent to Hutton's writings and they emerged from another tradition – from the natural history incorporating organic and sedimentary diversity and from the notion that biogeohistory could be extracted from the jumbled and fragmentary record of the rocks in the upper Earth's crust. An essential part of the story, indeed, is in the failed attempts to construct a timescale on a lithological-mineralogical succession.

We have, then, as the preconditions for biostratigraphy, the recognition of fossils for what they are, a strengthening notion that faunas and floras have changed through time and that some life forms have become extinct, and the freeing of men's minds of the impossible constraints of Earth's age measured in a few thousand years. There had to develop an historicist attitude, meaning an interest in (geo)history, in unique historical events (not general historical laws). Oldroyd (1979) made a strong case for the emergence of historical geology as an integral part of a major intellectual shift at the end of the eighteenth century. Scholars became more interested in the human past, the prehistorical past and the geological past, in contrast to the philosophers and scientists of the immediately preceding Enlightenment, who were thoroughly ahistorical in their concern with how the universe and its components, from atom to organism, fit together and function. For the early development of historical geology we must look to the 'neptunists' who produced the first global sucession: Primary crystalline rocks; Transition (mainly greywackes); Secondary, or older Floetz: younger Floetz, or Alluvial. The plutonist–neptunist controversy (Hallam, 1989) collapsed during a shift from neptunist stratigraphy, based on lithology and mineralogy, to the Phanerozoic stratigraphy that we know, based on palaeontology.'

Laudan (1987, 1989) demonstrated the place of taxonomy in the development of historical geology, including that great shift from mineralogy-dominated to strata-dominated Earth science. Linnaeus's success in the early eighteenth century in bringing unprecedented order into botany and zoology suggested that the mineral kingdom could be ordered on much the same precepts – that the world comprised natural kinds, or species; that they could be identified and classified according to their essential characters; that natural kinds could be grouped into a hierarchy of higher taxa. But Linnaeus concluded that the 'calamity' of mineralogy, the impossible difficulties in individuating them as for plants, sprang from their generation from 'irregularly sportive nature' instead of from the egg. Attempts to employ Linnaean principles in mineralogy collapsed by the later eighteenth century. One outcome was a shift in attention to rocks, hitherto merely aggregates of minerals but now needing their own classification, and A. G. Werner replaced mineralogy with mode and time of formation as their essential characteristics. Hence the rise of the *formation* as the basic systematic unit, explained by Cuvier and Brongniart as 'a group of beds of the same or different nature, but formed at the same epoch' (Laudan, 1989). Hence too a new emphasis on superposition – position in the succession – as the prime criterion for identification. Omalius d'Halloy distinguished two modes for dividing the 'country': geologically, according to the epoch of formation, and mineralogically, according to its mineralogy and chemistry (Laudan, 1989). Formations, *individual, historical entities*, catered for those who asked chronological questions (and built the geological timescale). A classification of rocks and rock-types was for those asking causal questions. Instead of cross-fertilization, there was a strong divergence by stratigraphy from mineralogy and petrology.

It is appropriate to interpolate here the advent of the disciplines linking biogeohistory with human history – prehistory and prehistoric archaeology. Butterfield (1949) concluded his crisp survey of the origins of modern science with a consideration of progress and evolution: '... the mind gradually came to see geology, pre-history and history in due succession to one another'. This mindshift occurred mostly in the eighteenth century – when else? – and it was as late as the nineteenth century when prehistory as a respectable, evidence-based discipline replaced the legendary prehistory of earlier times. The transition took place for three reasons, as outlined by Daniel (1962) in *The idea of prehistory*, within the timespan bracketed by Paley's *Works* (1802) and Darwin's *On the Origin of Species* ... (1859 [1964]). First, there was the development of geology, meaning an appreciation of both process geology and historical geology. Second, there was the discovery of early 'man'. The association in undisturbed strata of stone tools – 'stone objects now known to be cultural fossils of man' – with animal remains of species not known to exist now came to be accepted as a true – i.e. uncontaminated – association at about the same time that organic extinction itself could no longer be gainsaid. It is difficult to exaggerate the existential impact of the realization that humans once coexisted with animals now extinct, just as extinction itself had all sorts of cultural implications of its own. The third reason for the acceptance of prehistory was the demonstration of archaeological *succession* by Christian Thomsen in Copenhagen, from 1815 onwards – the three ages of man, the Stone, Bronze and Iron Ages.

Thus we have a remarkable flowering of history, *sensu lato*. Von Engelhardt (1982) reminded us that the late-eighteenth-century controversy between 'neptunism' and 'plutonism', that 'dismal squabble' in Arthur Holmes' words, actually was a collision of two different styles of Earth science. It was from among the neptunists that the historical disciplines arose – stratigraphy, palaeontology, palaeogeography, and so on. From plutonism came the geodynamic strategies concerned with Earth process – those phenomena rooted in the laws of physics and chemistry (which is not to imply, of course, that historical sciences are lawless). Laudan (1982) has drawn attention to the working out of the same tensions in the history of the Earth sciences in distinguishing between natural history and natural philosophy. Natural history is concerned with reconstructing the history of the earth and cannot avoid description, taxonomy, and relationship, whereas in natural philosophy we are engaged in understanding the processes of geological change (McGowran, 1986a). One way of generalizing the history of the discipline is as shifts in emphasis from one mode to the other. We might cite Lyell's uniformity and process in the 1830s, the geochemistry of radiochronology in the 1900s, the mobile Earth's crust in the 1960s, and the history of the oceans in the 1980s.

But is it history?

There is, or was, a view that historical knowledge derives from some internal process – living, human experience and human understanding – whereas science is merely apprehended externally. That view goes back at least as far as Dilthey in 1883 and is to be found in R. G. Collingwood's distinction between history proper and the pseudohistories geology, palaeontology and astronomy (Marwick, 1989). Whilst they all have narrative structure and relics, history is purposive but pseudohistory is non-purposive. This stifling view need not detain us. Among palaeontologists, Schindewolf (Reif, 1993) was especially forceful in asserting that that not only do Earth, life and mankind have histories, but these histories have fundamental properties essential to the hisrorical process in common – irreversibility and cycles, the latter comprising rapid origination and diversification followed by elaboration, stasis and extinction. Marwick listed nine points of difference between science and history but they refer mostly to functional and physicalist science, not to the comparative science that concerns us most here. (Mayr (e.g. 1982, 1988, 1997) has argued for decades that the historical sciences occupy a truly intermediate position between physical sciences and the history-type humanities.) Marwick's two points-in-common are concerned (i) with discovery, with bringing into being new knowledge about the world (*sensu lato*) and with solving problems; and (ii) with using systematic methods involving rigorous checks and the presentation of evidence as well as conclusions – harmless enough, surely, and non-controversial. More to the point here is Carr's conclusion that there is not a sharp break between the [historical] sciences and the non-sciences; Carr (1961) was 'not convinced that the chasm which separates the historian from the geologist is any deeper or more unbridgeable than the chasm which separates the geologist from the physicist' – and that too can hardly be controversial except to those impaled on the notion that physical and functional science is all of science.

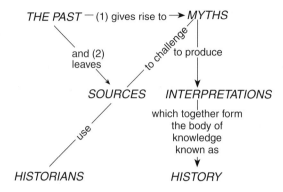

Figure 8.2 Historical analysis of the past (from Marwick, 1989). Can we replace 'history' with 'biogeohistory' and still be saying something useful?

Marwick (1989) attempted (Fig. 8.2) to capture the essence of what historians do. It includes the key ingredient 'myth', defined as 'a version of the past which usually has some element of truth in it, but which distorts what actually happened in support of some vested interest'. Can this diagram encapsulate biogeohistory similarly? At first sight, the importance of myths arising out of our own – human – history and needing to be challenged and sometimes to be overthrown might seem to be a crucial difference.

But on reflection ... consider the grandest of the several grand themes in Mayr's (1982) *The Growth of Biological Thought – Diversity, Evolution and Inheritance*. That theme is the tension reverberating down through the millennia between the worldview of Plato and the worldview of Aristotle. There is Plato, the geometer with little time for observations in the world of natural history, the essentialist for whom the world consists of a limited number of fixed and unchanging forms and an emphasis on discontinuity, constancy and typology. To Mayr, the influence of Plato on biology is nothing short of a disaster; the history of biology is to a significant extent the story of the overthrow of Platonic essentialism which is totally incompatible with evolutionism. The original Platonism – geometry, timeless unchanging realities, general principles of the cosmos – is the first of many instances when biology was severely harmed by the influence of mathematico-physical sciences. And then there is Aristotle, the 'first great naturalist of whom we know', an excellent observer who also asked not just *how*-questions but also *why*-questions (Mayr, 1997). It was Aristotle's notion of gradation from non-living to living nature and within the latter – e.g. the lower marine radial invertebrates resemble plants more than animals – which gave rise in due course to the *scala naturae*, thus penultimately to the emergence of evolutionary thinking, and ultimately to the overthrow of Platonic essentialism by Darwinian evolutionary and populational thinking. (Which is not, as Mayr made clear, to imply that Aristotle was an evolutionist – his concept of a static world of unlimited duration precluded that.)

It is not at all far-fetched in my opinion to compare the myths of human history with the tensions between essentialism or creationism and population-ism or evolutionism, between natural history and natural philosophy, or between time's arrow and time's cycle. We are not talking here simply about conjectures and refutations, about falsification. Instead, we have clashing worldviews. Teggart (1925) went so far as to assert that the split between the 'historical' study of events and the 'scientific' study of processes operating through time goes back to the seventeenth century. There is more than a trace here of what Rudwick (1982) distinguished as cognitive styles. Of four cognitive styles, the first two account for most Earth scientists (his examples were early mid nineteenth century). There is the *abstract* style, strongly causal in its analysis, conscious of an analogy between vast space and deep time and treating time dimensionally, open to external ideas (e.g. physics) and alert for reductionist or simplifying theories, and only weakly historical. There is the *concrete* style which focuses on the concrete order of strata, which raises controversies about bound-aries between the geological systems and their subdivisions because they are perceived as natural, not merely conventional, which is more concerned with order than causation and with the ordering of geological complexity, not sim-plifying it. (Rudwick also saw an agnostic style which refers to a few extreme empiricists, and a binary style covering the extremists of another stripe – creationists and violent catastrophists.) He preferred 'style' to 'tradition' or to 'paradigm'; 'style' does not prejudge or imply coherence in agreed theories. Certainly it is no longer sufficient to describe developments in this science in terms of simple opposition such as catastrophism versus uniformitarianism, and these 'styles' are an heuristic advance.

Perhaps, then, we could replace 'history' in Figure 8.2 with 'biohistory' and 'geohistory' and still be saying something useful.

There is one more matter in this matrix of geology and its relationship to history – the meaning of 'historicism'. First, two other items of jargon crop out from time to time – 'idiographic' and 'nomothetic', which come from the German historian Windelband in the 1890s (Mayr, 1997). From the glossary in Marwick (1989): 'ideographic' refers to 'the approach to history which argues that history is entirely different from the sciences and should follow purely pragmatic approaches of its own'; whereas the 'nomothetic' approach 'tries to assimilate [history] to the natural sciences by postulating general laws and the need for theory'. In simplistic terms, one might describe systematic and bio-stratigraphic palaeontology, and the search for causation and cycles in mass extinctions, for example, in those respective terms. In 'historicism' and

'historicist', we see history 'as an absolutely central discipline because it postulates that everything is explained by its past development, while at the same
time insisting that each age has unique characteristics, and a unique value of
its own'. That definition is necessary because Popper (1957) in *The Poverty of
Historicism* shifted the meaning pejoratively to grand-scale theorizing in history –
'metahistory' to Marwick – which in our terms would be analogous to the
inexorable unfolding of evolutionary lineages to some foreordained end –
finalism and orthogenesis, for example, which are no longer cogent (Simpson,
1949). That meaning of historicism spoils a good word for the historical
approach to sciences where such an approach is essential. Ferguson (1997)
observed that Popper's anti-'historicism' (which he would call a form of anti-
'determinism') was aimed not at causal explanations in general, but at those
that depended on general statements or deductive certainties. Hacohen (2000)
pointed out that Papper's antipathy was to historical prediction, and that he
distinguished pro-naturalist and anti-naturalist historicist doctrines, the former
assuming that the natural sciences applied to the social sciences, the latter not.

Gould (1980) heralded the promise of palaeobiology as a nomothetic science
based in evolutionary theory. Two extracts put biostratigraphy and historical
science firmly in their places.

> Invertebrate paleontology has cast its institutional allegiance with
> geology – more by historical accident than by current logic. When it
> operates as a geological discipline, paleontology has tended to be an
> empirical tool for stratigraphic ordering and environmental
> reconstruction. As a service industry, its practitioners have been
> schooled as minutely detailed, but restricted experts in the niceties of
> taxonomy for particular groups in particular times. We may affirm the
> absolute necessity of comprehensive geological training for success in
> palaeontology, but also admit that strictly non-biological approaches
> have not infused our profession with the excitement of ideas. To
> particularize Kant's dictum: with all biology and no geology,
> palaeontology is empty; but with geology alone, it is blind.
>
> In a classic case of 'methinks the lady doth protest too much', historical
> scientists have given away their disquiet about the validity of their
> discipline by discoursing at interminable length about the nature,
> meaning, and methods of their enterprise. Historicity – the necessity of
> working with complex and unique events in time – constitutes the
> central dilemma of these sciences. For if science is the search for
> common pattern in repeated phenomena, what can one do with the fall
> of Rome or the extinction of *Pharkidonotus percarinatus*? Psychologists long

ago coined a pair of opposite terms to express this dilemma, and they have been widely used in other historical sciences, including history itself (Nagel 1952). Among paleontologists, however, they are virtually unknown. *Ideographic* refers to the description of unique, unrepeated events; *nomothetic* to the lawlike properties reflected in repeated events. Science is nomothetic insofar as its descriptions include particulars of given times and individual objects only as boundary conditions, not as intrinsic referents in the laws themselves' (emphasis added).

Hoffmann and Reif (1988) took two polarizings – Mayr's functional and proximate vis-à-vis evolutionary and ultimate, and Windelband's ideographic vis-à-vis nomothetic – a stage further. They identified a tripartite configuration of biological enquiry – three kinds of problems, or *Fragestellungen*, demanding address by the biologist.

In *Fragestellung* One, one is concerned with the description of a biological entity, meaning its structure, function, or pattern. That embraces biological levels from molecule to biosphere and must include webs on interactions, organism–environment interaction, and patterns of change through time at the appropriate timescale. It would include an analysis of events giving a biostratigraphic zonation. Hoffmann and Reif invoked hermeneutics as the only method appropriate for the initial task of identifying what there is, in nature, to be analysed, described, and explained. Hermeneutics addresses the problem of determining a strategy for explaining a pattern; 'for if a system's function can be plausibly explained by a process, the system can be arguably regarded as a natural entity rather than merely as an artificial construct'.

In *Fragestellung* Two, one asks about the general theory theory controlling those entities, such as the theory of organic evolution by natural selection. The nomothetic approach belongs here; it is applicable in those disciplines dealing with 'some universal physical, chemical or biological aspects of biological phenomena'.

In *Fragestellung* Three, one asks ideographically about the historical process responsible for the existence and the structure of this entity or for the pattern of its change. The difference between historical description and historical explanation is exactly the same as the difference between chronicle and history – it is at this end of the spectrum of the biological sciences that the historical sciences and history are in a continuum. It is at this end that ecology, biosystematics, macroevolution, and biostratigraphy are found, where there is little promise of finding universal evolutionary law, where our tasks are to explain unique events and patterns.

Hoffmann and Reif emphasized that all of this does not constitute a one-way road leading from the descriptive to the ideographic to the nomothetic approaches. Instead, we have here a strong hermeneutic interdependence, a reciprocal illumination between theory and observation: we proceed from an initial pre-understanding, to an analysis, to an understanding to a further analysis, to an improved understanding, and so on, along an hermeneutic spiral.

'Consilience of inductions'

The weakest point in most philosophers' systems is the place of historical science in their scenarios of science and its doing. Among their biological achievements, four luminaries can count some incisive discussion of the fact that not all science is circumscribed by the laws and immanent properties of the physical sciences: Simpson (1964, 1970), Mayr (1976, 1982, 1988, 1997), Gould (1986, 1987), Ghiselin (1974, 1997). We can use examples from correlation at geological timescales to make that point.

Consider the elegant scenario of Cenozoic change, usually described as global climatic deterioration, in Figure 8.3. We have long known that the Eocene was a time of global warmth. Uniformitarian evidence for that generalization consists of the occurrence of crocodiles, palms, molluscs, corals at higher latitudes than are attained by their modern counterparts (e.g. Lyell, 1867). Using the same kind of strategy – extrapolation back from modern biogeographic faunal distribution, including what was known of the environmental requirements of neritic molluscs – Durham (1950) could reconstruct marine isotherms inferred from faunal shifts through geological time for the western margin of North America, and he demonstrated that those isotherms have been retreating equatorwards since the Eocene Epoch. Hence: long-term cooling. Similarly, Dorf (1955) could employ terrestrial floras as 'thermometers of the ages' to give a generalized curve for 40°–50° N latitude in the western USA, showing likewise that the thermometer has been dropping since Eocene times. Both strategies are strong but both are open to debate about the reliability or ambiguity of fossil leaves or of shells as indicators of past climatic states. We add Shackleton's (1985) generalized oceanic bottomwater curve of $\delta^{18}O$ values from benthic foraminifera. That sort of data too could be and was criticized variously as susceptible to diagenetic alteration, vital effects during biocalcification, salinity changes in the reservoir – criticisms as for the fossil curves, and in the same uniformitarian vein. And yet the three curves have a powerful mutual similarity! If quite disparate data from the terrestrial, neritic and pelagic realms of the biosphere show such a good mutual match through geological time, then the chances that we are seeing real climatic changes are suddenly much better than they were for each of the three datasets in isolation. The mutual reinforcement of shells, leaves and isotopes is stronger than the sum of the parts. Adding a curve of

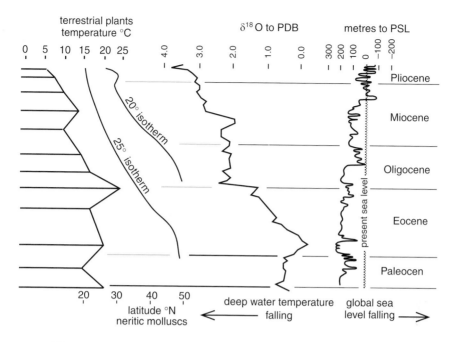

Figure 8.3 Matching trends in the Cenozoic, suggesting strong mutual reinforcement of entirely separate lines of enquiry: palaeontological in mutually exclusive environments (neritic molluscs and terrestrial plants), geochemical (oxygen isotopes from deep-ocean foraminifera) and sequence-stratigraphic (putatively global sea levels reconstructed through time). By *consilience of induction* the theory of a fall in temperatures and sea levels from the early Eocene to the present is much stronger than it would be for any one of the items by itself (modified from McGowran and Li, 1994; sources in the text).

putative global sea level (Haq *et al.*, 1987) to the array, we see a strengthening by further persuasive correlations. Sea level has fallen as temperature has fallen, from the same high point in the early Eocene. (There is a nagging mismatch where the major cooling leads the biggest fall in sea level by the duration of the early Oligocene – also heuristic because it demanded resolution.)

That lesson in mutual reinforcement by chronological correlation is at Cenozoic timescales. We can amplify the scenario for the Miocene series (Fig. 8.4) to make the same point. McGowran and Li (1996) correlated two kinds of curve independently to the revised chronology by Berggren *et al.* (1995a). One is the oceanic isotopic evidence for the Mi glaciations at 10^6 years scale, rising then falling on the trend at 10^7 years scale (Wright and Miller, 1993); and the other is the Exxon sea level curve, also at both 10^7 and 10^6 years scale (Haq *et al.*, 1987). The match at both scales is remarkable – coolings should fit sequence boundaries, either because of glacio-eustatic lowering of sea level, or

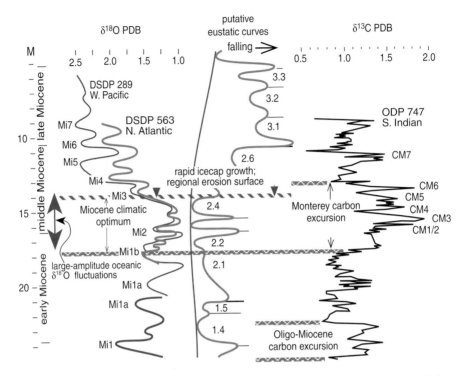

Figure 8.4 The Miocene oscillation and climatic optimum: independent correlations in the Miocene (geochronology is omitted except for Ma, but is shown in Figure 1.22). The oceanic δ¹⁸O curves were drawn by Wright *et al.* (1992) by filtering a cloud of points; the interpreted Mi third-order glacials have been confirmed by Oslick *et al.* (1994). The putative eustatic curves are from Haq *et al.* (1987) with third-order sequence notation. The δ¹³C curve with Oligo-Miocene and Monterey carbon excursions and higher-frequency peaks (CM) is from Wright and Miller (1992) (see also Hodell and Woodruff, 1994, Fig. 15). Large-amplitude δ¹⁸O fluctuations ∼17–14 Ma: Flower and Kennett (1995). Note that the zeniths of the three curves and the large-amplitude δ¹⁸O fluctuations all fall within the Miocene climatic optimum. Modified from McGowran and Li (1996).

because lower sea levels enhance continental-type climates. Thus the mismatches can be targeted for scrutiny and a fine-tuning of the correlations.

In both of these examples it is the *chronological correlation of disparate data* that counts for most in mutual reinforcement. This strategy has a noble ancestry and yet was curiously unacknowledged (Gould, 1986). Darwin, 'so keenly aware of both the strengths and limits of history, argued that iterated pattern, based on types of evidence so numerous and so diverse that no other coordinating interpretation could stand – even though any item, taken separately, could not provide conclusive proof – must be the criterion for evolutionary inference' (Gould, 1986). Darwin invoked this strategy repeatedly when his compound theory of evolution was challenged: all the difficulties and legitimate objections notwithstanding, how could a false theory explain so many classes of facts? – not just many facts, but many classes of facts: that is, widely differing lines of evidence? (Ruse, 1986.)

The early-nineteenth-century philosopher of science William Whewell is often quoted on the topic of classifications or systems in this passage from his *Philosophy of the Inductive Sciences* (1840): 'The maxim by which all Systems profess- ing to be natural must be tested is this: – *That the arrangement obtained from one set of characters coincides with the arrangement obtained from another set*' (his emphasis) (quoted in Ruse, 1986). Whewell called this strategy of coordinating different lines of evidence to form a pattern, 'consilience of inductions'. Hodge (1989) compared Whewellian consilience with the antecedent *vera causa* of the physic- alist philosophers, in which one contrasts true, known, real or existing causes with hypothetical, imaginary, unknown, conjectural or supposed causes: the difference being that the former, the *vera causa*, can draw evidence from inde- pendent sources. Hodge seems to suggest that the *vera causa* encompasses Whewell's strategy of consilience and that Darwin's theory of natural selection (*not* his more comprehensive theory/theories of evolution) was within the tradi- tion of (physics-dominated) science.

Magnus (1996) outlined the early-twentieth-century controversy over the role of genetic isolation in evolution, a controversy between the experimentalists (mutation theorists) and the naturalists. He demonstrated that the naturalists articulated and defended a strong notion of how good science ought to proceed, the most important *epistemological virtue* being 'to integrate as many lines of evidence as possible, preferably pointing to the same conclusion'. This virtue of *consilience* required considering a broad range of lines of evidence, in strong contrast to the narrowing demanded by the experimentalists' supreme virtue, *replicability*.

In his *Consilience: the Unity of Knowledge*, Wilson (1998) invoked Whewell's consilience to link together the broadest array of evidence of all, nothing less than all of human knowledge.

'Consilience of inductions' captures the spirit and the strategy of biostrati- graphy. How do the strata and their recorded events fit together as a pattern in space and time, and what does that reconstructed pattern tell us of biogeohis- tory? This is not an orderly one-way street from chronological correlations to history, but rather a turbulent two-way, hermeneutic thoroughfare populated more and more by lines of evidence demanding to be heard. Hodge (1989) asked, apropos of the philosophers of science quarrying the history of science, the two 'inescapable questions': 'Are they getting the history right? And, if not, how can they get the philosophy right?' Likewise, we must ask: Have we got the

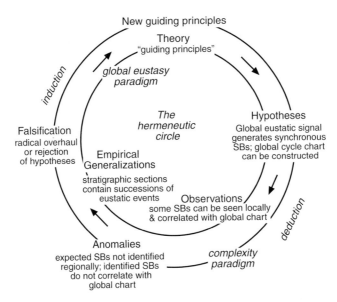

Figure 8.5 The hermeneutic circle, based on Miall and Miall (2001, Fig. 1) who applied it to the global-eustasy model (the inner circle here) with anomalies, falsification, and new guiding principles peeling off and vanishing. Miall and Miall contrasted the 'global-eustasy paradigm' most unfavourably with the 'complexity paradigm' (the outer circle added in this rendering) depicted as a more correct and fruitful exemplar of the Kuhnian paradigm (see their Table 1).

correlations right? And, if not, how can we ever get the geology or biology right? What could be more epistemologically virtuous that that? And that is the central theme of this book.

Another voice: hermeneutics rides again

From sustained criticism of sequence stratigraphy, Miall (1997) progressed to an analysis of the movement in sociological terms (Miall and Miall, 2001, 2002, 2004; Miall, 2004). Instead of the hermeneutic spiral, mentioned above, we have here the hermeneutic circle (Fig. 8.5). The template is well-seasoned – alternation of induction and deduction; analysis in such terms as observation, classification, generalization, explanation and verification; and scientific paradigms in the sense of Kuhn (1962). An hermeneutic circle including stratigraphic cycles and sequences ought also to include rigorous empirical testing, especially chronological correlation. Instead (in this view), we have not one paradigm but two in competition, capturing the situation not just of the past quarter-century but of the past century and a half. Thus there are the global-eustasy paradigm, as set out in the founding document of sequence stratigraphy (Vail *et al.*, 1977), and the complexity paradigm (Miall and Miall, 2001, especially their Table 1). The former sees simple chronostratigraphic surfaces in seismic pattern and simple signals of global eustasy in the stratigraphic record, whereas the latter believes none of this: there are no simple, globally correlatable reflection surfaces, or parallel unconformities, or signals of global eustasy. Instead, the picture is complex and subtle, requiring meticulous, comprehensive, precise, quantified, inductive methods. The global-eustasy paradigm is driven by top-down notions of and search for cycles and the 'pulse of the Earth'; the complexity paradigm by bottom-up, rigorous empiricism. The authors venture further still, bolstering these grand generalizations with behavioural conclusions such as mutual isolation in e.g. language and citation – an isolation that has sharpened during the rise and rise of sequence stratigraphy but actually well into its second century. This analysis is overtly polarizing, but buried in the 2001 paper is a note that the two paradigms are useful end-members of a broad spectrum of responses and approaches to the controversy over global eustasy – a bland exemplar of the Kuhnian vision of paradigms and revolutions. Certainly this writer would not be comfortable in either camp.

Miall's objections to sequence stratigraphy are numerous but two stand out: (i) is to the proposal that the bounding unconformities are mostly global in extent (pattern) and were generated by repeated eustatic changes in sea level; and (ii) is to the ensuing notion and construction of a global cycle chart which could be used as a 'universal correlation template'. The initial provocation was the statement by Vail *et al.* (1977) that geologists should combine their efforts to improve regional charts, which would feed into an ever-improving global chart, which thereby 'can become a more accurate and meaningful standard for Phanerozoic time' [than the available timescales]. The great fear for Miall and Miall is that 'deductive models' will contaminate the cold purity of rigorously, inductively built scales – inserting sequence surfaces into a scale presumably results in driving out the real data. Continuing the acknowledged polarizing in this discourse, a thought experiment (2001, Fig. 2) contrasted two approaches to correlating the same set of stratigraphic events. In the global-eustasy approach a dated sequence boundary acts as a powerful attractor to six other points and probable differences are ignored. In the empirical approach ages with error bars are determined independently.

The feared contamination has had a quarter-century to eventuate since the Vail manifesto. A persisting criticism of the global charts has been the withholding of the essential evidence for dating the key surfaces (not the authors' fault); even so, scrutiny has not merely been in-house within one or other of the perceived paradigms. Two major revisions of the integrated Cenozoic timescale have appeared (Berggren *et al.*, 1985a–c, 1995a, b). Two putatively global sequence charts have also appeared (Haq *et al.*, 1987, 1988; Hardenbol *et al.*, 1998), the first calibrated to a custom-built timescale and the second to the Berggren *et al.* (1995a) scale (the IMBS). Third, a major outgroup has developed in the form of oceanic chemostratigraphy, including oxygen-isotopic proxies for third-order glaciations. Fourth, Cenozoic sequences have been tested especially on the New Jersey continental margin (for the third and fourth points: Miller *et al.*, 1987–2003). Fifth, cyclostratigraphy has refined and sharpened key boundaries and events (e.g. the Oligocene–Miocene boundary: Shackleton *et al.*, 2000; Zachos *et al.*, 2001a). These accomplishments too have been constrained by the Berggren *et al.* timescale, itself being tested cyclostratigraphically.

All of this signifies progress – numerous known problems and unknown surprises notwithstanding, pollution of the timescale is not one of them. Consilience and Popperian conjecture–refutation are a more congenial philosophical environment for this historical science than Kuhnian paradigms and puzzles (itself fraying among some sociologists: Fuller, 2003). I agree that SB and MFS surfaces should not be intrinsic to geochronology – but I do not share the Miall–Wilson (1998) view that sequence-surfaces have no promise in correlation.

Early biostratigraphy: theory-free?

Eldredge and Gould (1977) observed that the early development of biostratigraphy was a prime case of progress preceding understanding. Comparable progress could never have happened in radiometric age determination, for example, where the physical theory had to be established before the possibility of its application to the ages of rocks and of the planet could arise. Eldredge and Gould compared biostratigraphy and Linnaean systematics in this respect. Both depended on the accumulation of data of the natural-historical type; both depended on the perception of pattern in nature; and neither needed sophisticated physical theory or even instrumentation to progress. (The construction of canals and other works of civil engineering stimulated physical stratigraphy as a surveying aid, but that is another story (Winchester, 2001).)

As Laudan (1982) expressed it: 'One can have an adequate understanding of the laws governing geological change without having a history of the earth. Similarly, the reconstruction of episodes of earth history does not ensure that we comprehend the processes of geological change.' Specifically in biostratigraphy, a pattern of the distribution of fossils in strata is established in one locality, incorporated in a composite regional succession, and confirmed elsewhere, all the while with the making of corrections and the filling of gaps. There is corroboration, there is falsification, there is scientific progress – all without a *necessary* theory of speciation or extinction or even, as in the graptolites or, later, the conodonts, without a clear understanding or consensus on the nature or the affinities of the organism. Pattern could be developed without the insights provided by an understood process or a correct taxonomic affinity.

There was, then, a delay in time of five or six decades between the acceptance of an Earth history, recording immense spans of time and, more to the point, decipherable, and the emergence of an acceptable theory of the fossil succession. A cause for Cuvier's revolutions resided in the tectonics of Elie de Beaumont, for by then it was clear that mountain building with its deformation and metamorphism was not confined to the early times of Earth history and could even postdate Cuvier's Paris Basin sections (Rudwick, 1972; Laudan, 1989). But where tectonism and the advances and retreats of the sea could explain extinctions, they did not explain the origin of species in the absence of an accepted notion of evolution. Here is Rudwick on early biostratigraphy: '... whatever the priorities, the fact is, that in the years after Cuvier and Brongniart's memoir was published much greater attention was given to the fossil contents of individual strata, and correlation by fossils was found to be a principle of great practical value. It remained however, an essentially empirical principle; and since a more biological approach to fossils tended at the same time to emphasize the geological implications of characteristic faunas, there was room for debate on the extent to which fossils could also be regarded as characteristic of the relative ages of strata.'

But there was theory, and one such was Lyell's (1833): '... we are apprehensive lest zoological periods in Geology, like artificial divisions in other branches of Natural History, should acquire too much importance, from being supposed to be founded on some great interruptions in regular series of events in the organic world, whereas, like the genera and orders in botany and zoology, we ought to regard them as invented for the convenience of systematic arrangement, always expecting to discover intermediate graduations between the boundary lines that we have drawn.'

Rudwick (1978, 1990, 1998) and Berggren (1998) discussed how Lyell's terms Eocene, Miocene, Pliocene were to define relatively short, isolated 'moments', randomly preserved, and based on a biological theory of the continuous piecemeal formation and extinction of species at a uniform rate; the 'moments' were to be dated by determining the percentage of extant molluscs in each group of strata (Eocene <3% extant species of molluscs, Miocene >8%, Older Pliocene 35–50%, Younger Pliocene 90–95%). The French malacologist Gérard Deshayes had already deduced a threefold temporal division of the Tertiary strata of Europe before he began his collaboration with Lyell (Berggren, 1998). Rudwick used a clock to illustrate Lyell's geological chronometer. Figure 8.6 illustrates

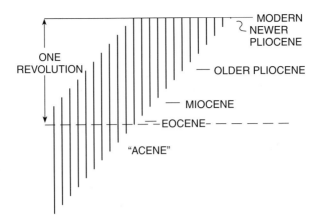

Figure 8.6 A visual representation of Lyell's notions of Eocene, Miocene and Pliocene as mere horizons in the (mostly unrecorded) geological past (McGowran, 1986a).

Lyell's revolution, meaning one complete turnover of a fauna; his piecemeal formation and extinction are shown as species of the same longevity, mutually offset by the same amount. This theory of organic succession is a steady-state model of biohistory, essentially maintaining the status quo (Simpson, 1970) and a shortlived attempt by Lyell to reconcile biohistory with geohistory under the aegis of configurational uniformitarianism – ceaseless change in a stable system, all adding up to a grand steady state. Species were lost as 'part of the constant and regular course of nature' and were replaced as new species arrived on the scene one by one. It is one of three steady-state theories, the others being (i) the anti-historicist, creationist assertion that species are created once and for all and have not varied significantly; and (ii) Lamarck's evolutionary theory of a continuum of organisms, up through which the mass of organisms moves, with no extinction and no species arising *de novo* (Simpson, 1970).

Contrasting with the ultra-uniform nature of Lyell's world was the world of d'Orbigny's stages (Chapter 7). In the 1840s d'Orbigny used the stage as a natural, worldwide unit overriding, by inclusion, local lithological or palaeontological units. Stage boundaries were based on an annihilation of an assemblage of life-forms and its replacement by another assemblage. (d'Orbigny is famous for his catastrophism; but as Hancock (1977) observed, his theory was more violent than his practice. His stages survived remarkably well, theory notwithstanding (Arkell, 1933).) But by the end of the century: 'The old doctrine of physical cataclysms attended by universal destruction of life has passed beyond serious consideration' (Chamberlin, 1898). For by then Darwin's *On the Origin of Species* had been published and evolution as a historical fact of life on Earth had been absorbed into the collective human intellect. Thus there was an understanding of why chronological correlation by fossils worked; the *Origin* appeared at about the time that Oppel and Quenstedt demonstrated the detailed, consistent, and consistently confirmed succession of fossils through different lithologies in the Jurassic of western Europe.

Darwin's theory was the greatest advance in biogeohistory between the perception of Earth history – *contingency* plus *process*– at the beginning of the nineteenth century and the beginnings of radiometric dating at the beginning of the twentieth. But we can ask: what did it do for biostratigraphy? Although Sir Archibald Geikie wrote of 'the flood of light which has been thrown upon geological history by the theory of evolution', in fact he gave more emphasis to Darwin's insistence on the imperfection of the geological record as one of the major advances of the later nineteenth century; 'the intervals which elapsed between consecutive formations may sometimes have been of far longer duration than the formations themselves' (Geikie, 1905). For Geikie, a uniformitarian record of deep geological time and of an always law-abiding nature was more significant than evolutionism – than either the directionalism or the genealogy of the organic world. Invertebrate palaeontology continued in business as usual in its great tasks of correlation and age determine and in biofacies analysis and environmental interpretation. It contributed little to the disciplines of evolutionary theory and phyletic pattern reconstruction (Bretsky, 1971) and it seems that the lack of influence was mutual.

Species and biostratigraphy

Biostratigraphy worked because it was based on a process that produced unique events. The succession through geological time of unique events gave the direction and the framework that repeatable, or reversible, events – geomagnetic field reversals, sea-level changes, climatic changes, epiboles – do not. The process is organic evolution and the unique events are speciations and extinctions. Each species emerges from its immediate ancestor and, once it has gone extinct, that is it: extinction is forever. Thus each species on this planet exists for one and only one lifetime and nowhere else. In its birth, its life and its death, each species is a cosmic singularity: '. . . given the fundamental make-up of the universe, gold is a permanent possibility. I do not see how anyone who thinks species are the things that evolve can maintain that they are in any sense "eternal". They come into existence and pass away and, once extinct, can never come into existence again' (Hull, 1983). And so, the presence of *Pseudodictyomitra pseudomacrocephala* in a stratum means that that layer of sediment accumulated in a marine environment during some part of a geologically brief interval of time in the middle Cretaceous (Albian–Cenomanian) when that radiolarian species existed – neither before nor after that slice of time. There are several statements in the preceding sentences that demand explication.

Why species? Scale, tiering and hierarchy

Hierarchy provides a framework to constrain and discipline enquiry within and across the range of geological timescales. Mayr (1982) distinguished two kinds in biology. One is the *constitutive hierarchy* in which members at a lower level are physically combined into new units or collectives (organs) at a higher level with unitary functions and emergent properties. Hence the series macro-molecule, cellular organelle, cell, tissue, organ, and so on up the levels of inclu-siveness. The units at each level do not have independent existences. In contrast stands the *aggregational hierarchy*, the best known example of which is the Linnaean hierarchy of taxonomic categories. Although units at each level are nested in units at the level above, this is strictly an arrangement of convenience. The basic units are physically independent and remain so even while they are being organized into collectives at the next highest level. Salthe (1985) referred to the *triadic system* comprising the hierarchical rank of interest (for whatever purpose), called the *focal level*, plus the ranks immediately above and below.

There are organizing structures in biology other than hierarchies, and Valentine and May (1996) call some of them *positional structures*. Examples include matrilinear family trees through human generations, evolutionary trees and cladograms. Still others attempt to rank processes and observations in terms of expanding timescales, as in the *tiers of the biosphere* (Gould, 1985; Bennett, 1990) which focus on timescales and on the question of whether we can extrapolate smoothly from top to bottom and back again. Table 6.3 summarizes a tiering from ecological time (Tier I), within which organisms, populations, and communities function, and natural selection operates, up to the tier with chronofaunas etc. (Tier IV), and it could be taken up to the four-billion-year-old biosphere ('Gaia herself'). To view the world in this way (which Valentine and May would exclude from the class of hierarchies) goes hand in hand with a painstaking awareness of timescales. If we enter into a dialogue about evolution or the nature of biostra-tigraphic correlation wherein one party is thinking at, say, level I scales and the other party is thinking at levels III and IV, then the possibilities for confusion and non-communication are ripe indeed. Meanwhile, it is important to consider the central position of timescales in the *explanations* – causes, processes – of what palaeobiologists and biostratigraphers reconstruct as *patterns* from the fossil record. Three paragraphs from Hoffman (1983) explicated the situation:

> There is no science without theories, without conjectures and refuta-tions. There is no science unless the question *how* is followed by *why*. It is

probably for this reason that paleobiologists have been so frequently and fervently engaged on modern disputes on theoretical and evolutionary biology, and also why they have so desperately searched for methods and/or approaches that might give them a chance to contribute, either theoretically or empirically, to those discussions. As an example one might cite a considerable number of palaeobiological studies pretending to have documented mortality patterns and life-history strategies, character displacement, and other morphological responses of populations, patterns of speciation and genetic evolution, patterns of colonization and ecological succession, and ecological equilibrium of ecosystems.

Most of those studies, however, have failed. There now seems to be a consensus among paleobiologists that biological processes operating on the ecological timescale have all too often been projected uncritically onto the fossil record. Consequently, paleobiologists have found themselves with the question, why be a paleobiologist, and with an even more urgent one, what to do as a paleobiologist. This then is their fundamental dilemma.

No doubt, geological, or evolutionary, time is the only dimension of biological processes that is accessible to paleobiologists but not to biologists. It should, therefore, present the major constituent of any uniquely paleobiological field of research. Paleobiologists, indeed, emphasize more and more commonly the reality of changes in structure of the biosphere in geological time, and the irreducibility of macroevolutionary processes to population-genetical and evolutionary-ecological principles. This emphasis forms the core of the modern paleobiological research programs.

Hoffman's statement is equally pertinent to fields interacting with palaeo-biology, such as biostratigraphy. How many textbooks have introduced the topic of biostratigraphy anachronously, with a weak discussion of the spread of populations at human timescales and of populational variation and the theory of allopatric speciation at ecological scales?

In terms of hierarchy and tiering, concentrating on the level of species means some attention to demes, below, and to monophyletic taxa, above. Why is this so? It became apparent in the analysis of Cretaceous and Cenozoic biostratigraphy in Chapters 1 and 2 that the workable zones, the spacing of biostratigraphic events, are of time spans in the low-single-digit millions of years. That is at level III in Table 6.3. The extinction events at level IV provide a few of the more apparent markers in the division of the Phanerozoic Eonothem (Phillips, 1860).

The recent advances in developing a numerical timescale in the Milankovitch band (Chapter 3) are located at level II.

On the 'species problem'

Species, then, the units of biodiversity (Claridge *et al.*, 1997), are the central units in the theory and practice of biostratigraphy. But what are species? Two books published in the mid 1980s exemplified contrasting responses to the persisting problems in answering that innocuous question (Ridley, 1986; Rosenberg, 1985). Two more books, published in the late 1990s, exemplify an even wider divergence (Ghiselin, 1997; Mahner and Bunge, 1997). A collection in between and a subsequent discussion displayed the range of views across the biological and philosophical issues pertaining to species (Ereshefsky, 1992, 2001). The earlier pair first: Ridley (1986), an evolutionary biologist, saw a difference in perceived importance and hence in emphasis between the era of the 'new systematics' (Huxley, 1940), in which attention was focused mostly on species, subspecies and populations, and the subsequent controversies over classification (the three schools known as evolutionary taxonomy, numerical phenetics and cladistics: Mayr and Ashlock, 1991) in which attention is captured more by the supraspecific reaches of the taxonomic hierarchy, where differing and conflicting concepts of species do not seem to make much difference. Interested though he was in the relationship of classification to organic evolution, Ridley chose to omit any treatment of the species concept. A philosopher, Rosenberg (1985), put the species problem squarely in the forefront of evolutionary biology. There is less agreement about what a species is now than there was in Darwin's time and before, and '... there can be no more serious cause for concern about the foundations of a discipline, its future prospects, and its current claims to knowledge than the admission that its key notion remains without an agreed theoretical significance over a century after the central theory in the discipline was framed' (Rosenberg, 1985). The theory is Darwinian evolution and the key notion, species, remains elusive as to explicit definition, sufficient definition, and necessary conditions for inclusion. Rosenberg gave the twofold presuppositions of traditional systematics:

1) There is a single correct description of the basic types of flora and fauna in the world.
2) This single correct description is not merely compatible with the rest of science but actively coheres with it, as reflected in the conviction that the explanatory power of the uniquely correct taxonomy of terrestrial flora and fauna can be grounded in the rest of biology.

But if it is only the species that is a suitable basic category for systematics, and if there is no theoretically grounded definition or characterization of the species, 'then there is no basic category for – no single correct description of – the basic types of flora and fauna'. It follows, Rosenberg pressed on, that scientific systematics, our way of managing organic diversity, is not possible.

There have been several definitions of species in recent decades. That there are competing definitions and that there is not a consensus turn on several points: (i) there really are different ways of perceiving the world; (ii) specializing in different groups (animals or plants, protists or procaryotes) is sure to produce different views on species' definitions; as will (iii) dealing with actively breeding populations, or with museum specimens, or with fragmentary fossil remains; (iv) different scientists use species for all sorts of reasons, including biostratigraphy, and there is a strongly pragmatic component in different perceptions of the species.

Table 8.1 assembles a wide array of species concepts. The broad arrangement of the concepts owes much to Endler (1989) who distinguished at least four differences in the aims of those groping for a concept: (i) taxonomic/evolutionary, (ii) theoretical/operational, (iii) contemporaneous/clade, (iv) reproductive/cohesive. They are helpful for clarification without being mutually exclusive or always clearcut, and so too are there numerous overlaps among the concepts listed in Table 8.1. 'Taxonomic' species concepts differ from 'evolutionary' species concepts in being strongly empirical (or operational) and in making no assumptions about their formation. None of the five overlapping taxonomic concepts has figured in the recent competitive search for the most comprehensive concept. All of those are in the category of the evolutionary species concepts, which Endler divides thus: 'the evolutionary species concept takes an evolutionarily instantaneous view of species, whereas the clade concept treats a species as a group connected by common descent as well as by common biology'.

In the biological species concept of Mayr (1942) definition was biological, permitting inferences about gene flow, disjunct gene pools, and so on; recognition on the other hand and for by far the most part was by morphological inference – there was mostly reliable correlation between genotypic disjunctness and gaps in observable characters, and there is no reason not to extend the morphological inference to asexual species (e.g. Mayr, 1963). Mayr later added the criterion of ecological autonomy because no species has completed the process of speciation until it can pass the ultimate test of coexistence with its nearest relatives. The isolation (biological, sensu Mayr) and recognition species concepts, collectively *reproductive*, have drawbacks: 'There seem to be real biological and evolutionarily independent entities present in

Table 8.1 *A digest of species concepts*

I Taxonomic species concepts	
Practical species	In determining whether a form should be ranked as a species or a variety, the opinion of naturalists having sound judgement and wide experience seems the only guide to follow (Darwin, 1859). A systematic unit considered to be a species by a competent systematist (preferably a specialist) (Mayr, 1942). Tate Regan's aphorism: a species is whatever a competent systematist says it is (Ghiselin, 1997).
Morphological species I	Static, with no reference to spatiotemporal changes, monotypic, all individuals approximating fairly closely to a single norm of variation, defined entirely on morphological characters, almost always clearly separated from nearest relatives (Cain, 1954).
Morphological species II	Based on specimens which are extreme variants of a single morphological series; not allowed in neontology, in palaeontology of use only to stratigraphers; category probably should be abolished (Cain, 1954).
Typological species	Treats species as random aggregates of individuals that share the essential properties of the type (Plato's *eidos*) of the species; variation is the imperfect manifestation of the idea implicit in each species (Mayr, 1969).
Nominalistic species	No essence in common, only the name is in common; there are only individuals, and no kingdoms or classes or genera or species – all of which are man-made abstractions (Mayr, 1969, 1982)
II Evolutionary species concepts	
	A mixture of empirical and theoretical concepts; includes assumptions about process of formation; broadly divided into *contemporaneous species concepts* and *clade species concepts*.
Contemporaneous species	Isolation and recognition species [collectively: reproductive species concept]; cohesion species abjuring reproduction; more useful for and emphasizes anagenesis.
Biological species [= *isolation species*]	A group of actually or potentially interbreeding natural populations reproductively isolated from other such groups (Mayr, 1942). A reproductive community of populations (reproductively isolated from others) occupying a specific niche in nature (Mayr, 1982).
Recognition species	The most extensive units in the natural economy such that reproductive competition occurs among their parts (Ghiselin, 1974). The most inclusive population of individual, biparental organisms which are a common fertilization system.
Cohesion species	Concept concentrates on what factors cause similarity and what processes maintain similarity of morphology, biology, ecology, behaviour and genetics (Templeton, 1989)
Evolutionary species	A lineage (an ancestral-descendant sequence of populations) evolving separately from others and with its own unitary role and tendencies (Simpson, 1961).

Table 8.1 (cont.)

	A single lineage of ancestor-dependant populations which maintains its identity from other such lineages and which has its own evolutionary tendencies and historical fate (Wiley, 1981).
	A population or group of populations sharing a common evolutionary fate through time (Templeton, 1989).
Ecological species	A lineage occupying an adaptive zone minimally different from that of any other lineage in its range evolving separately from all lineages outside its range (van Valen, 1976).
III Clade species concepts	
	A clade definition is claimed to be more useful for and emphasizes speciation.
Palaeospecies	Forms that seem to be good species at any one time may become indefinable because they are successive stages in a single evolutionary line (gens) and intergrade smoothly with each other, hence they are arbitrary sections of the gens (Cain, 1954).
Phylogenetic species	Evolutionary unit that is an irreducible cluster of organisms, diagnosably distinct from other clusters, and a discrete entity in space and time and capable of being compared one group to the next (Cracraft, 1987).
	The least inclusive taxon recognizable in a classification, into which organisms are grouped on evidence of monophyly (usually synapomorphies), and ranked as a species because it is the smallest important lineage deemed worthy of recognition (Mishler and Brandon, 1987).
Cladistic species	The group of species between two speciation events or between one speciation event and one extinction event or [living] descended from a speciation event; taking the individuality of species seriously requires subordinating the biological species concept, to the cladistic species concept (Ridley, 1986).
Monophyletic species	Minimal, morphologically diagnosable groups with at least one shared derived character (Smith, 1994).
IV Two recent attempts	
	These attempts to define the taxonomic category 'species' exemplify the philosophical split between species-as-individual and species-as-class.
From author of the 'radical solution' – SAI	Biological species are populations within which there is, but between which there is not, sufficient cohesive capacity to preclude indefinite divergence ('in some ways less bad than the available alternatives', Ghiselin, 1997).
From strong anti-bionominalists	A species is a *biospecies* if, and only if, (i) it is a natural kind (rather than an arbitrary collection); and (ii) all of its members are organisms (past, present, or future) (Mahner and Bunge, 1997).

nature ... or ... species are logical individuals with beginnings and ends, yet reproduction may have little to do with their boundaries. What causes them and what maintains them if sexual reproduction is neither necessary nor sufficient?' (Endler, 1989). Hence the cohesion species concept, which transcends reproduction in bringing together all the factors that cause similarity and sustain it (Templeton, 1989), emphasizing genetic and phenotypic cohesion within an entity.

That brings us to the evolutionary species itself, defined by Simpson (1961), modified by Wiley (1981) and with strong similarities to the cohesion concept. In fact, Endler (1989) pointed out several factors that connect the cohesion, evolutionary, and ecological species concepts. He identified the 'contemporaneous' species concepts as emphasizing and being more useful in anagenesis, whereas the 'clade' species concepts emphasize and are more useful in cladogenesis. There is not great disagreement among the proponents of these various definitions about one highly relevant thing: that the species exists 'out there already'. There is a strong sense of a shared research program of going out and finding, distinguishing, placing vis-à-vis its nearest phylogenetic relations, and naming a taxon that is a real entity with an evolutionary history.

But the search continues for a definition of the *category* species as distinguished from the individual *taxon* species. Cracraft (1987) criticized the biological species on several grounds, primarily in its role in the confounding of the roles of the species in taxonomy and in evolutionary theory respectively. He advanced four necessary characteristics of entities such as species: *reality*, which is existence even though maybe unobservable; *individuation*, meaning demarcation in space and time, giving discreteness (even if fuzzy at the boundaries in some cases); *irreducibility*, so that a species cannot be subdivided into the same kinds of entities; and *comparability* – you have to compare like with like. Cracraft promoted species defined in terms of evolutionary units. But Mishler and Brandon (1987) had a 'superior' phylogenetic species concept (expanded a little from Table 8.1):

A species is the least inclusive taxon recognized in a classification, into which organisms are grouped because of evidence of monophyly (usually, but not restricted to, the presence of synapomorphies), that is ranked as a species because it is the smallest 'important' lineage deemed worthy of formal recognition, where 'important' refers to the action of those processes that are dominant in producing and maintaining lineages in a particular case.

Mishler and Brandon (1987) emphasized two necessary distinctions. One distinction is between the two components in species concepts: *grouping*, the criteria for placing organisms together in a taxon; and *ranking*, the criteria for determining the cutoff point at which the taxon is designated a taxon. The distinction recalls the twofold practical problems relating to the species taxon as stated by Mayr (1982): (1) the assignment of individual variants ('phena'; see Mayr, 1969) to the appropriate species taxon, and (2) the delimitation of taxa against each other, particularly the decision which populations of a single variable aggregate of populations in time and space to include in a single species. The second distinction has to do with pluralism, which means keeping an open mind about causal agents. In the case of species concepts, pluralism denies that a universal species concept exists, but it splits into two variants (Mishler and Brandon, 1987). In one, there are several possible species classifications for a given situation depending on the needs and interests of the different systematists – e.g. biogeographers, economic botanists, conservationists. This pluralism was advocated by Kitcher (1984). The other brand refers to the sought-for general-purpose classification that exists, but which requires some tolerance of different criteria for different situations (Gould, 1973; Gould and Lewontin, 1979). The Kitcher brand of pluralism has more than a whiff of arbitrariness and nominalism – it implies that species are what suit us and suit our theories, not what exist 'out there' to be discovered and recognized and characterized. The other variety is more or less the reverse, and is more persuasive if you believe that species exist.

Species as individuals

The problem of species concepts seems to be whether we can capture the category 'species' – all species, animal and plant, sexual and asexual, protistan, procaryotic – in a definition. If we cannot do that, say the philosophers, then statements about evolution are not anchored securely to a firm consensual answer to the question: *what, then, is it that evolves – organisms? populations? species? lineages?* Enter the controversy over whether species are individuals or classes.

There are two fundamentally different logical categories lurking under the name 'species'. One is the *taxon*, 'a concrete zoological or botanical object' (Mayr, 1982), and the other is the *category*. The species category is 'the class, the members of which are species taxa'. It is the taxon that concerns us here. Here is Mayr (1982) in a typically concise statement on the biological species:

'There are three aspects of the biological species that required the adoption of new concepts. The first is to envision species not as types but as populations (or groups of populations), that is, to shift from essentialism to population thinking. The second is to define species not in terms of degree of difference but by distinctness, that is, by the reproductive gap. And third, to define species not by intrinsic properties but by their relation to other co-existing species, a relation expressed both behaviorally (noninterbreeding) and ecologically (not fatally competing). When these three conceptual changes are adopted, it becomes obvious that the species concept is meaningful only in the nondimensional situation: multidimensional considerations are important in the delimitation of species taxa but not in the development of the conceptual yardstick. It also becomes evident that the concept is called biological not because it deals with biological taxa but because the definition is biological, being quite inapplicable to species of inanimate objects; and that one must not confuse matters relating to the species taxon with matters relating to the concept of the species category.'

In 1987 Mayr complained that evolutionary taxonomists from the time of the 'new systematics' – the 1930s – to 1970 accepted that species are real entities, and thus that they rejected the class concept of the species, whilst at the same time this was overlooked by 'the philosophers', who continued to confuse class with individual. It was not until Ghiselin (1974) forcefully promoted the view that species taxa are not classes but individuals that the philosophers took notice (Ghiselin, 1997). (To Hull (1988, 1999) Mayr was rather too sweeping in his assessment of progressive taxonomists and reactionary philosophers.) The ontology of species-as-individual provides a clearer notion of the species and its problems than we had before, even as the discussion of the species concept goes on.

Species are real. Species exist. They have beginnings, lifetimes (or histories) and endings. They are 'spatiotemporally localized, well-organized, cohesive at any one time, and continuous through time' (Hull, 1987). They are not abstractions, they are not merely erected to the taste of an experienced taxonomist, they are not classes. They really are 'out there', challenging us to find and delimit them and to understand them. Mishler and Brandon (1988) 'unpacked' the notion of the individuality of species into four subparts: (i) spatial boundaries, because all known evolutionary processes produce entities (not just species) thus restricted: (ii) temporal boundaries – a taxon arises but once and goes extinct forever; (iii) integration, which refers to active interaction among parts of an entity; and (iv) cohesion wherein an entity behaves as a whole with respect to some process. The first two subconcepts are the *outcomes* of processes – i.e. patterns or configurations – whilst the others are the *actions* of processes.

Smith (1994) skilfully balanced the biology of species, their usefulness and practicality especially in the fossil record, and taxa including higher taxa. Species concepts boil down to two kinds: (i) those that emphasize biological processes and lead to process-related definitions, and (ii) those that emphasize operational means of definition and lead to pattern-based definitions. As many have observed, species recognition and differentiation have always been predominantly on diagnosable traits, and how a species is recognized as a species should be kept separate from hypotheses of how it came to be – especially in the case of fossils. Since we have to use morphological criteria, our species concept should be pattern-based. Smith discussed three kinds of species concept using morphological evidence ('morphospecies'): the concepts respectively of the phenetic species, phylogenetic species and monophyletic species. The phylogenetic species always is one of two kinds: minimally monophyletic with at least one shared derived (apomorphic) character, and sister taxa lacking diagnostic apomorphic characters (plesiomorphic). Thus the three kinds of morphospecies stand in an ascending series, because phenetic techniques are a good place to begin identifying 'minimal morphologically diagnosable groups' (phena) which will approximate phylogenetic species, after which the problem becomes the separation of the 'basal monophyletic taxa', called *species*, from the rest, which are not monophyletic and needing more information or study at a higher resolution ('plesiomorphic grades', called *metaspecies*).

Smith argued that if species participate in evolutionary processes which require continuity, then they cannot be spatiotemporally bounded entities. If on the other hand they are such entities, then they must be taxa and the historical products of evolutionary processes, not the effectors of such processes. Fossil species must be taxa, discovered by the same comparative methods of clustering and differentiation as higher taxa, i.e. not different in kind from higher taxa, but 'minimal' units. Fossil species fall into two categories: those based on single populations from one locality or horizon 'which can safely be assumed to be interbreeding populations' (ignoring postmortem condensation, etc.); and those built by using observed similarity to group populations from different localities or horizons. The fundamental split is between those categories – between single populations probably representing 'biologically cohesive participants in evolutionary processes', and groups of spatiotemporally dispersed populations representing taxa, 'the historical by-products of evolutionary processes' subject to the concepts of paraphyly and monophyly just as higher taxa are.

We come to the second pair of books, which could hardly be further apart on the metaphysical – ontological – nature of the biospecies. Ghiselin (1997) constructed an entire book around the ontological thesis of species-as-individual (SAI), his 'radical solution to the species problem' (1974). The difference between class and individual is altogether fundamental and the following criteria were advanced for recognizing individuals: individuals, unlike classes, are concrete; also unlike classes, they have no defining properties; again, they have no instances. Individuals are spatiotemporally restricted with a beginning and an end; they are incorporated in other individuals whereas classes are included in other classes; they can participate in processes; and they are not referred to as such by the laws of nature; in all of these they stand in stark contrast to classes.

Ghiselin approved Mayr's sustained criticisms of essentialism and his assertion that Darwin invented a whole new way of thinking by replacing typological thinking with population thinking (especially Mayr, 1959); indeed, he (Ghiselin) would go further – Darwin actually reversed the traditional priority of classes and individuals. Not surprisingly, perhaps, the SAI thesis has resonated widely among biologists engaged in controversies of their own – cladistics and punctuated equilibria especially.

Not so with certain philosophers. Ruse (1986, 1998) attempted to use arguments of consilience – if species are classes or natural kinds, then the various kinds of species, taken as different lines of evidence, will coincide, producing a strong structure that is unavailable to SAI. Ghiselin (1997) rejected this on the grounds that the different definitions simply do not pick out the same units. Further, in a neat inversion of Ruse's argument, he pointed out that three quite different arguments for SAI were developed independently by three philosophers – consilience in action, indeed.

Hostile to SAI were the authors of the most comprehensive treatise on biophilosophy (Mahner and Bunge, 1997). They use for SAI the appellation *bionominalism*, thus: nominalism, one of the three philosophical schools to do with taxonomy, was a 'primitive' reaction to platonic idealism, or platonism, another of the schools (the third, conceptualism, a sort of compromise, is the 'sole viable methodology and philosophy of taxonomy'). Nominalism comes in two variants – traditional, in which organisms exist, species being conventional names adopted pragmatically (Buffon, Lamarck, Darwin); and contemporary (neonominalism), itself in two forms. In weak neonominalism species are neither names nor concepts but concrete individuals. In strong neonominalism all taxa are historical entities, the reality of organic evolution precluding the conception of taxa as classes or kinds like e.g. the chemical elements. This strand is restricted to biology, hence bionominalism. For Mahner and Bunge, taxa are natural kinds, kinds are classes, and classes are collections. Since taxa are natural kinds, and since species are taxa, species are natural kinds. Since natural kinds are conceptual objects not material ones, species are constructs (yet not arbitrary or idle ones). Because the biological processes of ontogeny and evolution make biospecies different, the authors concede that species are biological kinds. The notion of a species as a class is *logically prior* to three other notions: of a reproductive community as a concrete system composed of organisms; of the notion of species-as-lineage, likewise; and thirdly of the notion of the phylogenetic species. All the properties and behaviours attributed to and evidence for SAI are attributable to the organism, not the taxon. Most cuttingly: evolutionary innovations are the properties of organisms, and organisms are the speciating entities. Most dismissively: '... we thus maintain that bionominalism is one of the major misconceptions in the current philosophy of biology.' Most quotably: 'The notion of a class or species is an indispensable logico-semantic concept, and cannot therefore be ontologized or reified without committing a category mistake.'

Just as Socrates attempted to define friendship and failed, so too is it likely that the category including those real, existing, spatiotemporally bounded entities called 'species' together with numerous other concepts will continue to elude a single comprehensive definition. The title of a review by Hull (1997) is eloquent: 'The ideal species concept – and why we can't get it.'

Gould on hierarchy and individuals

Gould (2002) structured his huge monograph around the 'Darwinian tripod', the three legs of what he called conceptual Darwinism and each of which he extended by prolonged argument. The three legs of the tripod and their extensions are as follows. (i) *Agency* – Darwinian organismal selection versus a hierarchy of units of selection. (ii) *Efficacy* – two approaches, functional (Darwinian externalist selection as virtually the sole creative force in evolutionary change) versus structuralist (internal constraints, inherited and ontogenetic). (iii) *Scope* – Darwinian extrapolationism, microevolutionary styles of selection sufficient to account for all evolution versus tiers of overriding causation – punctuated equilibrium (PE) at species level undoes microevolutionary anagenesis whilst catastrophic mass extinction in turn derails punctuated equilibrium.

To sustain his argument Gould paid particular attention to the species-level theory of punctuated equilibrium and to the place of the species in a hierarchy of individuals and their selection. There are three primary claims in PE: (i) scale: PE is about species in geological time, not below and not above; (ii) evolutionary stasis over long periods of time, meaning millions not thousands of years; and (iii) origination in geological moments, meaning on a 1% criterion, say, 40 000 years for a species' duration of 4 myr (For arguments Contra PE, see Hoffman (1989; and Levinton (2001); Prothero (1992) is a good example.) The macroevolutionary implications are that PE 'secures the hierarchical expansion of selectionist theory to the level of species', and PE defines species as basic units or atoms of macroevolution – they are stable 'things' (*Darwinian individuals*) not arbitrary segments of anagenetic continua. This individuation of species includes species selection as a crucial process and speciation by branching not anagenesis as equally crucial. All of this establishes the basis for an independent theoretical domain of macroevolution. In contrast to some of the discussion about species as individuals (above) Gould approached organic evolution and its products as comprising an hierarchy of Darwinian individuals – '*hierarchy and the six-fold way*', as set out in Table 8.2.

Table 8.2 *Gould's (2002) argument of non-fractal and non-extratrapolational hierarchy in the individuated organic world, Darwinian individuals being produced by tiered evolutionary processes. Speciational and transformational evolution are both variational, i.e. 'Darwinian', but hierarchical (Mayr, 1992)*

Level	Darwinian individual		Non–fractal tiering
VI	Clade–individual	Third Tier	Catastrophic mass extinction derails punctuated equilibrium
V	Species–individual	Second Tier	Punctuated equilibrium undoes anagenesis; differential success within clades [Speciational evolution]
IV	Deme–individual	First Tier	Anagenesis within populations in ecological time [Transformational evolution; phyletic gradualism]
III	Organism–individual		
II	Cell–individual		
I	Gene–individual		

Mayr (1992, 2002) agreed with the importance of hierarchy in organic evolution. Darwinian evolution is variational evolution at the level of population – but a similar variational evolution occurs at the level of species.

> Transformational evolution of species (phyletic gradualism) is not nearly as important in evolution as the production of a rich diversity of species and the establishment of evolutionary advance by selection among these species. In other words, speciational evolution is Darwinian evolution at a higher hierarchical level. The importance of this insight can hardly be exaggerated. In the modern spirit of collaboration between palaeontology and genomics in macroevolution, Gregory (2004) has argued that genome evaluation can only be understood from a hierarchical approach, 'thereby providing an intriguing conceptual link between the most reductionistic [genomes] and the most expensive [fossils] subjects of evolutionary study'.

In another view of the biological hierarchy, Simons (2002) and Leem (2003) discuss the notion that the same fundamental process of selection operates at all temporal scales from the ecological to the geological.

Questions for biostratigraphy

Table 8.3 summarizes several biological questions as they might pertain to biostratigraphy, and as discussed in various parts of this study.

Table 8.3 *Categories of comparative biology with some important or controversial questions and their relevance to biostratigraphy*

Biohistorical categories	Questions	Relevance to biostratigraphy
Nature of species	Are species natural kinds or individuals?	That species are individuals and cannot be natural kinds is the basis for biostratigraphy's unique attributes for ordering bioevents, strata and geohistorical events.
Nature of index taxa	Are index taxa discrete branches of the phyletic bush, convenient slices of lineages, or merely convenient morphotypes?	'Stage of evolution' avoids the issue; pragmatic typology avoids the issue; biostratigraphy is robustly indifferent and employs old-style morphotypes.
Definition of basic units	Are species concepts pluralist or monist?	Microfossil taxa are overwhelmingly pluralist, by default.
Development of patterns	Is evolutionary transformation concentrated at speciation, or is it gradual into and out of speciation?	'Actual working biostratigraphers are all punctuationists.'
Analysis of pattern	Is cladistic analysis of phylogeny the one way to the truth and the light?	The suggested correlation use of sister taxa in correlation notwithstanding, the relevance of cladistics to real biostratigraphy is tenuous at best.
Analysis of process I	Are the ecological polarizations r- and K-selection usefully extrapolated into geological timescales?	The effect hypothesis may help explain why good index fossils are good index fossils; r- and K-selection are useful notions in chronofaunas.
Analysis of process II	Does species selection exist?	This notion is not obviously relevant to biostratigraphy.
Analysis of process III	Are hierarchy and tiering relevant to analysis of spatiotemporal distribution?	Tiering is highly relevant to biostratigraphy; hierarchy less so.

The change in worldview

The style shift in biogeohistory

Chapter 1 concluded with the state of affairs in biostratigraphy as they had reached during the times of the gestation of the *International Stratigraphic Guide* – and the ISSC Reports preceding it – what I called 'classical biostratigraphy' (McGowran, 1986a). Among the prominent features of the *Guide*, two in particular displayed its philosophy. First was the stern anatomizing of the stratigraphic record into the celebrated trinity of *lithostratigraphic*, *biostratigraphic* and *chronostratigraphic* units. Second was a strong sense of a conscious effort to avoid imposing on the readership the authors' perceptions of the nature of the stratigraphic record, of how to read the chronicles and history entombed in it (McGowran, 1986a). We have referred often enough here to the 'punctuated' and 'gradualist' strands threading down through the decades in the development of our discipline, but it is relevant to capture the state of stratigraphy in the 1940s, 1950s, and into the 1960s, and to compare it to today's, for the changes that have occurred in the interim bear heavily upon our perceptions of biostratigraphy. 'Gradualism' may be defined as referring to geological processes – and then biological, as well – at slow rates and in small increments; 'catastrophism' at fast rates or instantaneously, and in large increments (Simpson, 1970). There has been long-ripening confusion in the tendency to associate gradualism with naturalism and catastrophism with preternaturalism – associations exploited tendentiously by the advocate-generalist Lyell in pressing the actualistic case for gradualism (e.g. Gould, 1987; in more measured terms: Rudwick, 1998).

The philosophy of the editor of the *Guide*, H. D. Hedberg, certainly was gradualist during the formative years (see especially Hedberg, 1948, 1961). It was characterized too by an open and opportunist view of all the available and potential evidence for stratigraphic correlation, classification and age determination. Thus fossils are but one of the sources of such evidence and this, more than any other sustained assertion, provoked disagreement among the various schools or traditions on the other side of the Atlantic. For it was fossils and fossils alone in their powers for correlation and age determination, that built the timescale in the early nineteenth century. It is striking, rereading Hedberg's elegantly persuasive prose, how persistently he downplayed the importance of biostratigraphy as being just one of numerous tools of correlation. By the 1940s, already:

> In the case of these smaller units (stages) we are approaching a limit to the degree of resolution possible in the use of evolutionary sequence of fossils, and at this scale the percentage error in the carrying of time-stratigraphic divisions by means of fossils has increased so much that this method is not infrequently exceeded in value by evidence from other stratigraphic features – lithology, mineralogy, electrical character, chemical composition, relation to diastrophism, relation to climate, etc. Since the application of time-stratigraphic concepts to the sediments of brief time spans is nevertheless extremely useful and

important to us, we are justified in employing any or all available lines of evidence for dividing and extending such minor time-stratigraphic units and we should not restrict criteria to any one group of fossils or any one facies of fossils, nor yet to fossils in general.

And we cannot, as some would insist (wrote Hedberg), recognize stage boundaries by fossils alone and by natural breaks in the evolutionary stratigraphic succession of fossils – for such breaks most likely will be at facies changes and unconformities; and it would be 'very amazing indeed' if the various main groups – vertebrates, plants, molluscs, foraminifera – had the same natural breaks [my Figures 1.3 and 1.4 would be a beautiful picture]. Although the thread of evolutionary change runs through the fossil record, making it particularly useful for the major geochronological divisions, 'the smaller the units which are under consideration the more predominant become facies changes and the less discernible the evolutionary changes'.

But this downplaying of the significance of fossils was part of a broader strategy of refuting all natural breaks, turning points, disjunctions or caesuras in the biogeohistorical record as acceptable worldwide markers: in that way did Hedberg establish the need for an opportunistic chronostratigraphy beholden to no class of potential tools of correlation. Consider changes in sea level (Hedberg, 1961): 'It seems evident that local vertical movements of the solid crust both on the continents and in the ocean basins have been so great and so variable geographically in relation to time as to leave much less order in the world-wide rock record of marine transgressions and regressions than some theorists might hope to see. Moreover, there is no reason to expect that the sediments of one transgression will have differed distinguishably from those of another. Gignoux (1936, p. 494–495) has brought out in excellent manner the caution with which one must look at even so widely accepted a transgression as that of the late Cretaceous.'

Not in organic evolution, nor in the notion of synchronous and episodic global diastrophism, usually associated with the name of Hans Stille, nor in sea-level or climatic change, has it 'yet been demonstrated that world-wide "natural breaks" in the character and continuity of our strata exist at the scale of the presently accepted geologic systems, nor has it been demonstrated that the evidence at the boundaries of the present systems is such as to allow them to be considered as the "natural" world-wide division points of the chronostratigraphic scale'. Our systems are only arbitrary chronostratigraphic units in a continuum – the Lyellian view of the 1830s. The achievement of international agreement on stratotypes is the critically important aim, not the discovery of a natural classification.

Much of Hedberg's sustained exposition of these principles was aimed at the very different tradition expressed by the Stratigraphic Commission of the USSR. That body began a declaration in 1960 (as in Hedberg, 1961) with the basic aim of finding the natural breaks in biohistory and in geohistory; that aim is holistic and the three-part division is unacceptable. The record in the rocks does indeed demonstrate periodicity in diastrophism and eustasy – in the whole gamut of recorded phenomena, in fact. To Hedberg, the Russian course was deductive – in contrast to his own inductive course – and had retained some of the influence of the old catastrophism. Whether this view was correct or not was not the point; it was a 'beautiful picture' of stratigraphy which may indeed, one day, prove to be correct, but in the meantime the classification of strata should be built up objectively and with no preconceptions as to the nature of the biogeohistory which they record. The geological timescale was 'natural' only to the extent that it was developed in western Europe and exported from there, and would have been quite different if the reverse had happened. Above all, urged Hedberg, 'let us beware of forcing nature into pigeonholes that she herself never knew to exist'.

The proposition that there were no worldwide breaks to be found in the record of diastrophism was entirely in accord with the orthodoxy of the 1950s and 1960s. Hedberg cited approvingly the most influential advocacy of the time, that of Gilluly (1949) on the distribution of mountain building in geologic time, and others including Arkell (1956): 'So far as our knowledge goes at present, it does not point to any master plan of universal, periodic, or synchronized orogenic and epeirogenic movements. The events were episodic, sporadic, not periodic. There was no "pulse of the earth".' As noted in Chapter 4, the prevailing orthodoxy in palaeontological/evolutionary theory was what later became dubbed as gradualistic – the macromutational theories of genetics and the saltational theories of palaeontology from earlier in the century were deemed to have failed, along with the coeval theories of diastrophism. All in all, it was a uniformitarian world, a world in which geological and biological rates of change varied widely in space and in time, to be sure, but also a world with no place for the sudden jumps associated with the term 'catastrophism'.

The search for a pulse in the record of organic evolution, for a pulse in the crust of the Earth, and for a correlation between the two which would suggest causation, was discussed extensively by Simpson (1944, 1953, 1970) who, in these matters at any rate, was philosophically close to Hedberg and Gilluly. Just as Gilluly, especially, found to be largely spurious the 'neocatastrophic' notion of tectonic episodes, identified first in Europe and followed across the planet as geologically brief 'revolutions', so did Simpson find similarly among the parallel theories of palaeontology. It is reasonable that geological events will have affected biohistory because, after all, they provide the setting. A notion of synchrony between major geohistorical and biohistorical episodes is as old as the discipline itself, and theories of a causal relationship continued to be supported by most geologists until relatively recently and still had advocates, according to Simpson in 1970. Simpson's main target was the theory of Schindewolf in which both extinction and origination were sudden, essentially synchronous over short periods of time, and due to some factor acting intermittently and distinct from the causes of less pronounced extinctions and originations (a forerunner of the distinction between mass extinction and background extinction). Sudden extinctions could be due to supernovae; originations to macromutations, the degree of whose impact determined the level of the taxon produced. Simpson had little difficulty demonstrating that the actual fossil record did not show such abrupt changes – for example, the initial appearances in the fossil record of major groups of mammals, originating ostensibly at the Cretaceous–Palaeogene boundary, in fact were strewn through no less than fifteen million years. He took some pains to spell out that there were indeed times of particularly marked changes in biotas that were very widespread, and that they were usually composed of high extinction rates followed by high rates of origination. But that is a 'modified, relatively mild and gradualistic form of revolutionism' which is consistent with our present knowledge of biohistory where neocatastrophism patently is not comparably consistent; and the association of biohistorical with geohistorical revolutions is, at best, unproven.

Thus, by the 1960s there was a powerful consensus among opinion leaders in stratigraphy, tectonics, palaeontology and evolutionary biology that the record in the rocks is episodic preservationally but gradualistic in its message and can be accommodated in Lyellian uniformitarianism. We perceive things differently, now: 'Since then, the pendulum has moved. Ocean-floor spreading and continental geodynamics are episodic (Trümpy, 1973; Schwan, 1980). Global climate changes from one state to the next by rapid shifts (Berger, 1982). Similarly, the stratigraphic record of broad changes in global sea level (most recently, from the Cretaceous high to the Neogene low) is punctuated by numerous, globally synchronous depositional sequences that reflect rapid eustatic sea-level changes – now recognized in one example as averaging only about 2 myrs in duration as far back as the Carboniferous (Ross and Ross, 1985). The Phanerozoic fossil record is punctuated by mass extinction. Finally, there are alternative models of evolutionary change at geological timescales: species changing gradually and usually slowly, or species usually not changing, but, instead, either splitting, or going extinct, or surviving in stasis (Eldredge and Gould, 1972)' (McGowran, 1986a). Punctuational change was vindicated (Gould, 1984).

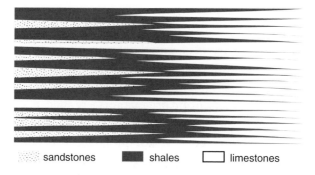

Figure 8.7 The facies sketch by Jukes (1861, Fig. 20) – clearly antecedent to the sand/shale ratios and other diachronous patterns in the mid twentieth century.

The question then is this: Hedberg went to great pains to establish a certain world view as a framework within which to advocate an agnostic but tripartite classification of the stratigraphic record. If our prevailing worldview has changed, as I think the majority would agree that it has, then we should ask whether our approach to stratigraphic classification should change accordingly. Can changing perceptions of Earth history be separated easily or clearly from changing perceptions of the correlations and calibrations in which that history must be grounded?

What happened to diachrony?

The facies concept is based in the fact that adjacent sedimentary environments will accumulate coeval but distinct suites of strata and fossils. As conditions change, the overall pattern of environments will shift in response. Ergo, the lithostratigraphic boundaries will be time-transgressive, i.e. diachronous (Fig. 8.7), and generations of students have learned that 'layer-cake' stratigraphy with time-parallel formation boundaries is invalid. Such a pattern is mostly unquestioned and, indeed, is a departure-point for seismic stratigraphy. How far though can we insinuate this inoffensive notion into the grand patterns of biogeohistory? A classically diachronous pattern is found in the Permian–Pennsylvanian cyclothems, in which the facies at a given location shift from non-marine, including coals, to marine, including offshore lime muds, and back again as the sea is inferred to advance and retreat (e.g. Moore, 1949). In lignitic phases in certain Cenozoic regimes, on the other hand, there is a close chronological relationship between the accumulation of coals and marine transgressions (Steininger *et al.*, 1989; Holdgate and Gallagher, 1997) which is counterintuitive, if the cyclothemic model is the style. Large-scale diachrony in the pelagic realm enjoyed a brief popularity when oceanic drilling confirmed the theory of seafloor spreading, whereby the interface between oceanic crust and overlying sediments increased in age from spreading ridge to abyssal plain. It followed that a given point on the seafloor would pass from a region of calcite accumulation to a region of clay accumulation as the site passed through the calcite compensation surface – hence, the contact between calcareous ooze and brown clay must be diachronous. With embellishments, as the site migrated through (under) the equatorial high-productivity belt with its depressed CCS and silceous belt, ocean-scale diachrony became an attractively inverted illustration of Walther's law: whereas environments sweep back and forth over a 'passive' site of accumulation in, classically, a delta, in the oceanic case an 'active' site migrates through the different sedimentary environments (Heezen *et al.*, 1973; Heezen and MacGregor, 1973; Hesse *et al.*, 1974). The notion reached the textbooks (Press and Siever, 1978) but did not prosper because it become clear, as the DSDP progressed and the discipline of palaeoceanography became established, that seafloor spreading in cm/year was a very much slower phenomenon than was the shift in watermasses as part of global climatic change. In a more modest example of diachronous stratigraphic response to crustal spreading, Pimm *et al.* (1974) showed that an upward-deepening, terrestrial-neritic-oceanic succession of facies was generated on the Ninetyeast Ridge with facies boundaries younging southwards as the plate and ridge migrated northwards (Fig. 8.8).

The Cenozoic climatic pattern of overall global cooling and overall retreat of isotherms towards the equator has stimulated suggestions of large-scale diachronism in biotas. One example is the retreat southwards by north-polar terrestrial floras (Hickey *et al.*, 1983); another is the northward–shift of an Antarctic neritic community since the Eocene (Zinnsmeister and Feldmann, 1984). It is reasonable to say of all of these examples of large-scale and long-term diachrony that they have not been particularly heuristic. Diachrony in all things is at root the null hypothesis of gradualism in deep time.

Diachrony has been a central theme in our perceptual shift from lithostratigraphy – gradualistic, preoccupied with diachrony between facies and formations and fearful of the dreaded 'layer-cake stratigraphy' – to sequence stratigraphy – punctuated, respectful of unconformities as being highly informative and not merely destructive of the stratigraphic record. Consider the contrasting renderings of the mid-Cenozoic record in southern Australia in Figure 8.9. Unenamoured of traditional biozoning in the southern mid latitudes and not motivated to overhaul the regional chronostratigraphic stages, I came to perceive a natural division of the record based on rapid regional transgressions (McGowran, 1989b). This view of the record led naturally enough to seeing a strong packaging of strata and their biofacies, as mentioned in previous

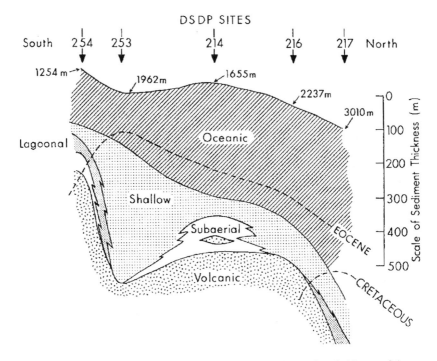

Figure 8.8 Stratigraphic diachrony generated during early subsidence of the Ninetyeast Ridge (Pimm *et al.*, 1974; Luyendyk, 1977, with permission). Pimm *et al.* emphasized that (a) the Ninetyeast Ridge subsided from subaerial to bathyal depths as part of a north-spreading plate, therefore (b) upward-deepening facies are older to the north and lithofacies boundaries are strongly diachronous. This early-1970s, pre-synthem-awareness synthesis still has some plausibility, but can be deconstructed similarly to neritic and terrestrial lithostratigraphy (Fig. 8.9).

chapters. Like numerous such reconstructions in stratigraphy, the stratigraphic diagrams for the Port Campbell, Gambier and Murray Basins display lithofacies and formation – even member – boundaries persisting for millions and in one case tens of millions of years. Lithofacies are bounded by *shazam lines* (Brett, 2001) – those oblique zigzags between named lithostratigraphic units 'that are more often a line of ignorance than a bit of information'. Such diagrams are very different from allostratigraphic packages defined by unconformities and within which facies patterns are plausible. Contrast this prolonged diachrony with the coal–limestone package – cycle or synthem or sequence: choose your terminology – surely a more heuristic and challenging stratigraphic model? The coal–limestone package focuses attention on unconformities as powerful unifiers of marine–non-marine strata. Likewise, the Gambier Limestone is shown to be packaged in a way that is consistent with the notion of global

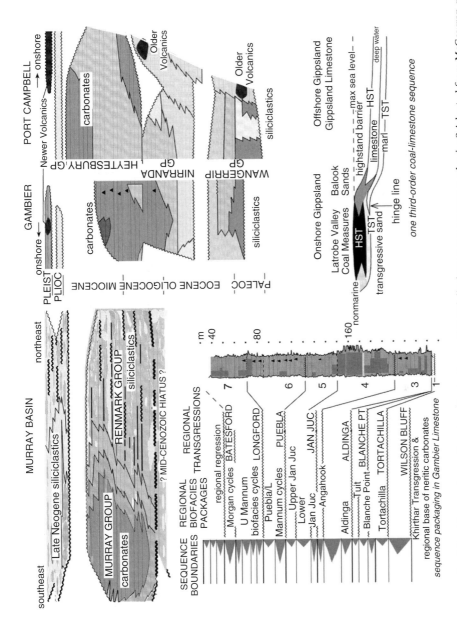

Figure 8.9 The Oligo-Miocene in southeastern Australia – prolonged diachrony or sequence-packaging? Adapted from McGowran *et al.* (2004), with permission. (Upper left) Lithostratigraphy of the Murray Group carbonates and Renmark Group siliciclastics (thick lines are lignites), as visualized by Brown and Stephenson (1991). Although the vertical scale is time (shared to left and right), intragroup unconformities and

Figure 8.9 (*cont.*)

the '?mid-Tertiary hiatus' display no hiatuses – a measure of the very broad time constraints in those facies and the gradualist assumptions of the compilers. Likewise, the Renmark Group includes the siliciclastics below the the '?mid-Tertiary hiatus.' Palaeocene–Eocene siliciclastics omitted. (Upper right) Lithostratigraphy of the Gambier and Port Campbell Basin. Data from various sources (Li *et al.*, 2000). The clutter of lithostratigraphic nomenclature has been omitted. Note the exuberant use of the 'Shazam!' symbol to indicate inferred, prolonged diachrony and interfingering. (Lower left) The Gambier Limestone depicted instead as an integral part of the pattern of rapid regional transgressions and regional biofacies packages (both sets given names after exemplifying rock units). Major lithologies of the Gambier Limestone in a borehole; rock symbols are limestone, dolomite and chert. The stratal units 1 to 7 represent allostratigraphic sediment packages at the third order separated by surfaces corresponding to glacio-eustatic events; sequences from Hardenbol *et al.* (1998). Third-order sequences and stratigraphic units were based on erosional surfaces, abundance changes in planktonic and deepwater benthic species, and clustered assemblages (details of age and biofacies in Li *et al.*, 2000). (Lower right) The third-order coal–limestone cycle or synthem (Gallagher and Holdgate, 1996). This is a good model for the brown coal deposits in southern Australia (Eocene–Miocene); the bar/hinge (the Balook Sands here) is prominent in all major accumulations. That this is a useful model has been corroborated by correlations and comparisons between coals and limestones and between both a curve of global ice volume, and coals vis-à-vis large benthic foraminifera in identified sequences (not shown). The point of this compilation is to contrast the exuberant perception of diachrony in stratigraphic reconstruction with the (perhaps comparably enthusiastic) perceptions of allostratigraphic pattern. The contrast indicates a replacement of the gradualistic by punctuational *Zeitgeist* during the rise of punctuated evolution and sequence stratigraphy in parallel in biohistory and geohistory, respectively.

sequences – a matching that could never arise from aprioristic diachrony in lithofacies and sedimentology.

Return of the layer cake?

Brett (2001) has outlined this shift in perception. At the mid-century high point in facies and diachrony, 'all laterally traceable nonvolcanic epeiric marine sedimentary rock units must be presumed to be diachronous' (Shaw, 1964). At the same time, there was little enthusiasm for allostratigraphic units below the level of the Sloss supersequences. Now, and not least due to the efforts of Brett and his colleagues, we emphasize lateral persistence in sedimentary surfaces, unique beds, and fossil horizons, often through changes in facies. The sequence boundaries and condensed sections are all part of this resurgence in awareness and re-emphasis of patterns – patterns that have been known for as long as facies themselves.

In a major product of this insight, Walther's Law of facies – their lateral and vertical equivalence where not disrupted by unconformity – is now constrained in its applicability to within third-order sequences but not between them (Holland, 1999). The diachrony sketched in Fig. 1.2 is constrained within one sequence.

Holland contrasted the 'new stratigraphy' more than favourably with 'traditional stratigraphy'. In this view traditional stratigraphy, focused on the Hedberg Triad of rocks, fossils and time (lithostratigraphy, biostratigraphy, and chronostratigraphy), progressively disenfranchised stratigraphy from its birthright, which is nothing less than the study of Earth history. Disenfranchisement happened in two ways: by an overemphasis on a coarse, arbitrary and non-genetic codification of the data that divorced the data from Earth history; and by emphasizing the distinctness of rock, fossil and time units, thereby obscuring their interrelationships and ceding the important scientific questions to sedimentology (which focuses at the small scale, so that the big questions went unasked altogether). Holland's point in terms of Harland's (1975) phenomenon category and framework category (Chapter 7) is that traditional stratigraphy got mired in the codes and rules of the framework category to the detriment of the real science in the phenomenon category. The New Stratigraphy mostly comprises sequence stratigraphy, in which the rock record is divided into genetically related packages bounded by unconformities. It asks the questions of eustasy and tectonics that were stifled (with exceptions) by the traditional bookkeeping of stratigraphy and its nomenclature. Among other virtues, the new stratigraphy provides a physical framework for interpreting the fossil record palaeobiologically. Holland concluded by urging that palaeobiology add an alliance with the new stratigraphy to its fruitful collaborations with such disciplines as geochemistry, geochronology and ecology.

Finale: the three central concerns of biostratigraphy

The environmental roots of modern biostratigraphy were fourfold. From the realm of the deep oceans (including their remnants in mountain belts) came the best fossil record of the late Phanerozoic microplankton, their phylogenetic reconstructions and phylozones, their calibration and cross-correlation within the IMBS, and their chemostratigraphic profiles. From the epicontinental seas of the Palaeozoic and Mesozoic Eras came various notions in ecostratigraphy (e.g. cohering palaeocommunities and coordinated stasis) and high-resolution whole-faunal analysis and event stratigraphy. From the Cenozoic ocean-marginal sedimentary wedges came the strongest impulses for sequence biostratigraphy. From terrestrial vertebrate assemblages came palaeobiological communities and chronofaunas, remaining more closely in touch with biological theories of organic evolution and ecology than did the other strands. There is now more mixing and communication than ever before among the workers in the three environmental realms and among palaeontologists, neontologists and various kinds of phyical-chemical stratigraphers; and that is truly an excellent thing.

There are three quite clear, though interlocking, interdisciplinary concerns central to the immediate future of our subject. They are:

i.   Marking off the passage of geological time by the irreversible events produced by organic evolution.

ii. Sequence biostratigraphy: reconciling classical biostratigraphy with biofacies.

iii. The palaeobiology of biostratigraphy: natural units in ecostratigraphy and macroevolution.

The first of these concerns is what I have called 'classical biostratigraphy'. Although the resolution of formally defined, planktonic biozones has not increased very much in recent years, their accuracy is still increasing in the wider context of the GPTS and IMBS and cyclostratigraphic calibration. We still lack a rigorous ordination of bioevents between the major skeletonized microplanktonic groups – calcareous, siliceous, and organic-walled, and phyto- and zoo-planktonic – and we lack too a goodly number of rigorous correlations of bioevents with the GPTS. This comment in no way qualifies our appreciation of recent advances, for it has been made already from time to time (e.g. Berggren and Miller, 1988; Berggren et al., 1995a). The integrated Cenozoic timescale will continue to strengthen as more oceanic sections are drilled and recovered intact, as first-order cross-correlation of bioevents among themselves and with geomagnetics is improved, as calibration improves by argon–argon dating of key events, and as fine-tuning progresses especially cyclostratigraphically (astrochronologically) but also by techniques such as graphic correlation.

It is sometimes said (verbally; I have nothing to cite, but I have heard it for decades) that the ongoing construction of the Cenozoic timescale, immensely important though it is, is not science. Thus grants have been available for the research into all facets of the work that feed into the integrated scale, but not for its actual construction. On the other side, the palaeobiology of the fossil record attracts support, but the correlation, age determination and chronological resolution upon which it depends utterly seems to be less attractive. This is the kind of nonsense that slips by when we forget that our science balances process with history – that, alone, one is sterile and the other is empty. The timescale is improved by the testing and correction or corroboration of ordination and correlation which continue between the technological advances. This alone makes the endeavour necessary, but more important are the demands put upon biostratigraphy and geochronology by the asking of biogeohistorical questions – for that history has some chance of being right if the correlations and age determinations are right, but it has no chance at all if they are wrong. Let there be no misunderstanding that this endeavour is little more than some sort of bookkeeping; every advance opens new questions and refocuses old ones in every corner of stratigraphy.

Because the IMBS reciprocally illuminates the science and technology of the oceans, there is a standing challenge to exploit its strength in the numerous Cenozoic problems of the other exogenic realms – terrestrial and neritic – most obviously in correlations in depositional sequences. This is the second major concern of biostratigraphy and the one that most clearly demands an integration of irreversible bioevents, speciations and extinctions, with the various kinds of reversible event that have been lumped together as biofacies-in-stratigraphy, or ecostratigraphy. One immediate challenge is to identify and date the key surfaces in regional depositional sequences, the sequence boundaries and flooding surfaces. From this we can test the timing of the putatively global sea-level curve, but we can, for example, tease out the chronological parallels between the extratropical carbonate realms of southern Australia and New Zealand in the south and Paratethys in the north, or search for generalizations on biofacies signals of other global chemofacies, such as upwelling and carbon burial in the Monterey event of the Early Miocene. The challenge extends into the non-marine realm and the tying of terrestrial biostratigraphic systems (vertebrates, palynomorphs) into the integrated web of geochronology, and it extends into the most difficult geohistorical problem of all, the history of weathering and the development of the regolith. The challenge extends further ito find and exploit the sections where cyclostratigraphy can cross the environmental realms at the levels of accuracy and precision that are fast developing in the oceanic realm.

Invertebrate palaeontology (and the later-christened micropalaeontology) diverged sharply from vertebrate palaeontology after the first Darwinian revolution, only research programmes on the vertebrates remaining firmly within the realm of evolution (Bretsky, 1979). Although this situation has changed during the recent resurgence in the field of macroevolution and the fossil record (Gould, 2002), the palaeobiological potential of the Cenozoic biostratigraphic record is immense. The tight integration of timescales and correlations, palaeoceanographic and palaeoclimatic scenarios, depositional sequences, and microfossil record has two broad aspects: systematics and taxonomy and phylogeny on the one hand and the ecostratigraphic extraction of communities and community cohesion from assemblages on the other. This is our third challenge.

For Fortey (1993) in resolutely inductivist mode, biostratigraphy had its real conceptual breakthrough in Lapworth's analysis, in the 1870s, of Early Palaeozoic complexities; all since has been fine tuning of what is already known. This stability and progressive refinement is a virtue in welcome contrast to Popperian hypothetico-deductive science. Fortey's is a highly constrained view of biostratigraphy which may sit well in the Early Phanerozoic, but not in its later reaches. To Simmons (1998) biostratigraphy seems to have 'turned the corner in its fortunes by becoming pragmatic and integrated'. For Holland (1999) the conceptual revolution in stratigraphy opens vistas of opportunity for palaeobiology, with biostratigraphy clearly at front-centre. Biostratigraphy lives!

References

Abele, C., Gloe, C. S., Hocking, J. B. *et al.* (1976). Tertiary. In *Geology of Victoria*, ed. J. G. Douglas & J. G. Ferguson, pp. 177–274. Melbourne: Geological Society of Australia Special Publication **5**.

Abreu, V. S. & Haddad, G. A. (1998). Glacioeustatic fluctuations: the mechanism linking stable isotope events and sequence stratigraphy from the Early Oligocene to middle Miocene. In *Mesozoic and Cenozoic Sequence Stratigraphy of European Basins*, ed. P.-C. de Graciansky, J. Hardenbol, T. Jacquin & P. R. Vail, Society of Sedimentary Geology Special Publication No. **60**, 245–60.

Abreu, V. S., Hardenbol, J., Haddad, G. A. *et al.* (1998). Oxygen isotope synthesis: a Cretaceous ice-house? In *Mesozoic and Cenozoic Sequence Stratigraphy of European Basins*, ed. P.-C. de Graciansky, J. Hardenbol, T. Jacquin & P. R. Vail. Society of Sedimentary Geology, Special Publication No. **60**, 75–80.

Adams, C. G. (1965). The foraminifera and stratigraphy of the Melinau Limestone, Sarawak, and its importance in Tertiary correlation. *Quarterly Journal of the Geological Society of London*, **121**, 283–338.

(1967). Tertiary foraminifera in the Tethys, American and Indo-Pacific provinces. *Systematics Association Special Publication*, **7**, 195–217.

(1970). A reconsideration of the East Indies Letter Classification of the Tertiary. *Bulletin British Museum (Natural History), London*, **19**, 87–137.

(1983). Speciation, phylogenesis, tectonism, climate and eustasy: factors in the evolution of Cenozoic larger foraminiferal bioprovinces. In *Evolution Time and Space: The Emergence of the Biosphere*, edited R. W. Sims, J. H. Price & P. E. S. Whalley. London: Academic Press, Systematics Association Special Volume **23**, 255–289.

(1984). Neogene larger foraminifera, evolutionary and geological events in the context of datum planes. In *Pacific Neogene Datum Planes*, ed. N. Ikebe & R. Tsuchi, pp. 47–67. Tokyo: University of Tokyo Press.

(1992). Larger foraminifera and the dating of Neogene events. In *Pacific Neogene: Environment, Evolution, and Events*, ed. R. Tsuchi & J. C Ingle, pp. 221–235. Tokyo: University of Tokyo Press.

Adams, C. G., Butterlin, J. & Samanta, B. K. (1986). Larger foraminifera and events at the Eocene–Oligocene boundary in the Indo-West Pacific region. In *Terminal Eocene Events*, ed. Ch. Pomerol & I. Premoli Silva, pp. 237–252. New York: Elsevier Science.

Adams, C. G., Lee, D E. & Rosen, B. R. (1990). Conflicting isotopic and biotic evidence for tropical sea-surface temperatures during the Tertiary. *Palaeogeography, Palaeoclimatology, Palaeoecology*, **77**, 289–313.

Ager, D. V. (1984). The stratigraphical code and what it implies. In *Catastrophes and Earth History*, ed. W. A. Berggren & J. A. van Couvering, pp. 91–100. Princeton: Princeton University Press.

Agterberg, F. P. & Gradstein, F. M. (1999). The RASC method for ranking and scaling of biostratigraphic events. *Earth-Science Reviews*, **46**, 1–25.

Allan, R. S. (1933). On the System and Stage names applied to subdivisions of the Tertiary Strata of New Zealand. *Transactions New Zealand Institute*. **63**, 81–108.

(1966). The unity of stratigraphy. *New Zealand Journal of Geology & Geophysics*, **9**, 491–494.

Allmon, W. D. & Bottjer D. J. (2001). *Evolutionary Paleoecology: The Ecological Context of Macroevolutionary Change*. New York: Columbia University Press.

Alroy, J. (1992). Conjunction among taxonomic distributions and the Miocene mammalian biochronology of the Great Plains. *Paleobiology*, **18**, 326–343.

(1994). Appearance event ordination: a new biochronological method. *Paleobiology*, **20**, 191–207.

(1998a). Diachrony of mammalian appearance events: implications for biochronology. *Geology*, **26**, 23–27.

(1998b). Equilibrial diversity dynamics in North American mammals. In *Biodiversity Dynamics: Turnover of Populations, Taxa and Communities*, ed. M. L. McKinney & J. Drake, Columbia UP, pp. 232–287.

(2000). New methods for quantifying macroevolutionary patterns and processes. *Paleobiology*, **26**, 707–733.

Alroy, J., Koch, P. L. & Zachos, J. C. (2000). In *Deep Time: Paleobiology's Perspective*, ed. D. H. Erwin & S. L. Wing, *Palaeobiology*, Supplement to vol. **26**(4), 259–288.

Alvarez, W., Arthur, M. A., Fischer, A. G. *et al.* (1977). Upper Cretaceous–Paleocene magnetic stratigraphy at Gubbio, Italy. V. Type section for the Late Cretaceous–Paleocene geomagnetic reversal scale. *Geological Society of America, Bulletin*, **88**, 383–389.

Alvarez, L. W., Alvarez, W., Asaro, F. & Michel, H. V. (1980). Extraterrestrial cause for the Cretaceous–Tertiary extinction. *Science*, **208**, 1095–1118.

Andreasen, D. J. & Ravelo, A. C. (1997). Tropical Pacific Ocean thermocline depth reconstruction for the last glacial maximum. *Paleoceanography*, **12**, 395–413.

Applin, E. R., Ellisor, A. E. & Kniker, H. T. (1925). Subsurface stratigraphy of the coastal plain of Texas and Louisiana. *American Association of Petroleum Geologists, Bulletin*, **9**, 79–122.

Arkell, W. J. (1933). *The Jurassic System in Great Britain*. Oxford: Clarendon Press.

(1956). Comments on stratigraphic procedure and terminology. *American Journal of Science*, **254**, 457–467.

Armentrout, J. M. (1996). High resolution sequence biostratigraphy: examples from the Gulf of Mexico Plio-Pleistocene. In *High Resolution Sequence Stratigraphy: Innovations and Applications*, ed. J. A. Howell & J. F. Aitken, pp. 65–86. The Geological Society Special Publication No. 104.

Arnold, A. J. (1983). Phyletic evalution in the *Globorotalia crassaformis* (Galloway and Wissler) lineage: a preliminary report. *Palaeobiology*, **9**, 390–8.

Arnold, A. J. & Parker, W. C. (1999). Biogeography of planktonic foraminifera. In *Modern Foraminifera*, ed. B. K. Sen Gupta, pp. 103–122. Dordrecht: Kluwer Academic Publishers.

Aubry, M.-P. (1995). From chronology to stratigraphy: interpreting the lower and middle Eocene stratigraphic record in the Atlantic Ocean. In *Geochronology Time Scales and Global Stratigraphic Correlation*, ed. W. A. Berggren, D. V. Kent, M.-P. Aubry & J. Hardenbol. Tulsa, SEPM Special Publication **54**, 213–274.

   (2000). Where should the Global Stratotype Section and Point (GSSP) for the Paleocene–Eocene boundary be located? Comment définir la Limite Paléocène–Eocène? *Bulletin de la Societé géologique de France*, **171**.

Aubry, M.-P. & Berggren, W. A. (2000). The homeless GSSP: the dilemma of the Paleocene/Eocene boundary. *Tertiary Research*, **20**, 107–112.

Aubry, M.-P., Berggren, W. A., Kent, D. V. *et al.* (1988). Paleogene geochronology: an integrated approach. *Paleoceanography*, **3**, 707–742.

Aubry, M.-P., Berggren, W. A., Stott, L. & Sinha, A. (1996). The upper Paleocene-lower Eocene stratigraphic record and the Paleocene/Eocene boundary carbon isotope excursions. In *Correlation of the Early Paleogene in northwest Europe*, ed. R. W. O'B Knox, R. M. Corfield & R. E. Dunay, p. 353–380. London: Geological Society, Special Publication No. **101**.

Aubry, M.-P., Berggren, W. A., Van Couvering, J. A. & Steininger, F. (1999). Problems in chronostratigraphy: stages, series, unit and boundary stratotypes, global stratotype section and point and tarnished golden spikes. *Earth-Science Reviews*, **46**, 99–148.

Aubry, M.-P., Cramer, B. S., Miller, K. G., Wright, J. D., Kent, D. V. & Olsson, R. K. (2000). Late Paleocene event chronology: unconformities, not diachrony. *Bulletin de la Societé géologique de France*, **171**, 367–378.

Aubry, M.-P., Lucas, S. G. & Berggren, W. A., eds (1998). *Late Paleocene–Early Eocene Climatic and Biotic Events in the Marine and Terrestrial Records*. New York: Columbia University Press, 513 pp.

Australasian Petroleum Company (1961). The geological results of petroleum exploration in western Papua. *Journal of the Geological Society of Australia*, **8**, 1–133.

Ayala, F. (1983). Microevolution and macroevolution. In *Evolution from Molecules to Men*, ed. D. S. Bendall, pp. 387–402. Cambridge: Cambridge University Press.

Backman, J. & and Raffi, I. (1997). Calibration of Miocene nannofossil events to orbitally tuned cyclostratigraphies from Ceara Rise. *Proceedings of the Ocean Drilling Program, Scientific Results*, **154**, 83–99. College Station, Texas.

Bakker, R. (1986). *The Dinosaur Heresies*. London: Longman Scientific & Technical.

Baldi, T. (1980). The early history of the Paratethys. *Földtani Közlöny*, **110**, 1–18.

Bambach, R. K. & Bennington, J. B. (1996). Statistical testing for paleocommunity recurrence: are similar fossil assemblages ever the same? In New perspectives of faunal stability in the fossil record, ed. L. C., Ivany & K. M. Schopf, *Palaeogeography, Palaeoclimatology, Palaeoecology*, **127**, 103–133.

Bandy, O. L. (1972). Origin and development of *Globorotalia (Turborotalia) pachyderma* (Ehrenberg). *Micropaleontology*, **18**, 294–318.

Bandy, O. L., Frerichs, W. E. & Vincent, E. (1967). Origin, development and geologic significance of *Neogloboquadrina* Bandy, Frerichs and Vincent, gen. nov. *Contributions Cushman Foundation Foraminiferal Research*, **18**, 152–157.

Banner, F. T. & Blow, W. H. (1959). The classification and stratigraphical distribution of the Globigerinaceae. *Palaeontology*, **2**, 1–27.

  (1965). Progress in the planktonic foraminiferal biostratigraphy of the Neogene. *Nature*, **208**, 1164–1166.

Barbin, V. (1988). In *The Eocene–Oligocene Boundary in the Marche–Umbria Basin (Italy)*, ed. I. Premoli Silva, R. Coccioni & A. Montanari, p. 163–171. International Subcommission on Paleogene Stratigraphy, International Union of Geological Sciences, Eocene–Oligocene meeting, Ancona (Italy), Special Publication.

Barry, J. C., Morgan, M. E., Flynn, L. J. *et al.* (2002). Faunal and environmental change in the Late Miocene Siwaliks of northern Pakistan. *Paleobiology Memoirs*, Suppl. to Vol. **28**(2), Memoir 3, 1–72.

Bé, A. W. H. (1977). An ecological, zoogeographic and taxonomic review of recent planktonic foraminifera. In *Oceanic Micropalaeontology*, ed. A. T. S. Ramsay, pp. 1–100. London: Academic Press.

  (1982). Biology of plankton foraminifera. In *Foraminifera: Notes for a Short Course*, ed. T. W. Broadhead, *Studies in Ecology*, **6**, 51–92.

Bé, A. W. H. & Tolderlund, D. S. (1971). Distribution and ecology of living planktonic foraminifera in surface waters of the Atlantic and Indian Oceans. In, *Micropaleontology of the Oceans*, ed. B. M. Funnell & W. R. Riedel, pp. 105–150. Cambridge: Cambridge University Press.

Behrensmeyer, A. K., Todd, N. E., Potts, R. & McBrinn, G. E. (1997). Late Pliocene faunal turnover in the Turkana Basin, Kenya and Ethiopia. *Science*, **278**, 1589–1594.

Bennett, K. D. (1990). Milankovitch cycles and their effects on species in ecological and evolutionary time. *Paleobiology*, **16**, 11–21.

  (1997). *Evolution and Ecology: The Pace of Life*. Cambridge: Cambridge University Press.

Benson, R. A. (1995). Editorial: is the death of an ocean falling through a stratigraphic crack? *Paleoceanography*, **10**, 1–3.

Benson, R. A. & Hodell, D. A. (1994). Comment on 'A critical re-evaluation of the Miocene–Pliocene boundary as defined in the Mediterranean' by F. J. Hilgen and C. G. Langereis. *Earth & Planetary Science Letters*, **124**, 245–250.

Benton, M. J. & Pearson, P. N. (2001). Speciation in the fossil record. *Trends in Ecology & Evolution*, **16**, 405–411.

Berger, W. H. (1970). Biogenous deep-sea sediments: fractionation by deep-sea circulation. *Geological Society of America Bulletin*, **81**, 1385–1402.

(1974). Deep-sea sedimentation. In *The Geology of Continental Margins*, ed. C. A. Burk & C. L. Drake, pp. 213–241. New York: Springer-Verlag.

(1982). Deep-sea stratigraphy: Cenozoic climatic steps and the search for chemo-climatic feedback. In *Cyclic and Event Stratification*, ed. G. Einsele & A. Seilacher, pp. 121–157. Berlin: Springer-Verlag.

Berger, A. & Loutre, M. F. (1991). Insolation values for the climate of the last 10 million years. *Quaternary Science Reviews*, **10**, 297–317.

Berger, W. H. & Vincent, E. (1981). Chemostratigraphy and biostratigraphic correlation: exercises in systemic stratigraphy. *Proceedings 26th International Geological Congress, Oceanologica Acta 4 (Supplement)*, 115–127.

(1986). Deep-sea carbonates: reading the carbon-isotope signal. *Geologische Rundschau*, **75**, 249–269.

Berggren, W. A. (1960). Paleogene biostratigraphy and planktonic foraminifera of the SW Soviet Union: an analysis of recent Soviet publications. *Stockholm Contributions in Geology*, **6**(5), 63–125.

(1964). The Maestrichtian, Danian and Montian Stages and the Cretaceous–Tertiary boundary. Stockholm Contributions in Geology, **11**, 103–176.

(1968). Phylogenetic and taxonomic problems of some Tertiary planktonic foraminiferal lineages. *Tulane Studies in Geology*, **6**, 1–22.

(1969a). Cenozoic chronostratigraphy, planktonic foraminiferal zonation and the radiometric time scale. *Nature*, **224**, 1072–1075.

(1969b). Rates of evolution in some Cenozoic planktonic foraminifera. *Micropaleontology*, **15**, 351–365.

(1971a). Multiple phylogenetic zonations of the Cenozoic based on planktonic foraminifera. In *Proceedings of the II Planktonics Conference, Roma 1970*, ed. A. Farinacci, 41–56. Roma: Edizioni Technoscienza.

(1971b). Tertiary boundaries and correlations. In *The Micropaleontology of the Oceans*, ed. B. F. Funnell & W. R. Riedel, pp. 693–808. Cambridge: Cambridge University Press.

(1978). Biochronology. In *Contributions to the Geological Time Scale*, ed. G. V. Cohee, M. F. Glaessner & H. D. Hedberg. American Association of Petroleum Geologists, Studies in Geology, **6**, 39–55.

(1984). Neogene planktonic foraminiferal biostratigraphy and biogeography: Atlantic, Mediterranean, and Indo-Pacific regions. In *Pacific Neogene Datum Planes: Contributions to Biostratigraphy and Chronology*, ed. N. Ikebe & R. Tsuchi, pp. 111–161. University of Tokyo Press.

(1998). The Cenozoic Era: Lyellian (chrono)stratigraphy and nomenclatural reform at the milennium. In *Lyell: The Past is the Key to the Present*, ed. D. J. Blundell & A. C. Scott. Geological Society Special Publication No. **143**, 111–132.

Berggren, W. A. & Aubry, M.-P. (1996). A late Paleocene–early Eocene NW European and North Sea magnetobiochronological correlation network. In *Correlation of the Early Paleogene in Northwest Europe*, ed. R. W. O'B. Knox, R. M. Corfield & R. E. Dunay, pp. 309–352. London: Geological Society, Special Publication No. 101.

Berggren, W. A. & Miller, K. G. (1988). Paleogene tropical planktonic foraminiferal biostratigraphy and magnetobiochronology. *Micropaleontology*, **34**, 362–380.

Berggren, W. A. & Norris, R. D. (1997). Biostratigraphy, phylogeny and systematics of Paleocene trochospiral foraminifera. *Micropaleontology*, **43**, Supplement 1, 1–116.

Berggren, W. A. & Prothero, D. R. (1992). Eocene–Oligocene climatic and biotic evolution: an overview. In *Eocene–Oligocene Climatic and Biotic Evolution*, ed. D. R. Prothero & W. A. Berggren, pp. 1–28. Princeton: Princeton University Press.

Berggren, W. A. & van Couvering, J. A. (1978). Biochronology. In *Contributions to the Geological Time Scale*, ed. G. V. Cohee, M. F. Glaessner & H. D. Hedberg. American Association of Petroleum Geologists, Studies in Geology, **6**, 39–55.

Berggren, W. A., Hilgen, F. J., Langereis, C. G. *et al.* (1995b). Late Neogene chronology: New perspectives in high-resolution stratigraphy. *Geological Society of America, Bulletin*, **107**(11), 1272–1287.

Berggren, W. A., Kent, D. V. & Flynn, J. J. (1985). Paleogene geochronology and chrono-stratigraphy. In *The Chronology of the Geological Record*, ed. N. J. Snelling. Geological Society of London, Memoir **10**, 141–195.

Berggren, W. A., Kent, D. V., Flynn, J. J. & Van Couvering, J. A. (1985). Cenozoic geo-chronology. *Geological Society of America, Bulletin*, **96**, 1407–1418.

Berggren, W. A., Kent, D. V., Swisher, C. C. III & Aubry, M.-P. (1995). A revised Cenozoic geochronology and chronostratigraphy. In *Geochronology Time Scales and Global Stratigraphic Correlation*, ed. W. A. Berggren, D. V. Kent, M.-P. Aubry & J. Hardenbol. Tulsa, SEPM (Society of Sedimentary Geology) Special Publication **54**, 129–212.

Berggren, W. A., Kent, D. V. & Van Couvering, J. A. (1985). Neogene geochronology and chronostratigraphy. In *The Chronology of the Geological Record*, ed. N. J. Snelling. Geological Society of London Memoir **10**, 211–260.

Berry, W. B. N. (1968). *Growth of a Prehistoric Time Scale Based on Organic Evolution*. San Francisco: W. H. Freeman & Co.

(1977). Graptolite biostratigraphy: a wedding of classical principles and current concepts. In *Concepts and Methods of Biostratigraphy*, ed. E. G. Kauffmann & J. E. Hazel, pp. 321–338. Stroudsberg, PA: Dowden, Hutchison & Ross, Inc.

Bignot, G. (1985). *Elements of Micropalaeontology*. London: Graham & Trotman Ltd.

Billups, K., Channell, J. E. T. & Zachos, J. C. (2002). Late Oligocene to early Miocene geochronology and paleoceanography from the subantarctic South Atlantic. *Paleoceanography*, **17**, 4–11.

Blackwelder, R. E. (1967). *Taxonomy: A Text and Reference Book*. New York: John Wiley.

Blackwelder, R. E. & Boyden, A. (1952). The nature of systematics. *Systematic Zoology*, **1**, 26–33.

Blow, W. H. (1956). Origin and evolution of the foraminiferal genus *Orbulina* d' Orbigny. *Micropaleontology*, **2**, 57–70.

(1959). Age, correlation and biostratigraphy of the Upper Tocuyo (San Lorenzo) and Pozón formations eastern Falcón, Venezuela. *Bulletin of American Paleontology*, **39**, 1–251.

(1969). Late middle Eocene to Recent planktonic foraminiferal biostratigraphy. In *Proceedings of the First International Conference on Planktonic Microfossils*, ed. P. Brönnimann & H. H. Renz. Leiden: E. J. Brill, vol. 1, pp. 199–421.

(1970). Validity of biostratigraphic correlations based on the Globigerinacea. *Micropaleontology*, **16**, 257–268.

(1979). The Cainozoic Globigerinida: a study of the morphology, taxonomy, evolutionary relationships and the stratigraphical distribution of some Globigerinida (mostly Globigerinacea). 3 volumes: I. text, part I and part 2, section 1, xvii + 752 pp., 118 figures. II. part II, section 2, ix + pp. 753–1413. III. Atlas, xxi pp., 264 pp. Leiden: E. J. Brill.

Blow, W. H. & Banner, F. T. (1962). Part Two: The mid-Tertiary (Upper Eocene to Aquitanian) Globigerinaceae. In, *Fundamentals of Mid-Tertiary Stratigraphical Correlation*, ed. F. E. Eames, F. T. Banner, W. H. Blow & W. J. Clarke, pp. 61–151. Cambridge: Cambridge University Press.

Bobe, R. & Eck, G. G. (2001). Responses of African bovids to Pliocene climatic change. *Paleobiology Memoirs*, **27**, Supplement to No. 2, 1–48.

Boersma, A., Premoli Silva, I. & Hallock, P. (1998). Trophic models for the well-mixed and poorly mixed warm oceans across the Paleocene/Eocene Epoch boundaries. In *Late Paleocene–Early Eocene Climatic and Biotic Events in the Marine and Terrestrial Records*, ed. M. -P. Aubry, S. G. Lucas & W. A. Berggren, 204–213. New York: Columbia University Press.

Bolli, H. M. (1950). The direction of coiling in the evolution of some Globorotaliidae. *Contributions of the Cushman Foundation of Foraminiferal Research*, **1**, 82–89.

(1957a). The genera *Globigerina* and *Globorotalia* in the Paleocene–lower Eocene Lizard Springs Formation of Trinidad, B. W. I. *United States National Museum, Bulletin*, **215**, 51–81.

(1957b). Planktonic foraminifera from the Eocene Navet and San Fernando Formations of Trinidad, B. W. I. *United States National Museum, Bulletin*, **215**, 155–172.

(1957c). Planktonic foraminifera from the Oligocene-Miocene Cipero and Lengua Formations of Trinidad, B. W. I. *United States National Museum, Bulletin*, **215**, 97–123.

(1966). Zonation of Cretaceous to Pliocene sediments based on planktonic foraminifera. *Boletin Informativo, Asociacion Venezolano de Geologia, Mineraria y Petroleo*, **9**, 3–32.

(1967). The subspecies of *Globorotalia fohsi* Cushman and Ellisor and the zones based on them. *Micropalaeontology*, **13**, 502–512.

(1971). The direction of coiling in planktonic foraminifera. In *The Micropalaeontology of Oceans*, ed. Funnell, B. M. & Riedel, W. R., pp. 639–648. Cambridge: Cambridge University Press.

Bolli, H. M. & Saunders, J. B. (1985). Oligocene to Holocene low-latitude planktic foraminifera. In *Plankton Stratigraphy*, ed. H. M. Bolli, J. B. Saunders & K. Perch-Nielsen, pp. 155–262. Cambridge, Cambridge University Press, 2 volumes.

Bolli, H. M., Loeblich, A. R., Jr. & Tappan, H. (1957). Planktonic foraminiferal families Hantkeninidae, Orbulinidae, Globorotaliidae, and Globotruncanidae. *United States National Museum, Bulletin*, **215**, 3–50.

Bolli, H. M., Saunders, J. B. & Perch-Nielsen, K., Editors (1985). *Plankton Stratigraphy*. Cambridge, Cambridge University Press, 2 volumes.

Bonuso, N., Newton, C. R., Brower, J. C. & Ivany, L. C. (2002). Does coordinated stasis yield taxonomic and ecologic stability?: Middle Devonian Hamilton Group of New York. *Geology*, **30**, 1055–1058.

Boucot, A. J. (1982). Ecostratigraphy. In, *Stratigraphy Quo Vadis?* ed. E. Seibold & J. H. Meulenkamp, American Association of Petroleum Geologists, Studies in Geology No. **16**, 55–60.

(1983). Does evolution take place in an ecological vacuum? *Journal of Paleontology*, **57**, 1–30.

(1990a). Modern paleontology: using biostratigraphy to the utmost. *Revista Española de Paleontología*, **5**, 63–70.

(1990b). Community evolution: its evolutionary and biostratigraphic significance. In *Paleocommunity Temporal Dynamics: The Long-term Development of Multispecies Assemblages*, ed. W. Miller III. The Paleontological Society Special Publication, **5**, 48–70.

(1994). The episodic, rather than periodic nature of extinction events. *Revista de la Sociedad Mexicana de Paleontologia*, **7**, 15–35.

Boudagher-Fadel, M. K. (2002). The stratigraphical relationship between planktonic and larger berthic foraminifera in the middle Miocene to lower Pliocene carbonate facies of Sulawesi, Indonesia. *Micropalaeontology*, **48**, 153–76.

Boudagher-Fadel, M. K. & Bonner, F. T. (1999). Revision of the stratigraphic significance of the Oligocere–Miocere letter Stages. *Reveue de Micropaléontologie*, **42**, 93–7.

Bralower, T. J., Leckie, R. M., Sliter, W. V. & Thierstein, H. R. (1995). An integrated Cretaceous microfossil biostratigraphy. In *Geochronology Time Scales and Global Stratigraphic Correlation*, ed. W. A. Berggren, D. V Kent, M.-P. Aubry & J. Hardenbol. SEPM (Society of Sedimentary Geology), Special Publication, **54**, 65–80.

Bramlette, M. N. & Martini, E. (1964). The great change in calcareous nannoplankton fossils between the Maastrichtian and Danian. *Micropaleontology*, **10**, 291–322.

Brasier, M. D. (1980). *Microfossils*. London: Allen & Unwin.

(1995). Fossil indicators of nutrient levels. 2. Evolution and extinction in relation to oligotrophy. In *Marine Palaeoenvironmental Analysis from Fossils*, ed. D. J. W. Bosence & P. A. Allison. London: The Geological Society, Special Publication No. **83**, 133–150.

Bretsky, P. W. (1979). History of paleontology: post-Darwinian. In *Encyclopedia of Paleontology*, ed. R. W. Fairbridge & D. Jablonski, pp. 384–395. Stroudsberg, PA: Dowden, Hutchison & Ross, Inc.

Brett, C. E. (1998). Sequence stratigraphy, paleoecology, and evolution: biotic clues and responses to sea-level fluctuations. *Palaios*, **13**: 241–262.

(2001). A slice of the 'layer cake': the paradox of 'frosting continuity'. *Palaios*,

Brett, C. E. & Baird, G. C. (1995). Coordinated stasis and evolutionary ecology of Silurian to middle Devonian faunas in the Appalachian Basin. In *New Approaches to Speciation in the Fossil Record*, ed. D. H. Erwin & R. L. Anstey. pp. 285–315, New York: Columbia University Press.

(1997). *Paleontological Events: Stratigraphic, Ecological, and Evolutionary Implications*. New York: Columbia University Press.

Brett, C. E., Ivany, L. C. & Schopf, K. M. (1996). Coordinated stasis: an overview. *Palaeogeography, Palaeoclimatology, Palaeoecology*, **127**, 1–20.

Brinkhuis, H. (1994). Late Eocene to Early Oligocene dinoflagellate cysts from the Priabonian type-area (northwest Italy): biostratigraphy and palaeoenvironmental interpretation. *Palaeogeography, Palaeoclimatology, Palaeoecology*, **107**, 121–163.

Brinkhuis, H. & Visscher, H. (1995). The upper boundary of the Eocene Series: a reappraisal based on dinoflagellate cyst biostratigraphy and sequence stratigraphy. In *Geochronology, Time Scales and Global Stratigraphic Correlation*, ed. W. A. Berggren, D. V. Kent, M.-P. Aubry & J. Hardenbol. Tulsa, SEPM Special Publication **54**, pp. 295–304.

Brönnimann, P., and Resig, J. (1971). A Neogene Globigerinacean biochronologic timescale of the southwestern Pacific. *Initial Reports of the Deep Sea Drilling Project*, **7**, 1235–1469.

Brotzen, F. (1959). On *Tylocidaris* species (Echinoidea) and the stratigraphy of the Danian of Sweden, with a bibliograp[hy of the Danian and the Paleocene. *Sveriges Geologiska Undersökning, Series C*, **54**, 1–81.

Brown, C. W. & Stephenson, S. (1991). *Geology of the Murray Basin*. Canberra: Bureau of Mineral Resources, Bulletin 235.

Buckland, W. (1837). *Geology and Mineralogy Considered with Reference to Natural Theology*. London: William Pickering, 2 volumes.

Buckman, S. S. (1902). The term 'Hemera'. *Geological Magazine*, n.s., **9**, 554–557.
    (1903). The term 'Hemera'. *Geological Magazine*, n.s., **10**, 95–96.

Buffetaut, E. (1987). *A Short History of Palaeontology*. London: Wolfboro.

Butler, R. F. & Lindsay, E. H. (1985). Mineralogy of magnetic minerals and revised magnetic polarity stratigraphy of continental sediments, San Juan Basin, New Mexico. *Journal of Geology*, **93**, 535–554.

Butterfield, H. (1949). *The Origins of Modern Science: 1300–1800*. London: Bell, 217 pp.
    (1951). *The Whig Interpretation of History*. London: G. Bell, 132 pp.

Buzas, M. A. & Culver, S. J. (1984). Species duration and evolution: Benthic foraminifera on the Atlantic continental margin of North America. *Science*, **225**, 829–830.
    (1989). Biogeographic and evolutionary patterns of continental margin foraminifera: *Paleobiology*, **15**(1), 11–19.
    (1998). Assembly, disassembly, and balance in marine paleocommunities. *Palaios*, **13**, 263–275.

Cain, A. J. (1954). *Animal Species and their Evolution*. London: Hutchinson's University Library, 190 pp.
    (1960). *Animal Species and their Evolution*. New York: Harper & Brothers.

Callomon, J. H. (1995). Time from fossils: S. S. Buckman and Jurassic high-resolution geochronology. In *Milestones in Geology*, ed. M. J. Le Bas, Geological Society, London, Memoir, **16**, 127–150.

Cande, S. C. & Kent, D. V. (1992). A new geomagnetic polarity time scale for the Late Cretaceous and Cenozoic. *Journal of Geophysical Research*, **97**, 13917–13951.

(1995). Revised calibration of the geomagnetic polarity timescale for the Late Cretaceous and Cenozoic. *Journal of Geophysical Research*, **100**, 6093–6095.

Carney, J. L. & Pierce, R. W. (1995). Graphic correlation and composite standard databases as tools for the exploration biostratigrapher. In *Graphic Correlation and the Composite Standard Approach*, ed. K. O. Mann, H. R. Lane, & J. R. Stein, pp. 23–44. SEPM (Society for Sedimentary Geology), Special Publication No. **53**.

Caron, M. & Homewood, P. (1983). Evolution of early planktic foraminifers. *Marine Micropaleontology*, **7**, 453–462.

Carpenter, W. B., Parker W. K. & Jones, T. R. (1862). *Introduction to the Study of the Foraminifera*. London: The Ray Society.

Carr, E. H. (1961). *What is History?* New York, N.Y: Random House, 209 pp.

Carter, A. N. (1958a). Pelagic foraminifera in the Tertiary of Victoria: *Geological Magazine*, **95**, 297–304.

(1958b). Tertiary foraminifera from the Aire District, Victoria: *Geological Survey of Victoria Bulletin* **55**, 1–76.

Carter, R. M. (1974). A New Zealand case-study of the need for local time-scales. *Lethaia*, **7**, 181–202.

Carter, R. M. and Naish, T. R. (1998). Have local stages outlived their usefulness for the New Zealand Pliocene–Pleistocene? *New Zealand Journal of Geology and Geophysics*, **41**, 271–279.

Castradori, D. (2002). A complete standard chronostratigraphic scale: how to turn a dream into reality? *Episodes*, **25**, 107–110.

Cavelier, C., Chateauneuf, J. J., Pomerol, C., Rabussier, D., Renard, M. & Vergnaud-Grassini, C. (1981). The geological events at the Eocene/Oligocene boundary. *Palaeogeography, Palaeoclimatology, Palaeoecology*, **36**, 223–248.

Chaisson, W. P. (2003). Vicarious living: Pliocene menardellids between an isthmus and an ice sheet. *Geology*, **31**, 1085–8.

Chamberlin, T. C. (1898). The ulterior basis of time divisions and the classification of geologic history. *Journal of Geology*, **6**, 449–462.

Chapman, M. R. (2000). The response of planktonic foraminifera to the Late Pliocene intensification of Northern Hemisphere glaciation. In *Biotic Response to Global Change: The Past 145 Million Years*, ed., S. J. Culver & P. F. Rawson, pp. 79–96. London: British Museum of Natural History & Cambridge: Cambridge University Press.

Chaproniere, G. C. H. (1975). Palaeoecology of Oligo-Miocene Larger Foraminiferida, Australia. *Alcheringa*, **1**, 37–58.

(1981). Australasian mid-Tertiary correlations, larger foraminiferal associations and their bearing on the East Indian Letter Classification. *BMR Journal of Australian Geology and Geophysics*, **6**, 145–151. Canberra: Bureau of Mineral Resources (now Geoscience Australia).

(1984). Oligocene and Miocene larger Foraminiferida from Australia and New Zealand. *BMR Bulletin Australian Geology & Geophysics*, **188**, 1–98.

Chaproniere, G. C. H., Shafik, S., Truswell, E. M., Macphail, M. K. & Partridge, A. D. (1996). Ch. 2.10, Cainozoic (Chart 10). In *An Australian Phanerozoic Timescale*, ed. G. C. Young & J. R. Laurie, pp. 175–186. Oxford: Oxford University Press.

Cifelli, R. (1969). Radiation of Cenozoic planktonic foraminifera. *Systematic Zoology*, **18**, 154–168.

(1990). A history of the classification of the foraminifera (1826–1933). Part I, Foraminiferal classification from d'Orbigny to Galloway. *Cushman Foundation for Foraminiferal Research, Special Publication*, **27**, 1–88.

Cita, M. B. (1975). The Miocene–Pliocene boundary: history and definition. In *Late Neogene Epoch Boundaries*, ed. T Saito & L. H. Burckle. New York: Micropaleontology Press, Special Publication **1**, 1–30.

Claridge, M. F., Dawah, H. A. & Wilson, M. R., Editors (1997). *Species: The Units of Biodiversity*. London: Chapman and Hall.

Clarke, W. J. & Blow, W. H. (1969). The inter-relationships of some late Eocene, Oligocene and Miocene larger foraminifera and plankton biostratigraphic indices. In *Proceedings of the International Conference on Planktonic Microfossils (Geneva 1967)*, ed. P. Brönnimann & H. H. Renz. Leiden: E. J. Brill, vol. 2, 82–97.

Colin, J.-P. & Lethiers, F. (1988). The importance of ostracods in biostratigraphic analysis. In *Ostracoda in the Earth Sciences*, ed. P. De Deckker, J.-P. Colin & J.-P. Peypouquet, pp. 27–46. Amsterdam: Elsevier.

Conkin, B. M. & Conkin, J. E. (1984). *Stratigraphy: Foundations and Concepts*. Van Nostrand and Reinhold, New York, 335 pp.

Cooper, R. A., ed. (2004). The New Zealand geological timescale. *Institute of Geological and Nuclear Sciences*, Monograph **22**.

Cooper, R. A., Crampton, J. S., Raine, J. I. *et al.* (2001). Quantitative biostratigraphy of the Taranaki Basin, New Zealand: a deterministic and probabilistic approach. *American Association of Petroleum Geologists Bulletin*, **85**, 1469–1498.

Cooper, R. A., Crampton, J. S. & Uruski, C. I. (2000). *The Time-Calibrated Composite: A Powerful Tool in Basin Exploration*. Proceedings of the 2000 New Zealand Petroleum Conference, Christchurch, New Zealand, pp. 346–364. Wellington: Ministry of Commerce.

Corfield, R. M. & Cartlidge, J. E. (1991). Oceanographic and climatic implications of the Palaeocene carbon isotope maximum. *Terra Nova*, **4**, 443–455.

Corfield, R. M. & Norris, R. D. (1998). The oxygen and carbon isotopic context of the Paleocene/Eocene Epoch boundary. In *Late Paleocene–early Eocene Climatic and Biotic Events in the Marine and Terrestrial Records*, ed. M.-P. Aubry, S. G. Lucas & W. A. Berggren, 124–137. New York: Columbia University Press.

Cowie, J. W. (1986). Guidelines for boundary stratotypes. *Episodes*, **9**, 78–82.

(1990). Global boundary stratotypes: overview. In *Palaeobiology: A Synthesis*, ed. D. E. G. Briggs & P. R. Crowther, pp. 471–475. Oxford: Blackwell.

Cowie, J. W., Ziegler, W. & Remane, J. (1989). Stratigraphic commission accelerates progress. *Episodes*, **12**, 79–83.

Cox, A., Doell, R. R. & Dalrymple, G. B. (1964). Geomagnetic polarity epochs. *Science*, **143**: 351–352.

Cracraft, J. (1987). Species concepts and the ontology of evolution. *Biology and Philosophy*, **2**, 329–346.

Crespin, I. (1943). *The Stratigraphy of the Tertiary Marine Rocks in Gippsland*, Victoria: Palaeontological Bulletin, 101 p. (mimeo) (Department of Supply and Shipping, Australia).

Croneis, C. (1941). Micropaleontology, past and future. *American Association of Petroleum Geologists, Bulletin*, **35**, 1308–1355.

Cross, T. A. & Lessenger, M. A. (1988). Seismic stratigraphy. *Annual Reviews of Earth and Planetary Sciences*, **16**, 319–354.

Culver, S.J. & Buzas, M.A. (2000). Response of shallow water foraminiferal palaeocommunities to global and regional environmental change. In *Biotic Response to Global Change: The Past 145 Million Years*, ed. S. J. Culver, & P. F. Rawson, pp. 122–134. London: British Museum of Natural History & Cambridge: Cambridge University Press.

Cushman, J. A. (1927). Some new genera of the foraminifera. *Contributions Cushman Laboratory Foraminiferal Research*, **2**, 77–81.

Cushman, J. A. & Stainforth, R. M. (1945). The foraminifera of the Cipero Marl Formation of Trinidad, British West Indies. *Cushman Laboratory Foraminiferal Research Special Publication*, **14**, 1–74.

D'Hondt, S. (1991). Phylogenetic and stratigraphic analysis of earliest Paleocene triserial and biserial planktonic foraminifera. *Journal of Foraminiferal Research*, **21**, 168–181.

D'Hondt, S. & Zachos, J. C. (1993). On stable isotopic variation and earliest Paleocene planktic foraminifera. *Paleoceanography*, **8**, 527–547.

  (1998). Cretaceous foraminifera and the evolutionary history of planktic photosymbiosis. *Paleobiology*, **24**, 512–523.

Daniel, G. (1962). *The Idea of Prehistory*. London: Watts, 171 pp.

Darling, K. F., Kucera, M., Pudsey, C. J. et al. (2004). Molecular evidence links gyptic diversification in palar planktonic pratists to Quaternary chriate dynamics. *Proceedings National Academy of Sciences, USA*, **101**, 7657–66.

Darling, K. F., Wade, C. M., Kroon, D. & Leigh Brown, A. J. (1997). Planktonic foraminiferal molecular evolution and their polyphyletic origins from benthic taxa. *Marine Micropaleontology*, **30**, 251–266.

Darling, K. F., Wade, C. M., Kroon, D., Leigh Brown, A. J. & Bijma, J. (1999). The diversity and distribution of modern planktonic foraminiferal small subunit ribosomal RNA genotypes and their potential as tracers of past and present ocean circulations. *Paleoceanography*, **14**, 3–12.

Darling, K. F., Wade, C. M., Stewart, I. A. et al. (2000). Molecular evidence for genetic mixing of Arctic and Antarctic subpolar populations of planktonic foraminifers. *Nature*, **405**, 43–47.

Darragh, T. A. (1985). Molluscan biogeography and biostratigraphy of the Tertiary of southern Australia. *Alcheringa* **9**, 83–116.

Darwin, C. (1859, [1964]). *On the Origin of Species: A Facsimile of the 1st Edition with an Introduction by Ernst Mayr*. Cambridge, MA: Harvard University Press.

Davies, A. M. (1934). *Tertiary Faunas: A Text-Book for Oilfield Palaeontologists and Students of Geology*. London: Thomas Murby, 1934–35 [vol. 1, 1935], 2 volumes.

De Vargas, C., Bonzon, M., Rees, N., Pawlowski, J., Zaninetti, L. (2002). A molecular approach to biodiversity and biogeography in the planktonic foraminifer *Globigerinella siphonifera* (d'Orbigny). *Marine Micropaleontology*, **45**, 101–116.

de Vargas, C., Norris, R. D., Zaninetti, L., Gibb, S. W. & Pawlowski, J. (1999). Molecular evidence of cryptic speciation in planktonic foraminifers and their relation to oceanic provinces. *Proceeding of the National Academy of Sciences USA*, **96**, 2864–2868.

De Vargas, C., Renaud, S., Hillbrecht, H. & Pawlowski, J. (2001). Pleistocene adaptive radiation in *Eloborotalia transcatulinoides*: genetic, morphologic, and environmental evidence. *Paleobiology*, **27**, 104–2s.

Desmond, A. J. (1989). *The Politics of Evolution: Morphology, Medicine, and Reform in Radical London*. Chicago: University of Chicago Press, 503 pp.

Diener, C. (1925). *Grundzüge der Biostratigraphie*. Leipzig: Deuticke.

DiMichele, W. A. (1994). Ecological patterns in time and space. *Paleobiology*, **20**, 89–92.

DiMichele, W. A., Behrensmeyer, A. K., Olszewski, T. D. *et al.* (2004). Long-term stasis in ecological assemblages: evidence from the fossil record. *Annual Review of Ecology and Systematics*, **35**, 285–322.

Dobzhansky, T. (1937). *Genetics and the Origin of Species*. New York: Columbia University Press.

Donovan, D. T. (1966). *Stratigraphy: An Introduction to Principles*. London: Murby.

Dorf, E. (1955). Plants and the geological time scale. In *Crust of the Earth*, ed. A. Poldervaart. Geological Society of America, Special Paper, **62**, 575–592.

Dott, R. H. Jr, ed. (1992). *Eustasy: the historical ups and downs of a major historical concept*. Geological Society of America, Memoir, **180**.

Douglas, R. G. & Savin, S. M. (1978). Oxygen isotope evidence for the depth stratification of Tertiary and Cretaceous planktonic foraminifera. *Marine Micropaleontology* **3**, 175–196.

Dowsett, H. J. (1988). Diachrony of late Neogene microfossils in the southwest Pacific Ocean: application of the graphic correlation method. *Paleoceanography*, **3**, 209–222.

Doyle, P. & Bennett, M. R. (Editors) (1997). *Unlocking the Stratigraphical Record: Advances in Modern Stratigraphy*. New York: John Wiley & Sons.

Drooger, C. W. (1956). Transatlantic correltion of the Oligo-Miocene by means of foraminifera. *Micropaleontology*, **2**, 183–192.

(1963). Evolutionary trends in the Miogypsinidae. In *Evolutionary Trends in Foraminifera*, ed. G. H. R. von Koenigswald, J. D. Emeis, W. L. Buning & C. W. Wagner, pp. 315–349. Amsterdam: Elsevier Publishing Company.

(1966). Zonation of the Miocene by means of planktonic foraminifera. With additional comments by Z. Reiss, M. B. Cita, W. H. Blow, F. E. Eames, R. M. Stainforth & H. M. Bolli. In *Committee on Mediterranean Stratigraphy*, ed. C. W. Drooger, Z. Reiss, R. F. Rutsch & P. Marks. Proceedings of the third session in Berne, 8–13 June 1964. International Union of Geological Sciences Commission on Stratigraphy, pp. 40–50, Leiden: E. J. Brill.

(1974). The boundaries and limits of stratigraphy. *Koninklijke Nederl. Akad. Wetenschappen, Amsterdam, Proc. Series B*, **77**, 159–176.

(1993). *Radial Foraminifera: Morphometrics and Evolution*. Amsterdam, New York: North-Holland, 242 pp.

Durham, J. W. (1950). Cenozoic marine climates of the Pacific coast. *Geological Society of America, Bulletin*, **61**, 243–1264.

Eames, F. E. & Savage, R. J. G. (1975). *Tertiary Faunas: A Text-Book for Oilfield Palaeontologists and Students of Geology*, by A. Morley Davies, 2nd edition revised and brought up to date. London: Allen & Unwin, vols. 1, 2.

Eames, F. E., Banner, F. T., Blow, W. H. & Clarke, W. J. (1962). *Fundamentals of Mid-Tertiary Stratigraphical Correlation*. Cambridge: Cambridge University Press.

Edwards, L. E. (1982a). Quantitative biostratigraphy: the methods should suit the data. In *Quantitative Stratigraphic Correlation*, ed. J. M. Cubitt & R. A. Reyment, pp. 45–60. New York: John Wiley.

(1982b). Numerical and semi-objective biostratigraphy: review and predictions. *Proceedings of the Third North American Palaeontological Convention*, **1**, 147–152.

(1984). Insights on why the graphic correlation (Shaws method) works. *Journal of Geology*, **92**, 583–597.

(1989). Supplemented graphic correlation: A powerful tool for paleontologists and nonpaleontologists. *Palaios*, **4**, 127–143.

Eicher, D. L. (1976). *Geologic Time* (2nd edn). Englewood Cliffs: Prentice-Hall.

Eldredge, N. (1979). Alternative approaches to evolutionary theory. *Carnegie Museum Natural History Bulletin*, **13**, 7–19.

(1989). *Macroevolutionary Dynamics: Species, Niches and Adaptive Peaks*. New York: McGraw-Hill.

Eldredge, N. & Gould, S. J. (1972). Punctuated equilibria: an alternative to phyletic gradualism. In *Models in Paleobiology*, ed. T. J. M. Schopf, pp. 82–115. San Francisco: Freeman, Cooper.

(1977). Evolutionary models and biostratigraphic strategies. In *Concepts and Methods of Biostratigraphy*, ed. E. G. Kauffman & J. E. Hazel, pp. 3–22. Stroudsberg, PA: Dowden, Hutchison and Ross.

Emiliani, C. (1955). Pleistocene temperatures. *Journal of Geology*, **63**, 538–578.

(1969). A new paleontology. *Micropaleontology*, **15**, 265–300.

(1982). Extinctive evolution: extinctive and competitive evolution combine into a unified model of evolution. *Journal of Theoretical Biology*, **97**, 13–33.

Endler, J. (1989). In *Speciation and its Consequences*, ed. D. Otte & J. Endler. Sunderland, MA: Sinauer & Associates.

Ereshefsky, M., (ed.) (1992). *The Units of Evolution: Essays on the Nature of Species*. Cambridge, MA: The MIT Press.

(2001). *The Poverty of the Linnaean Hierarchy: A Philosophical Study of Biological Taxonomy*. Cambridge: Cambridge University Press.

Ericson, D. B. (1961). Pleistocene climatic record in some deep sea sediment cores. *Science*, **95**, 537–541.

Ericson, D. B. & Wollin, G. (1956). Micropaleontological and isotopic determinations of Pleistocene climates. *Micropaleontology*, **2**, 257–270.

(1968). Pleistocene climates and chronology in deep sea sediments. *Science*, **162**, 1227–1234.

Evernden, J. F., Savage, D. E., Curtis, G. H. & James, G. T. (1964). Potassium-argon dates and the Cenozoic mammalian chronology of North America. *American Journal of Science*, **62**, 145–198.

Fejfar, O. & Heinrich, W.-D. (1989). Muroid rodent biochronology of the Neogene and Quaternary. In *European Neogene Mammal Chronology*, ed. E. H. Lindsay, V. Fahlbusch & P. Mein, pp. 91–118. NATO Advanced Research Workshop on European Neogene Mammal chronology, Schloss Reisensburg. New York & London: Plenum Press.

Finlay, H. J. (1947). The foraminiferal evidence for Tertiary trans-Tasman correlation. *Transactions Royal Society New Zealand*, **76**, 327–352.

Finlay, H. J. & Marwick, J. A. (1940). The divisions of the upper Cretaceous and Tertiary in New Zealand. *Transactions of the Royal Society of New Zealand*, **70**, 77–135.

(1947). New divisions of the New Zealand upper Cretaceous and Tertiary. *New Zealand Journal of Science & Technology*, **B28**, 228–236.

Fornaciari, E. & Rio, D. (1996). Latest Oligocene to early Miocene quantitative calcareous nannofossil biostratigraphy in the Mediterranean region. *Micropaleontology*, **42**, 1–26.

Fuller, S. (2003). *Kuhn vs Popper: The Struggle for the Soul of Science*. Icon Books, UK.

Fischer, A. G. (1980). Gilbert: bedding rhythms and geochronology. In *The Scientific Ideas of G. K. Gilbert*, ed. E. Yochelson, pp. 93–104. Geological Society of America, Special Paper 183.

(1981). Climatic oscillations in the biosphere. In *Biotic Crises in Ecological and Evolutionary Time*, ed. M. H. Nitecki, pp. 103–121. New York: Academic Press.

(1984). The two Phanerozoic supercycles. In *Catastrophes and Earth History*, ed. W. A. Berggren & J. A. van Couvering, pp. 9–34. Princeton: Princeton University Press.

(1986). Climate rhythms recorded in strata. *Annual Review of Earth & Planetary Sciences*, **14**, 351–376.

Fischer, A. G. & Arthur, M. A. (1977). Secular variations in the pelagic realm. In *Deep-Water Carbonate Environments*, ed. H. Cook & P. Enos. Society of Economic Paleontologists and Mineralogists, Special Publication, **25**, 19–25.

Fischer, A. G. & Herbert, T. (1986). Stratification rhythms: Italo-American studies in the Umbrian facies. *Mem. Geol. Soc. Italia*, **31**, 45–51.

Fischer, A. G., Herbert, T. D., Napoleone, G., Premoli Silva, I. & Ripepe, R.(1991). Albian pelagic rhythms (Piobbico core). *Journal of Sedimentary Petrology*, **62**, 1146–1172.

Fisher, D. C. (1994). Stratocladistics: morphological and temporal patterns and their relation to phylogenetic process. In *Interpreting the Hierarchy of Nature: From Systematic Patterns to Evolutionary Process Theories*, ed. L. Grande & O. Rieppel. pp. 133–171, Orlando: Academic Press.

Flower, B. P. & Kennett, J. P. (1993) Relations between Monterey Formation deposition and global cooling. *Geology*, **88**, 10–11.

(1995) Middle Miocene deepwater paleoceanography in the southwest Pacific: relations with East Antarctic ice sheet development. *Paleoceanography*, **10**, 1095–1113.

Fordham, B. G. (1986). Miocene–Pleistocene foraminifers from DSDP Sites 208 and 77, and phylogeny and classification of Cenozoic species. *Evolutionary Monographs*, **6**, 200 pp.

Fortey, R. A. (1993). Charles Lapworth and the biostratigraphic paradigm. In *Milestones in Geology*, ed. M. J. Le Bas, Geological Society, London, Memoir, **16**, 93–104.

Frerichs, W. E. (1971). Evolution of planktonic foraminifera and paleotemperatures. *Journal of Paleontology*, **45**, 963–968.

Funnell, B. M. (1964). The Tertiary Period. In *The Phanerozoic time scale: a symposium*. *Quarterly Journal Geological Society London*, **120**(S), 179–191. [no editor named]

Gallagher, S. J. & Holdgate, G. R. (1996). *Sequence Stratigraphy and Biostratigraphy of the Onshore Gippsland Basin, S. E. Australia*. Australian Sedimentologists Group Field Guide Series No. 11, Geological Society of Australia, 70 pp.

Gardner, J. (1931). Relation of certain foreign faunas to Midway fauna of Texas. *American Association of Petroleum Geologists, Bulletin*, **15**, 149–160.

Gayon, J. (1990). Critics and criticisms of the modern synthesis: the viewpoint of a philosopher. *Evolutionary Biology*, **24**, 1–49.

Geikie, A. (1905). *The Founders of Geology*. London: Macmillan.

George, T. N. (1956). Biospecies, chronospecies and morphospecies. *Systematics Association Publication No.* **2**, 123–137.

George, T. N. *et al.* (1967). The stratigraphical code – Report of the Stratigraphical Code Subcommittee. *Geological Society of London, Proceedings*, **1638**, 75–87.

(1969). Recommendations on stratigraphical usage. *Geological Society of London, Proceedings*, **1638**, 139–166 (2nd revision of 1967 Report of the Stratigraphical Code Subcommittee).

Ghiselin, M. T. (1974). A radical solution to the species problem. *Systematic Zoology*, **23**, 536–544.

(1997). *Metaphysics and the Origin of Species*. Albany: State University of New York Press, 377 pp.

Gignoux, M. (1955). *Stratigraphic Geology* (English translation by G. G. Woodford of the 1950 French edition of 'Géologie Stratigraphique'). San Francisco: W. H. Freeman, 682 pp.

Gili, E., Skelton, P. W., Vicens, E. & Obrador, A. (1995). Corals to rudists – an environmentally induced assemblage succession. *Palaeogeography, Palaeoclimatology, Palaeoecology*, **119**, 127–136.

Gillispie, C. C. (1951). *Genesis and Geology: A Study in the Relations of Scientific Thought, Natural Theology, and Social Opinion in Great Britain, 1790–1850*. Cambridge, MA: Harvard University Press.

Gilluly, J. (1949). Distribution of mountain building in geologic time. *Geological Society of America, Bulletin*, **60**, 561–590.

Gingerich, P. D. (1979). Stratophenetic approach to phylogeny reconstruction in vertebrate paleontology. In *Phylogenetic Analysis and Paleontology*, ed. J. Cracraft & N. Eldredge, pp. 41–79. New York: Columbia University Press.

(1990). Stratophenetics. In *Paleobiology: A Synthesis*, ed. D. E. G. Briggs & P. R. Crowther, pp. 437–441. Oxford: Blackwells Scientific Publishing.

Glaessner, M. F. (1937). Planktonforaminiferen aus der Kreide und dem Eozän und ihre stratigraphische Bedeutung. *Studies in Micropaleontology*, **1**, 27–52.

(1943). Problems of stratigraphic correlation in the Indo-Pacific region. *Proceedings of the Royal Society of Victoria*, **55**, 41–80.

(1945). *Principles of Micropalaeontology*. Melbourne: Melbourne University Press, 296 pp.

(1951). Three foraminiferal zones in the Tertiary of Australia. *Geological Magazine*, **88**, 273–283.

(1953). Time-stratigraphy and the Miocene Epoch. *Geological Society of America, Bulletin*, **64**, 647–658.

(1955). Taxonomic, stratigraphic and ecologic studies of foraminifera and their interrelations. *Micropaleontology*, **1**, 3–8.

(1966). Problems of palaeontology. *Journal of the Geological Society of India*, **7**, 14–27.

(1967). Time scales and Tertiary correlations. In *Tertiary Correlations and Climatic Changes in the Pacific*, ed. K. Hatai, 11th Pacific Science Congress, Tokyo, 1966, Symposium No. **25**, 1–5.

Glaessner, M. F. & Wade, M. (1958). The St. Vincent Basin. In *Geology of South Australia*, ed. M. F. Glaessner & L. W. Parkin. *Journal of the Geological Society of Australia*, **5**(2), 115–126.

Goldstein, S. (1999). Foraminifera: a biological overview. In *Modern Foraminifera*, ed. B. K. Sen Gupta, pp. 37–56. Dordrecht: Kluwer Academic Publishers.

Gould, S. J. (1980). Is a new and general theory of evolution emerging? *Paleobiology*, **6**, 119–130.

(1982). Darwinism and the expansion of evolutionary theory. *Science*, **216**, 380–387.

(1984). Toward the vindication of punctuational change. In *Catastrophes and Earth History*, ed. W. A. Berggren & J. A. van Couvering, pp. 9–34. Princeton: Princeton University Press.

(1985). The paradox of the first tier: an agenda for paleobiology. *Paleobiology*, **11**, 2–12.

(1986). Evolution and the triumph of homology, or why history matters. *American Scientist*, Jan–Feb, 60–69.

(1987). *Time's Arrow, Time Cycle*. Cambridge, MA: Harvard University Press.

(1996). *Life's Grandeur: The Spread of Excellence from Plato to Darwin*. London: Jonathan Cape.

(2002). *The Structure of Evolutionary Theory*. Cambridge, MA, & London: Belknap Press of Harvard University Press.

Gould, S. J. & Eldredge, N. (1977). Punctuated equilibria; the tempo and mode of evolution reconsidered. *Paleobiology*, **3**, 115–151.

(1993). Punctuated equilibrium comes of age. *Nature*, **366**, 223–227.

Gould, S.J. & Lewontin, R.W. (1979). The spandrels of San Marco and the Panglossian paradigm: a critique of the adaptationist programme. *Proceedings of the Royal Society of London, Series B*, **205**, 581–598.

Grabau, A. W. (1940). *The Rhythm of the Ages*. Peking: Henri Vetch Publishers.

Gradstein, F. M. & Agterberg, F. P. (1982). Models of Cenozoic foraminiferal stratigraphy – northwestern Atlantic margin. In *Quantitative Stratigraphic Correlation*, ed. J. M. Cubitt & R. A. Reyment, pp. 119–173. New York: John Wiley.

(1985). Quantitative correlation in exploration micropaleontology. In *Quantitative Stratigraphy*, ed. F. M Gradstein, F. P. Agterberg, J. C. Brower & W. S. Schwarzacher, pp. 309–357. Paris: UNESCO & D. Reidel Publishing Company.

Gregory, T. R. (2004). Macroevalution, hierarchy theory, and the C-value sigma. *Paleobiology* **30**: 179–202.

Grünbaum, A. (1963). *Philosophical Problems of Space and Time*. New York: Alfred A. Knopf, 448 pp.

Guex, J. (1991). *Biochronological Correlations*. Berlin: Springer-Verlag.

Hacohen, M. H. (2000). *Karl Popper, the Formative Years 1902–1945: Politics and Philosophy in Interwar Vienna*. Cambridge University Press.

Hailwood, E. A. (1989). The role of magnetostratigraphy in the development of geological time scales. *Paleoceanography,* **4**, 1–18.

Hallam, A. (1983). Plate tectonics and evolution. In *Evolution from Molecules to Men*, ed. D. S. Bendall, pp. 367–386. Cambridge: Cambridge University Press.

(1989). *Great Geological Controversies*, Second edition. Oxford: Oxford University Press.

(1993). *Phanerozoic Sea Level Changes*. New York: Columbia University Press.

(1985). Why are larger foraminifera large? *Paleobiology*, **11**, 195–208.

(1987). Fluctuations in the trophic resource continuum: a factor in global diversity cycles? *Paleoceanography* **2**, 457–471.

(1999). Symbiont-bearing foraminifera. In *Modern Foraminifera*, ed. B. K. Sen Gupta, pp. 123–139. Dordrecht: Kluwer Academic Publishers.

Hallock, P., Premoli Silva, I. & Boersma, A. (1991). Similarities between planktonic and larger foraminiferal evolutionary trends through Paleogene paleoceanographic changes. *Palaeogeography, Palaeoclimatology, Palaeoecology*, **83**, 49–64.

Hancock, J. M. (1977). The historic development of concepts of biostratigraphic correlation. In *Concepts and Methods of Biostratigraphy*, ed. E. G. Kauffman & J. E. Hazel, pp. 3–22. Stroudsberg, PA: Dowden, Hutchison & Ross.

Hansen, T. A. (1988). Early Tertiary radiation of marine molluscs and the long-term effects of the Cretaceous–Tertiary extinction. *Paleobiology*, **14**, 37–51.

Haq, B. U. (1982). Climatic acme events in the sea and on land. In *Climate and Earth History*, pp. 126–132. Washington: National Academy Press.

Haq, B. U. & Boersma, A., Editors (1978). *Introduction to Marine Micropaleontology*. New York: Elsevier.

Haq, B. U., Hardenbol, J. & Vail, P. R. (1987). The chronology of fluctuating sea level since the Jurassic. *Science*, **235**, 1156–67.

(1988). Mesozoic and Cenozoic chronostratigraphy and cycles of sea-level change. In *Sea-Level Changes, an Integrated Approach*, ed. C. K. Wilgus, B. S. Hastings, C. G. S. C. Kendall, H. W. Posamentier, C. A. Ross, & J. C. van Wagoner, Society of Economic Paleontologists and Mineralogists, Special Publication, **42**, 71–108.

Hardenbol, J. & Berggren, W. A. (1978). A new Paleogene numerical time scale. In *Contributions to the Geological Time Scale*, ed. G. V. Cohee, M. F. Glaessner &

H. D. Hedberg. American Association of Petroleum Geologists, Studies in Geology, **6**, 216–234.

Hardenbol, J., Thierry, J., Farley, M. B., Jacquin, T., de Graciansky, P.-C. & Vail, P. R. (1998). Mesozoic and Cenozoic sequence chronostratigraphic framework of European basins. In *Mesozoic and Cenozoic Sequence Stratigraphy of European Basins*, ed. P.-C. de Graciansky, J. Hardenbol, T. Jacquin & P. R. Vail, SEPM (Society of Sedimentary Geology) Special Publication No. **60**, 3–13.

Harland, W. B. (1973). Stratigraphic classification, terminology and usage – essay review of 'Hedberg, H. D., Editor, 1972. An international guide to stratigraphic classification, terminology and usage, Introduction and summary'. *Geological Magazine*, **110**, 567–574.

(1975). The two geological time scales. *Nature*, **253**, 505–507.

(1992). Stratigraphic regulation and guidance: a critique of current tendencies in stratigraphic codes and guides. *Geological Society of America Bulletin*, **104**, 1231–1235.

Harland, W. B., Armstrong, R. L., Cox, A. V., Craig, L. E., Smith, A. G. & Smith, D. G. (1990). *A Geological Time Scale 1989*. Cambridge: Cambridge University Press, 163 pp.

Hart, M. B. (1980). A water depth model for the evolution of the planktonic Foraminiferida. *Nature*, **286**, 252–254.

(1990). Major evolutionary radiations of the planktonic foraminiferida. In *Major Evolutionary Radiations*, ed. P. D. Taylor & G. P. Larwood. Systematics Association Special Volume No. **42**, 59–72.

Harzhauser, M. & Piller, W. E. (2004). Integrated stratigraphy of the Sarmatian (upper Middle Miocene) in the western central Paratethys. *Stratigraphy*, **1**, 65–86.

Hay, W. W. (1972). Probabilistic stratigraphy. *Ecologae Geologicae Helvetiae*, **65**, 255–266.

Hay, W. W. & Mohler, H. P. (1969). Paleocene–Eocene calcareous nannoplankton and high-resolution biostratigraphy. In *Proceedings of the First International Conference on Planktonic Microfossils,* ed. P. Brönnimann & H. H. Renz, pp. 250–253. Leiden: E. J. Brill.

Hay, W. W. & Southam, J. R. (1978). Quantifying biostratigraphic correlation. *Annual Reviews of Earth and Planetary Sciences*, **6**, 353–375.

Haynes, J. (1981). *Foraminifera*. London: Macmillan.

Hays, J. D., Imbrie, J. & Shackleton, N. J. (1976). Variations in the earth's orbit: pacemaker of the ice ages. *Sciences*, **194**, 1121–1132.

Hazel, J. E. (1989). Chronostratigraphy of Upper Eocene microspherules. *Palaios*, **4**, 318–329.

(1993). Biostratigraphy. In *Fossil Prokaryotes and Protists*, ed. J. H. Lipps, pp. 44–50. Boston: Blackwell Scientific Publishers.

Heath, R. S. & McGowran, B. (1984). Neogene datum planes: foraminiferal successions in Australia with reference sections from the Ninety-east Ridge and the Ontong-Java Plateau. In *Pacific Neogene Datum Planes: Contributions to Biostratigraphy and Chronology*, ed. N. Ikebe & R. Tsuchi, pp. 187–192. Tokyo: University of Tokyo Press.

Hedberg, H. D. (1948). Time-stratigraphic classification of sedimentary rocks. *Geological Society of America, Bulletin* **59**, 447–462.

(1959). Toward harmony in stratigraphic classification. *American Journal of Science*, **257**, 674–683.

(1961). The stratigraphic panorama (an inquiry into the bases for age determination and age classification of the earth's rock strata). *Geological Society of America, Bulletin* **72**, 499–518.

ed. (1976). *International Stratigraphic Guide: A Guide to Stratigraphic Classification, Terminology and Procedure*. New York: John Wiley & Sons, 200 pp.

Heezen, B. C. & MacGregor, I. (1973). The evolution of the Pacific. *Scientific American*, **229**, 102–112.

Heezen, B. C., MacGregor, I., Foreman, H. P., Forristall, G. Z., Hekel, H., Hesse, Hoskins, R. H., Jones, E. J. W., Krasheninnikov, V., Okada, H. & Ruff, M. H. (1973). Diachronous deposits: a kinematic interpretation of the post-Jurassic sedimentary sequence on the Pacific plate. *Nature*, **241**, 25–32.

Heirtzler, J. R., Dickson, G. O., Herron, E. M., Pitman, W. C. & Le Pichon, X. (1968). Marine magnetic anomalies, geomagnetic field reversals, and motions of the ocean floor and continents. *Journal Geophysical Research*, **73**, 2119–2136.

Hemleben, Ch., Spindler, M. & Anderson, O. R. (1989). *Modern Planktonic Foraminifera*. New York: Springer-Verlag, 363 pp.

Herbert, T. D. (1999). Toward a composite orbital chronology for the late Cretaceous and early Palaeocene GPTS. *Philosophical Transactions Royal Society*, **357**, 1891–1905.

Herbert, T. D. & D'Hondt, S. L. (1990). Precessional climate cyclicity in late Cretaceous–early Tertiary marine sediments: a high resolution chronometer of Cretaceous–Tertiary boundary events. *Earth & Planetary Science Letters*, **99**, 263–275.

Herbert, T. D. & Fischer, A. G. (1986). Milankovitch climatic origin of mid-Cretaceous black shale rhythms in central Italy. *Nature*, **321**, 739–743.

Herbert, T. D., Gee, J. S. & DiDonna, S. (1999). Precessional climatic cycles in the late Cretaceous South Atlantic: Long-term consequences of high-frequency variations. In *Late Cretaceous Climates*, ed. E. Barrera & C. Johnson. Boulder: Geological Society of America Special Volume 322, 105–120.

Herbert, T. D., Premoli Silva, I., Erba, E. & Fischer, A. G. (1995). Orbital chronology of Cretaceous–Paleocene marine sediments. In *Geochronology Time Scales and Global Stratigraphic Correlation*, ed. W. A. Berggren, D. V Kent, M.-P. Aubry & J. Hardenbol. Tulsa: SEPM (Society of Sedimentary Geology) Special Publication **54**, 81–94.

Herbert, T. D., Stallard, R. F & Fischer, A. G. (1986). Anoxic events, productivity rhythms, and the orbital signature in a mid-Cretaceous pelagic core. *Paleoceanography*, **1**, 495–506.

Hess, J., Stott, L. D., Bender, M. L., Kennett, J. P. & Schilling, J. G. (1989). The Oligocene marine microfossil record: age assessments using strontium isotopes. *Paleoceanography*, **4**, 655–679.

Hesse, R., Foreman, H. P., Forristall, G. Z., Heezen, B. C., Hekel, H., Hoskins, R. H., Jones, E. J. W., Kaneps, A. G., Krasheninnikov, V., MacGregor, I. & Okada, H. (1974). Walther's facies rule in pelagic realm – a large-scale example from the Mesozoic–Cenozoic Pacific. *Zeitschrift Deutsches Geologisches Gesellschaft*, **125**, 151–172.

Hickey, L. J., West, R. M., Dawson, M. R. & Choi, D. K. (1983). Arctic terrestrial biota: paleomagnetic evidence of age disparity with mid–northern latitudes during the Late Cretaceous and Early Tertiary. *Science*, **221**, 153–1154.

Hilgen, F. J. (1991). Astronomical calibration of Gauss to Matuyama sapropels in the Mediterranean and implications for the geomagnetic polarity time scale. *Earth and Planetary Science Letters*, **104**, 226–244.

(1994). An astronomically calibrated (polarity) time scale for the Pliocene–Pleistocene: a brief review. *International Association of Sedimentologists, Special Publications*, **19**, 109–116.

Hilgen, F. J. & Langereis, C. G. (1993). A critical re-evaluation of the Miocene/Pliocene boundary as defined in the Mediterranean. *Earth & Planetary Science Letters*, **118**, 167–179.

(1994). Reply to comment on 'A critical re-evaluation of the Miocene–Pliocene boundary as defined in the Mediterranean', by R. H. Benson & D. A. Hodell. *Earth and Planetary Science Letters*, **118**, 124, 251–254.

Hilgen, F. J. & Krijgsman, W. (1999). Cyclostratigraphy and astrochronology of the Tripoli diatomite formation (pre-evaporite Messinian, Sicily, Italy). *Terra Nova*, **11**, 16–22.

Hilgen, F. J., Aziz, H. A., Krijgsman, W. *et al.* (1999). Present status of the astronomical (polarity) time-scale for the Mediterranean Neogene. *Philosophical Transactions of the Royal Society of London A*, **357**, 1931–1947.

Hilgen, F. J., Krijgsman, W., Langereis, C. G. & Lourens, L. J. (1997). Breakthrough made in dating of the geological record. *Eos*, **78**, 285–292.

Hilgen, F. J., Krijgsman, W., Raffi, I., Turco, E. & Zachariasse (2000). Integrated stratigraphy and astronomical calibration of the Serravallian–Tortonian boundary section at Monte Gibliscemi (Sicily, Italy). *Marine Micropaleontology*, **38**, 181–211.

Hilgen, F. J., Lourens, L. J., Berger, A. & Loutre, M. F. (1993). Evaluation of the astronomically calibrated time scale for the late Pliocene and earliest Pleistocene. *Paleoceanography*, **8**, 549–566.

Hills, S. J. & Thierstein, H. R. (1989). Plio-Pleistocene calcareous plankton biochronology. *Marine Micropaleontology*, **14**, 67–96.

Hodell, D. A. (1994). Editorial: progress and paradox in strontium isotope stratigraphy. *Paleoceanography*, **9**, 395.

Hodell, D. A., Curtis, J. H., Sierro, F. J. & Raymo, M. E. (2001). Correlation of late Miocene to early Pliocene sequences between the Mediterranean and North Atlantic. *Paleoceanography*, **16**, 164–178.

Hodell, D. A. & Kennett, J. P. (1986). Late Miocene–early Pliocene stratigraphy and paleoceanography of the South Atlantic and southwest Pacific Oceans: a synthesis. *Paleoceanography*, **1**, 285–311.

Hodell, D. A. & Vayavananda, A. (1993). Middle Miocene paleoceanography of the western equatorial Pacific (DSDP site 289) and the evolution of *Globorotalia (Fohsella)*. *Marine Micropaleontology*, **22**, 279–310.

Hodell, D. A. & Woodruff, F. (1994). Variations in the strontium isotopic ratio of sea water during the Miocene: stratigraphic and geochemical implications. *Paleoceanography*, **9**, 405–426.

Hodge, M. J. S. (1989). Darwin's theory and Darwin's argument. In *What the Philosophy of Biology Is: Essays Dedicated to David Hull*, ed. M. Ruse, pp. 163–182. Dordrecht: Kluwer Academic, 337 pp.

Hoffman, A. (1983). Paleobiology at the crossroads: a critique of some modern paleobiological research programs. In *Dimensions of Darwinism*, ed. M. Grene, pp. 241–271. Cambridge: Cambridge University Press.

   (1989). *Arguments on Evolution: A Paleontologist's Perspective*. New York, Oxford: Oxford University Press, 274 pp.

Hoffmann, A. & Reif, W.-E. (1988). On methodology of the biological sciences: from an evolutionary biological perspective. *Neues Jahrbuch Geologie & Paläontologie, Abhandlungen*, **177**, 185–211.

Holdgate, G. R & Gallagher, S. J. (1997). Microfossil paleoenvironments and sequence stratigraphy of Tertiary cool-water carbonates, onshore Gippsland Basin, S. E. Australia. In *Cool and Temperate Water Carbonates*, ed. N. James & J. D. A. Clarke. Tulsa: Special Publication of the Society of Economic Paleontologists and Mineralogists, **56**, 205–220.

Holland, C. H. (1978). Stratigraphical classification and all that. *Lethaia*, **11**, 85–90.

   (1990). Biostratigraphic units and the stratorype/golden spike concept. In *Paleobiology: A Synthesis*, ed. D. E G. Briggs & P. R. Crowther, 461–465. Oxford: Blackwells Scientific Publishing.

Holland, S. M. (1995). Depositional sequences, facies control and the distribution of fossils. In *Sequence Stratigraphy and Depositional Response to Eustatic, Tectonic and Climatic Forcing*, ed. B. U. Haq. pp. 1–23, Dordrecht: Kluwer Academic Publishers.

   (1999). The new stratigraphy and its promise for paleobiology. *Paleobiology*, **25**, 409–416.

   (2000). The quality of the fossil record: a sequence stratigraphic perspective. In *Deep time: Paleobiology's Perspective*, ed. D. H. Erwin & S. L. Wing, Paleobiology, Supplement to vol. **26**(4), 148–168.

Hornibrook, N. deB. (1958). New Zealand Foraminifera: Key species in stratigraphy – No.6. *New Zealand Journal Geology & Geophysics*, **1**, 653–676.

   (1965). A viewpoint on stages and zones. *New Zealand Journal of Geology & Geophysics*, **8**, 1195–1212.

   (1969). *Report on a Visit to the U.S.A. to Attend the I.U.G.S. Working Group for a Biostratigraphic Zonation of the Cretaceous and the Cenozoic*. Lower Hutt: New Zealand Geological Survey Report **42**, 1–11.

   (1971). Inherent instability of biostratigraphic zonal schemes. *New Zealand Journal of Geology & Geophysics*, **14**, 727–733.

   (1976). Jurassic, Cretaceous and Cenozoic stages divisions and zones used in New Zealand. *United Nations Mineral Resources Development Series*, **42**, 81–93.

Hornibrook, N. deB. & Edwards, A. R. (1971). Integrated planktonic foraminiferal and calcareous nannoplankton datum levels in the New Zealand Cenozoic. In *Proceedings of the II Planktonics Conference, Roma 1970*, ed. A. Farinacci, 649–657. Roma: Edizioni Technoscienza.

Hornibrook, N. deB., Brazier, R. C. & Strong, C. P. (1989). *Manual of New Zealand Permian to Pleistocene Foraminiferal Biostratigraphy*. Lower Hutt, New Zealand Geological Survey Paleontological Bulletin **56**, 175 pp.

Hottinger, L. (1981). *The Resolution Power of the Biostratigraphic Clock Based on Evolution and its Limits*. Barcelona: Universidad Barcelona, International Symposium on Concepts and Methods in Palaeontology, pp. 233-242.

  (1982). Larger foraminifera, giant cells with a historical background. *Naturwissenschaft,* **69**, 361-371.

  (1983). Processes determining the distribution of larger foraminifera in space and time. *Utrecht Micropaleontological Bulletins,* **30**, 239-253.

  (1990). Significance of diversity in shallow benthic foraminifera. *Atti del Quarto Simposio Ecologia e Paleoecologia Comunità Benthonice*, Sorrento, 1-5 November 1988. Museo Regionale di Scienze Natural – Torino, pp. 35-51.

  (1996). Sels nutritifs et biosédimentation. *Mémoir de la Societé géologique de France*, n.s., **169**, 99-107.

  (1997). Shallow benthic foraminiferal assemblages as signals for depth of their deposition and their limitations. *Bulletin de la Société géologique de France*, **168**, no. 4, 491-505.

  (1998). Shallow benthic foraminifera at the Paleocene-Eocene boundary. *Strata*, **9**, 61-64.

House, M. (1985). *A new approach to an absolute timescale from measurements of orbital cycles and sedimentary microrhythms*. *Nature*, **316**, 721-725.

Huber, B. T., Bijma, J. & Darling, K. (1997). Cryptic speciation in the living planktonic foraminifer *Globigerinella siphonifera* (d'Orbigny). *Paleobiology*, **23**, 33-62.

Hughes, N. F. (1989). *Fossils as Information: New Recording and Stratal Correlation Techniques*. Cambridge, New York: Cambridge University Press, 136 pp.

Hulburt, R. C. Jr (1993). Taxic evolution in North American Neogene horses (Subfamily Equinae): the rise and fall of an adaptive radiation. *Paleobiology*, **19**, 216-234.

Hull, D. (1983). Popper and Plato's metaphor. In *Advances in Cladistics*, Vol. **2**, ed. N. Platnick & V. A. Funk, pp. 177-189. New York: Columbia University Press.

  (1984). Historical entities and historical narratives. In *Minds Machines and Evolution*, ed. C. Hookway, pp. 17-42. Cambridge: Cambridge University Press.

  (1988). *Science as a Process: An Evolutionary Account of the Social and Conceptual Development of Science*. Chicago: University of Chicago Press, 583 pp.

  (1989). A function for actual examples in philosophy of science. In *What the Philosophy of Biology Is: Essays Dedicated to David Hull*, ed. M. Ruse, pp. 309-322. Dordrecht: Kluwer Academic Publishers.

Hull, D. L. (1997). The ideal species concept – and why we can't get it. In *Species: The Units of Biodiversity*, ed. M. F. Claridge, H. A. Dawah & M. R., pp. 357-380. Chapman & Hall.

Hunter, R. S. T., Arnold, A. J. & Parker, W. C. (1988). Evolution and homeomorphy in the development of the Paleocene *Planorotalites pseudomenardii* and the Miocene *Globorotalia (Globorotalia) margaritae* lineages. *Micropaleontology*, **31**, 181-192.

Huxley, J. S., ed. (1940). *The New Systematics*. Oxford: The Clarendon Press.

Huxley, T. (1862). The anniversary address. *Geological Society of London, Quarterly Journal*, **18**, xl–liv.

Iaccarino, S. (1985). Mediterranean Miocene and Pliocene planktic foraminifera. In *Plankton Stratigraphy*, ed. H. M. Bolli, J. B. Saunders & K. Perch-Nielsen, pp. 283–315. Cambridge: Cambridge University Press, 2 volumes.

Imbrie, J. & Imbrie, K. P. (1979). *Ice Ages: Solving the Mystery*. London: Macmillan.

Ingle, J. C. (1973). *Summary Comments on Neogene Biostratigraphy, Physical Stratigraphy, and Paleo-Oceanography in the Marginal Northeast Pacific Ocean. Initial Reports of the Deep Sea Drilling Project*, **18**, 949–960.

Israelsky, M. C. (1949). Oscillation chart. *American Association of Petroleum Geologists, Bulletin*, **33**, 92–98.

Ivany, L. C. (1999). So ... now what? Thoughts and ruminations about coordinated stasis. *Palaios*, **14**(4), 4 pp.

Ivany, L. C. & Schopf, K. M. (1996). New perspectives on faunal stability in the fossil record. *Palaeogeography, Palaeoecology, Palaeoclimatology, Special Issue*, **127**, 359 pp.

Jackson, J. B. C. and Johnson, P. A. (2000). Life in the last few million years. In *Deep Time: Paleobiology's Perspective*, ed. D. H. Erwin & S. L. Wing, *Paleobiology*, Supplement to vol. **26**(4), 221–235.

Jackson, J. B. C., Budd, A. F. & Pandolfi, J. M. (1996). The shifting balance of natural communities? In *Evolutionary Paleobiology: Essays in Honor of James W. Valentine*, ed. D. Jablonski, D. H. Erwin & J. H. Lipps, pp. 89–122. Chicago: University of Chicago Press.

Jenkins, D. G. (1960). Planktonic foraminifera from the Lakes Entrance oil shaft, Victoria, Australia. *Micropaleontology*, **6**, 345–371.

(1966a). Planktonic foraminiferal zones and new taxa from the Danian to lower Miocene of New Zealand. *New Zealand Journal of Geology & Geophysics*, **8**, 1088–1126.

(1966b). Planktonic foraminiferal datum planes in the Pacific and Trinidad Tertiary. *New Zealand Journal of Geology & Geophysics*, **9**, 424–427.

(1967). Planktonic foraminiferal zones and new taxa from the Lower Miocene to the Pleistocene of New Zealand. *New Zealand Journal of Geology & Geophysics*, **10**, 1064–1078.

(1968). Variations in the numbers of species and subspecies of planktonic Foraminiferida as an indicator of New Zealand Cenozoic paleotemperatures. *Paleogeography, Paleoclimatology, Paleoecology*, **5**, 309–313.

(1971). New Zealand Cenozoic planktonic foraminifera. Lower Hutt: *New Zealand Geological Survey Paleontological Bulletin*, **56**, 175 pp.

(1973). The present status and future progress in the study of Cenozoic planktonic foraminifera. *Revista Española de Micropaleontología*, **5**, 133–146.

(1985). Southern mid-latitude Paleocene to Holocene planktonic foraminifera. In *Plankton Stratigraphy*, ed. H. M. Bolli, J. B. Saunders & K. Perch-Nielsen, pp. 263–282. Cambridge: Cambridge University Press.

(1993). Cenozoic southern mid- and high-latitude biostratigraphy and chronostratigraphy based on planktonic foraminifera. In *The Antarctic Paleoenvironment: a Perspective on Global Change*, ed. J. P. Kennett & D. A. Warnke, pp. 125–144.

Washington, D.C.: Antarctic Research Series, Part Two, vol. **60**, American Geophysical Union.

Johnson, D. A. & Nigrini, C. (1985). Time-transgressive late Cenozoic radiolarian events of the equatorial Indo-Pacific. *Science*, **230**, 538–540.

JOIDES (Joint Oceanographic Institutes for Deep Earth Sampling) (1981). Report of the Conference on scientific ocean drilling, November 16–18, 1981. Washington: JOI Inc.

Jukes, J. B. (1862). *The Student's Manual of Geology*. Edinburgh: Adam & Charles Black, 763 pp.

Kauffman, E. G. (1969). Evolutionary rates and biostratigraphy. In *North American Paleontological Convention* (Chicago, 1969), Proc., Part F (Correlation by Fossils),

(1977). Evolutionary rates and biostratigraphy. In *Concepts and Methods of Biostratigraphy*, ed. E. G Kauffman & J. E. Hazel, pp. 109–142. Stroudsberg, PA: Dowden, Hutchison & Ross.

(1987). The uniformitarian albatross. *Palaios*, **2**, 531.

Kauffman, E. G., Elder, W. P. & Sageman, B. B. (1991). High-resolution correlation: a new tool in stratigraphy. In *Cycles and Events in Stratigraphy*, ed. G. Einsele *et. al.*, pp. 795–819. Berlin: Springer-Verlag.

Kellogg, T. B (1975). In *Investigation of Late Quaternary Paleoceanography and Paleoclimatology*, ed. R. M. Cline & J. D. Hays. Geological Society of America Memoir 145.

Kelly, D. C., Arnold, A. J. & Parker, W. C. (1996). Paedomorphosis and the origin of the planktonic foraminiferal genus *Morozovella* . *Paleobiology*, **22**, 266–281.

Kelly, D. C., Bralower T. J. & Zachos, J. C. (1998). Evolutionary consequences of the latest Paleocene thermal maximum for tropical planktonic foraminifera. *Palaeogeography, Palaeoclimatology, Palaeoecology*, **141**, 139–161.

Kemple, W. G., Sadler, P. M. & Strauss, D. J. (1995). Extending graphic correlation to N dimensions. In *Graphic Correlation and the Composite Standard Approach*, ed. K. O. Mann, H. R. Lane, & J. R. Stein, pp. 65–82. Tulsa: SEPM (Society for Sedimentary Geology), Special Publication No. 53.

Kennett, J. P. (1968). Latitudinal variation in *Globigerina pachyderma* (Ehrenberg) in surface sediments of the south-west Pacific Ocean. *Micropaleontology*, **14**, 305–319.

(1982). *Marine Geology*. Englewood Cliffs: Prentice-Hall, 813 pp.

Kennett, J. P. & Shackleton, N. J. (1976). Oxygen isotopic evidence for the development of the psychrosphere 38 Myr ago. *Nature*, **260**, 513–515.

Kennett, J. P. & Stott, L. D. (1991). Abrupt deep-sea warming, paleoceanographic changes and benthic extinctions at the end of the Paleocene. *Nature*, **353**, 225–229.

Kitcher, P. (1984). Species. *Philosophy of Science*, **51**, 303–333.

Kitts, D. B. (1966 (1977)). Geologic time. *Journal of Geology*, **74**, 127–146.

(1977). *The Structure of Geology*. Dallas: SMU Press.

Kleinpell, R. M. (1979). *Criteria in Correlation: Relevant Principles of Science*. Bakersfield, California: Pacific Section of the American Association of Petroleum Geologists, 44 pp.

Klitgord, K. D. & Schouten, H. (1986). Plate kinematics of the central Atlantic. In *The Geology of North America*, Vol. M, *The Western North Atlantic Region*, ed. P. R. Vogt & B. E. Tucholke, pp. 351–377. Boulder: Geological Society of America.

Knowlton, N. (1993). Sibling species in the sea. *Annual Review of Ecology and Systematics*, **24**, 189–216.

(2000). Molecular analysis of species boundaries in the sea. *Hydrobiologia*, **420**, 73–90.

Knox, R. W. O'B. (1996). Correlation of the early Paleogene in northwest Europe: an overview. In *Correlation of the Early Paleogene in Northwest Europe*, ed. R. W. O'B. Knox, R. M. Corfield & R. E. Dunay, London: Geological Society, Special Publication **101**, 1–11.

Koch, P. L., Zachos, J. C. & Gingerich, P. D. (1992). Correlations between isotope records in marine and continental reservoirs near the Paleocene/Eocene boundary. *Nature*, **385**, 319–322.

Kominz, M. A., Miller, K. G. & Browning, J. V. (1998). Long-term and short-term global Cenozoic sea-level estimates. *Geology*, **26**, 311–314.

Kouwenhoven, T. J., Seidenkrantz, M.-S. & van der Zwaan, G. J. (1999). Deep-water changes: the near-synchronous disappearance of a group of benthic foraminifera from the late Miocene Mediterranean. *Palaeogeography, Palaeoclimatology, Palaeoecology*, **152**, 259–281.

Krassilov, V. A. (1974). Causal biostratigraphy. *Lethaia*, **7**, 173–179.

(1978). Organic evolution and natural stratigraphical classification. *Lethaia*, **11**, 93–104.

Krijgsman, W., Gaboardi, S., Hilgen, F. J., Iaccarino, S., de Kaenel, E. & Van der Laan, E. (2004). Revised astrochronology for the Ain el Beida section (Atlantic Morocco): no glacio-eustatic control for the onset of the Messinian salinity crisis. *Stratigraphy*, **1**, 87–101.

Kucera, M. & Darling, K. F. (2002). Cryptic species of planktonic foraminifera: their effect on palaeoceanographic reconstructions. *Philosophical Transactions of the Royal Society. Series A*, **360**, 695–718.

Kucera, M. & Malmgren B. A. (1998). Differences between evolution of mean form and evolution of new morphotypes: an example from late Cretaceous planktonic foraminifera. *Paleobiology*, **24**, 49–63.

Laskar, J. (1999). The limits of Earth orbital calculations for geological time-scale use. *Philosophical Transactions of the Royal Society of London A*, **357**, 1735–1760.

Laskar, J., Joutel, F. & Boudin, F. (1993). Orbital, precessional, and insolation quantities for the Earth from 20 myr to 10 myr. *Astronomy & Astrophysics*, **270**, 522–533.

Laudan, R. (1976). *From Mineralogy to Geology: The Foundations of a Science, 1650–1830*. Chicago: University of Chicago Press.

(1982). Tensions in the concept of geology: natural history or natural philosophy? *Journal of the History of the Earth Sciences*, **1**, 7–13.

(1987). *From Mineralogy to Geology: The Foundations of a Science, 1650–1830*. Chicago: University of Chicago Press, 278 pp.

(1989). Individuals, species and the development of mineralogy and geology. In *What the Philosophy of Biology Is: Essays Dedicated to David Hull*, ed. M. Ruse, pp. 221–233. Dordrecht: Kluwer Academic Publishers.

Lazarus, D. B., (1983). Speciation in pelagic protists and its study in the planktonic microfossil record: a review. *Paleobiology*, **9**, 327–340.

Lazarus, D. B., Hilbrecht, H., Spencer–Cervato, C. & Thierstein, H. (1995). Sympatric speciation and phylogenetic change in Globorotalia truncatulinoides. *Paleobiology*, **21**, 28–51.

Lear, C. H., Elderfield, H. & Wilson, P. A. (2000). Cenozoic deep-sea temperatures and global ice volumes from Mg–Ca in benthic foraminiferal calcite. *Science*, **287**, 269–272.

Leckie, R. M. (1989). A paleoceanographic model for the early evolutionary history of planktonic foraminifera. *Palaeogeography, Palaeoclimatology, Palaeoecology*, **73**, 107–138.

Lee, M. S. Y. (2003). The geometric meaning of macroevolution. *Trends in Ecology and Evolution*, **18**, 263–266.

LeRoy, L. W. (1948). The foraminifer *Orbulina universa* d'Orbigny, a suggested middle Tertiary time indicator. *Journal of Palaeontology*, **22**, 500–508.

Lethiers, F. (1983). Les extensions stratigraphiques des espèces d'Ostracodes sur les plates-formes dévoniennes: le concept de répartitions sigmoidales. *Palaeogeography, Palaeoecology, Palaeoclimatology*, **43**, 299–312.

Leupold, W. & van der Vlerk, J. M. (1931). The Tertiary [of the East Indies]. Overdruk uit Leidsche Geol. Mededeelingen, Deel vol. (Festbundel K. Martin), 611–648.

Levinton, J. (2001). *Genetics, Paleontology, and Macroevolution*, Second Edition. Cambridge: Cambridge University Press.

Li, Q., & McGowran, B. (1994). Miocene upwelling events: foraminiferal evidence from southern Australia. *Australian Journal of Earth Sciences*, **41**, 593–603.

(1997). Miocene climatic oscillations recorded in the Lakes Entrance oil shaft, southeastern Australia: benthic foraminiferal response on a mid-latitude margin. *Micropaleontology*, **43**, 149–164.

(2000). The Miocene foraminifera from Lakes Entrance Oil Shaft, southeastern Australia. *Association of Australasian Palaeontologists, Memoir* **22**, 142 pp., 26 plates.

Li, Q., Davies, P. J., McGowran, B. & Van der Linden, T., (1999). Foraminiferal sequence biostratigraphy of the Oligo-Miocene Janjukian strata from Torquay, southeastern Australia. *Australian Journal of Earth Sciences* **46**, 261–273.

(2003). High-resolution foraminiferal ecostratigraphy: an experimental study on upper Oligocene cool-water carbonates from southeastern Australia. In, Olson, H. and Leckie, M. (eds), *Paleobiological, Geochemical, and other Proxies of Sea Level Change*, pp. 147–171, SEPM Special Volume **75** (Society of Sedimentary Geology).

Li, Q., McGowran, B., James, N. P. & Bone, Y., 1996. Foraminiferal biofacies on the mid-latitude Lincoln Shelf, South Australia: oceanographic and sedimentological implications. *Marine Geology*, **129**, 285–312.

Li, Q., McGowran, B. & White, M. R. (2000). Foraminiferal biostratigraphic and biofacies packages in the mid-Cainozoic Gambier Limestone, South Australia: reappraisal of foraminiferal evidence. *Australian Journal of Earth Sciences*, **47**, 955–970.

Li, Q., Simo, J. A., McGowran, B. & Holbourn, A., (2004). The eustatic and tectonic origin of Neogene unconformities from the Great Australian Bight. *Marine Geology*, **203**, 57–81.

Lindsay, E. H. (1989). The setting. In *European Neogene Mammal Chronology*, ed. E. H. Lindsay, V. Fahlbusch & P. Mein, pp. 1–15. NATO Advanced Research Workshop on European Neogene Mammal chronology, Schloss Reisenburg. New York & London: Plenum Press.

Lindsay, E. H. & Tedford, R. H. (1989). Development and application of land mammal ages in North America and Europe, a comparison. In *European Neogene Mammal Chronology*, ed. E. H. Lindsay, V. Fahlbusch & P. Mein, pp. 601–624. NATO Advanced Research Workshop on European Neogene Mammal chronology, Schloss Reisenburg. New York & London: Plenum Press.

Lindsay, E. H., Jacobs, L. L. & Butler, R. F. (1978). Biostratigraphy and magnetostratigraphy of Paleocene terrestrial deposits, San Juan Basin, New Mexico. *Geology*, **6**, 425–429.

Lindsay, E. H., Opdyke, N. D., Johnson, N. M. & Butler, R. F. (1987). Mammalian chronology and the magnetic polarity time scale. In *Cenozoic Mammals of North America: Geochronology and Biostratigraphy*, ed. M. O. Woodburne, 269–284. Berkeley: University of California Press.

Lindsay, J. M. (1967). Foraminifera and stratigraphy of the type section of Port Willunga Beds, Aldinga Bay, South Australia. *Transactions Royal Society of South Australia*, **91**, 93–110.

(1981). Tertiary stratigraphy and foraminifera of the Adelaide City area, St Vincent Basin, South Australia. Unpublished MSc Thesis, The University of Adelaide, Adelaide.

(1985). Aspects of South Australian foraminiferal biostratigraphy, with emphasis on studies of *Massilina* and *Subbotina*. In *Stratigraphy, Palaeontology, Malacology, Papers in Honour of Dr Nell Ludbrook*, ed. J. M. Lindsay, pp. 187–214. Adelaide Department of Mines and Energy, Special Publication 5, D. J. Woolman, Government Printer.

Lipps, J. H. (1970). Plankton evolution. *Evolution*, **24**, 1–22.

(1981). What, if anything, is micropaleontology? *Paleobiology*. **7**, 167–199.

ed. (1993). *Fossil Prokaryotes and Protists*. Boston: Blackwell Scientific Publishers.

Loeblich, A. R. & Tappan, H. (1957a). Correlation of the Gulf and Atlantic Coastal Plain Paleocene and Lower Eocene formations by means of planktonic foraminifera. *Journal of Paleontology*, **31**, 1109–1136.

(1957b). Planktonic foraminifera of Paleocene and early Eocene age from the Gulf and Atlantic Coastal Plain. In Loeblich, A. R., & Collaborators (1957), *Studies in foraminifera. US National Museum Bulletin* **215**, 173–198.

(1964a). Sarcodina, chiefly 'Thecamoebians' and Foraminiferida, in *Treatise on Invertebrate Paleontology*, ed. R. C. Moore. Boulder: Geological Society of America, Part C, vols. 1–2, 900 pp.

(1964b). Foraminiferal classification and evolution. *Journal of the Geological Society of India*, **5**, 5–40.

Loeblich, A. R. *et al.* (1957). Studies in foraminifera. *US National Museum Bulletin* **215**.

Lohmann, G. P. & Malmgren, B. A. (1983). Equatorward migration of *Globorotalia truncatulinoides* ecophenotypes through the late Pleistocene:gradual evolution or ocean change? *Paleobiology*, **9**, 414–421.

Loutit, T. S., ed. (1992). Sea Level Working Group (SLWG), Report. JOIDES Journal, **18**, 28-36.

Loutit, T. S. & Kennett, J. P. (1981). New Zealand and Australian Cenozoic cycles and global sea-level changes. *The American Association of Petroleum Geologists Bulletin*, **65**, 1586-1601.

Loutit, T. S., Hardenbol, J. & Vail, P. R. (1988). Condensed sections: the key to age determination and correlation of continental margin sequences. In *Sea-Level Changes, an Integrated Approach*, ed. C. K. Wilgus, B. S. Hastings, C. G. S. C. Kendall, H. W. Posamentier, C. A. Ross & J. C. van Wagoner. Society of Economic Paleontologists and Mineralogists, Memoir, **42**, 183-213.

Loutit, T. S., Romine, K. K. & Foster, C. B. (1997). Sequence biostratigraphy, petroleum exploration and *A. cinctum*. *The APPEA Journal* **1997**, 272-284.

Lovejoy, A. O. (1936). *The Great Chain of Being*. Cambridge, MA: Harvard University Press.

Lowman, S. W. (1949). Sedimentary facies in the Gulf coast. *The American Association of Petroleum Geologists Bulletin*, **39**, 1939-1947.

Ludbrook , N. H. (1971). Stratigraphy and correlation of marine sediments in the western part of the Gambier Embayment, In *The Otway Basin in Southeasten Australia*, ed. H. Wopfner, H. & Douglas J. G., pp. 4–6. Adelaide and Melbourne, Geological Surveys of South Australia & Victoria, Special Bulletin.

  (1973). *Distribution and Stratigraphic Utility of Cenozoic Molluscan Faunas in Southern Australia*. Tohoku University, Science Reports, 2ndseries (Geology), Special Volume No. **6** (Hatai Memorial Volume), pp. 241–261.

Ludbrook, N. H. & Lindsay, J. M. (1969). Tertiary foraminiferal zones in South Australia. In *Proceedings of the First International Conference on Planktonic Microfossils*, ed. P. Brönnimann & Renz, H. H., pp. 36–74, vol. 2. Leiden: E. J. Brill.

Luterbacher, H.-P. (1964). Studies in some *Globorotalia* from the Paleocene and lower Eocene of the central Apennines. *Ecologae Geologicae Helvetiae*, **57**, 631-730.

  (1998). Sequence stratigraphy and the limitations of biostratigraphy in the marine Paleogene strata of the Tremp Basin (central part of the southern Pyrenean foreland basins, Spain). In *Mesozoic and Cenozoic Sequence Stratigraphy of European Basins*, ed. P.-C. de Graciansky, J. Hardenbol, T. Jacquin, & P. R. Vail, SEPM (Society of Sedimentary Geology) Special Publication No. **60**, 303-310.

Luterbacher, H.-P. & Premoli Silva, I. (1964). Biostratigrafia del limite Cretaceo-Terziarionell' Appennino centrale. *Rivista Italiana Paleontologia Stratigrafia*, **70**, 67-128.

Luyendyk , B. (1977). Deepsea drilling on the Ninetyeast Ridge: synthesis and a tectonic model. In *Indian Ocean Geology and Biostratigraphy*, Studies Following Deep Sea Drilling Legs 22-29, ed. J. R. Heirtzler, H. M. Bolli, T. A. Davies, J. B. Saunders, & J. G. Sclater, Washington, D.C.: American Geophysical Union, p. 165-168.

Lyell, C. (1871). *Students Elements of Geology*. London: John Murray.

MacFadden, B. J. & Hulburt, R. Jr (1988). Explosive speciation at the base of the adaptive radiation of Miocene grazing horses. *Nature*, **336**, 466–468.

MacLeod, N. (1991). Punctuated anagenesis and the importance of stratigraphy to paleobiology. *Paleobiology*, **17**, 167–188.

(1993). The Maastrichtian-Danian radiation of triserial and biserial planktic foraminifera: testing phylogenetic and adaptational hypotheses in the (micro) fossil record. *Marine Micropaleontology*, **21**, 547–100.

(2001). The role of phylogeny in quantitative paleobiological data analysis. *Paleobiology* **27**, 226–240.

MacLeod, N. & Keller, G. (1991a). Hiatus distributions and mass extinctions at the Cretaceous–Tertiary boundary. *Geology*, **19**, 109–147.

(1991b). How complete are Cretaceous–Tertiary boundary sections? A chronostratigraphic estimate based on graphic correlation. *Geological Society of America, Bulletin*, **103**, 1439–1457.

MacLeod, N. & Sadler, P. (1995). Estimating the line of correlation. In *Graphic Correlation and the Composite Standard Approach*, ed. K. O. Mann, H. R. Lane, J. R. Stein, pp. 51–64. Tulsa: SEPM (Society for Sedimentary Geology), Special Publication No. 53.

Magnus, D. (1996). Heuristics and biases in evolutionary biology. *Biology and Philosophy*, **12**, 1–20.

Mahner, M. & Bunge, M. (1997). *Foundations of Biophilosophy*. Berlin & New York: Springer.

Malmgren, B. A. & Berggren, W. A. (1987). Evolutionary changes in some late Neogene planktonic foraminiferal lineages and their relationship to paleoceanographic changes. *Paleoceanography*, **2**, 445–456.

Malmgren, B. A. & Kennett, J. P. (1981). Phyletic gradualism in a late Cenozoic planktonic foraminiferal lineage; DSDP Site 284, southwest Pacific. *Paleobiology*, **7**, 230–240.

Malmgren, B. A., Berggren, W. A. & Lohmann, G. P. (1983). Evidence for punctuated gradualism in the late Neogene *Globorotalia tumida* lineage of planktonic foraminifera. *Paleobiology*, **9**, 377–388.

(1984). Species formation through punctuated gradualism in planktonic foraminifera. *Science*, **225**, 317–319.

Malmgren, B. A., Kucera, M. & Ekman, G. (1996). Evolutionary changes in supplementary apertural characteristics of the late Neogene *Sphaeroidinella dehiscens* lineage (planktonic foraminifera). *Palaios*, **11**, 192–206.

Mancini, E. & Tew, B. H. (1991). Relationships of Paleogene stage and planktonic foraminiferal zone boundaries to lithostratigraphic and allostratigraphic contacts in the eastern Gulf Coastal Plain. *Journal of Foraminiferal Research*, **21**, 48–66.

(1995). Geochronology, biostratigraphy and sequence stratigraphy of a marginal marine to marine shelf stratigraphic succession: Upper Paleocene and Lower Eocene, Wilcox Group, eastern Gulf Coastal Plain, U.S.A. In *Geochronology Time Scales and Global Stratigraphic Correlation*, ed. W. A. Berggren, D. V. Kent, M.-P. Aubry & J. Hardenbol. Tulsa, SEPM Special Publication **54**, 28–94, Mann, K. O. & Lane,

H. R., eds. (1995). Graphic Correlation. Tulsa: SEPM (Society of Sedimentary Geology) Special Publication, **53**.

Margulis, L. & Fester, R. (1991). *Symbiosis as a Source of Evolutionary Innovation*. Cambridge, MA: MIT Press.

Marshall, C. R. (1997). Confidence intervals on stratigraphic ranges with nonrandom distributions of fossil horizons. *Paleobiology*, **23**, 165–173.

Martin, R. E. (1991). Beyond biostratigraphy: micropaleontology in transition? *Palaios*, **6**, 437–438.

(1995). The once and future profession of micropaleontology. *Journal of Foraminiferal Research*, **25**, 372–373.

Martin, R. E. & Fletcher, R. R. (1995). Graphic correlation of Plio-Pleistocene sequence boundaries, Gulf of Mexico: oxygen isotopes, ice volume, and sea level. In *Graphic Correlation and the Composite Standard Approach*, ed. K. O. Mann, H. R. Lane & J. R. Stein, pp. 235–248. Tulsa: SEPM (Society for Sedimentary Geology), Special Publication No. 53.

Martin, R. E., Neff, E. D., Johnson, G. W. & Krantz, D. E. (1993). Biostratigraphic expression of Pleistocene sequence boundaries, Gulf of Mexico. *Palaios*, **8**, 155–171.

Marwick, A. (1989). *The Nature of History*. 3rd edition. Houndmills, Basingstoke: Macmillan, 442 pp.

Mayr, E. (1942). *Systematics and the Origin of Species from the Viewpoint of a Zoologist*. New York: Columbia University Press.

(1961). Cause and effect in biology. *Science* **134**, 1501–1506.

(1963). *Animal Species and Evolution*. Cambridge: Harvard University Press.

(1964). Introduction. In *C. Darwin, On the Origin of Species* [Facsimile of first edition, 1859, ed. E. Mayr], vii–xxvii. Cambridge, MA: Harvard University Press.

(1969). *Principles of Systematic Zoology*. New York: McGraw-Hill, 428 pp.

(1976). *Evolution and the Diversity of Life*. Cambridge: Harvard University Press.

(1982). *The Growth of Biological Thought: Diversity, Evolution, and Inheritance*. Cambridge, MA: Belknap Press, 974 pp.

(1988). *Toward a New Philosophy of Biology: Observations of an Evolutionist*. Cambridge, MA: Belknap Press of Harvard University Press, 564 p.

(1997). *This is Biology: The Science of the Living World*. Cambridge, MA: Belknap Press of Harvard University Press, 327 pp.

(1992). Speciational evolution or punctuated equilibria. In *The Dynamics of Evolution*, ed. A. Somit & S. Peterson, pp. 21–48. New York: Cornell University Press.

(2002). *What Evolution Is*. London: Phoenix.

Mayr, E. & Ashlock, P. D. (1991). *Principles of Systematic Zoology*, 2nd Edition. New York: McGraw-Hill, 475 pp.

McDougall, I. & Tarling, D. H. (1963). Dating of the polarity zones in the Hawaiian Islands. *Nature*, **200**, 171–172.

McGowan, J. A. (1986). The biogeography of pelagic ecosystems. In *Pelagic Biogeography*, ed. A. C. Pierrot–Bults *et al.*, 191–200. Paris: UNESCO.

McGowan, J. A. & Walker, P. W. (1993). Pelagic diversity patterns. In *Species Diversity in Ecological Communities*, ed. R. E. Ricklefs & D. Schluter, pp. 203–214. Chicago: University of Chicago Press.

McGowran, B. (1968a). Reclassification of Early Tertiary *Globorotalia*. *Micropaleontology*, **14**, 179–198.

(1968b). Late Cretaceous and Early Tertiary correlations in the Indo-Pacific region. In *Cretaceous-Tertiary Formations of South India*, ed. L. Rama Rao. Bangalore: Geological Society of India, Memoir **2**, 335–360.

(1971). On foraminiferal taxonomy. In *Proceedings of the II Planktonics Conference*, Roma 1970, ed. A. Farinacci, 813–820. Roma: Edizioni Technoscienza.

(1977). Maastrichtian to Eocene foraminiferal assemblages in the northern and eastern Indian Ocean region: correlations and historical patterns. In *Indian Ocean Geology and Biostratigraphy*, Studies Following Deep Sea Drilling Legs 22–29, ed. J. R. Heirtzler, H. M. Bolli, T. A. Davies, J. B. Saunders & J. G. Sclater, p. 417–458, Washington, D.C.: American Geophysical Union.

(1978a). Early Tertiary biostratigraphy in southern Australia: a progress report. In, *The Crespin Volume: Essays in Honour of Irene Crespin*, ed. D. J. Belford & V. Scheibnerova. Canberra: Bureau Mineral Resources Australia, *Geology & Geophysics, Bulletin*, **192**, 83–95.

(1978b). Stratigraphic record of early Tertiary oceanic and continental events in the Indian Ocean region. *Marine Geology*, **26**, 1–39.

(1979). The Australian Tertiary: foraminiferal overview. *Marine Micropaleontology*, **4**, 235–264.

(1986a). Beyond classical biostratigraphy. *PESA Journal*, **9**, 28–41.

(1986b). Cainozoic oceanic events: the Indo-Pacific biostratigraphic record. In *Global Stratigraphic Correlation of Mesozoic and Cainozoic Sediments*, ed. W. A. Berggren, *Palaeogeography, Palaeoclimatology, Palaeoecology*, 55, 247–265.

(1989a). The later Eocene transgressions in southern Australia. *Alcheringa*, **13**, 45–68.

(1989b). Silica burp in the Eocene ocean. *Geology*, **17**, 857–860.

(1990). Fifty million years ago. *American Scientist*, **78**, 30–39.

(1991). Evolution and environment in the early Palaeogene. In, *The World of Martin F. Glaessner*, ed. B. P. Radhakrishna. Geological Society of India, Memoir 20, 21–53.

McGowran, B., & Beecroft, A. (1985). *Guembelitria* in the early Tertiary of southern Australia and its palaeoceanographic significance. In *Stratigraphy, Palaeontology and Malacology*, Papers in Honour of N. H. Ludbrook, ed. J. M. Lindsay. South Australian Dept Mines & Energy, Special Publication, **5**, 247–261.

McGowran, B. & Li, Q. (1994). The Miocene oscillation in southern Australia. *Special Volume on Australian Vertebrate Evolution, Palaeontology, & Systematics, Records of the South Australian Museum*, **27**, 197–212.

(1996). Ecostratigraphy and sequence biostratigraphy, with a neritic foraminiferal example from the Miocene in southern Australia. *Historical Biology*, **11**, 137–169.

(2000). Evolutionary palaeoecology of Cainozoic foraminifera: Tethys, IndoPacific, southern Australia. *Historical Biology*, **15**, 3–28.

(2002). Sequence biostratigraphy and evolutionary palaeoecology: Foraminifera in the Cenozoic Era. *Memoirs of the Association of Australasian Palaeontologists*, **27**, 167–188.

McGowran, B., Archer, M., Bock, P. *et al.* (2000). Australasian palaeobiogeography: the Palaeogene and Neogene record. In *Palaeobiogeography of Australasian faunas and floras*, ed. A. J. Wright, G. C. Young, J. A. Talent & J. R. Laurie, pp. 405–470. Australasian Association of Palaeontologists, Memoir, 23.

McGowran, B., Holdgate, G. R., Li, Q. & Gallagher, S. J. (2004). Cenozoic statigraphic succession in southeastern Australia. *Australian Journal of Earth Sciences*, **51**, 459–496.

McGowran, B., Li, Q., Cann, J. Padley, D., Mckirdy, D. & Shafik, S. (1997b). Biogeographic impact of the Leeuwin Current in southern Australia since the late middle Eocene. *Palaeogeography, Palaeoclimatology, Palaeoecology*, **136**, 19–40.

McGowran, B. Li, Q. & Moss, G. (1997a). The Cenozoic neritic record in southern Australia: The biogeohistorical framework. In *Cool-water Carbonates in Space and Time*, ed. N. P. James & J. D. A. Clarke, Society of Economic Paleontologists & Mineralogists, Special Volume, **56**, 185–203.

McGowran, B., Lindsay, J. M. & Harris, W. K. (1971). Attempted reconciliation of Tertiary biostratigraphic systems, Otway Basin. In *The Otway Basin in Southeasten Australia*, ed. H. Wopfner, H. & Douglas J. G., pp. 27–81. Adelaide and Melbourne, Geological Surveys of South Australia & Victoria, Special Bulletin.

McGowran, B., Moss, G. & Beecroft, A. (1992). Late Eocene and early Oligocene in southern Australia: local neritic signals of global oceanic changes. In, *Eocene-Oligocene Climatic and Biotic Evolution*, ed. D. R. Prothero & W. A. Berggren, pp. 178–201. Princeton: Princeton University Press.

McLean, D. M. (1981). *Cretaceous–Tertiary extinctions and possible terrestrial and extraterrestrial causes*. Syllogeous, 39. Ottawa: National Museums of Canada.

Mein, P. (1975). Résultats du Groupe de Travail des Vertébrés. In *Report on Activity of R.C.M.S. Working Group*, ed. J. Senes. Regional Committee of Mediterranean Stratigraphy, 75–81.

(1981). Mammal zonations: introduction. In *Annales Géologiques des Pays Helléniques*, ed. G. Marinos & G. Symeonidis, pp. 83–88. Proceeding VIIth International Congress on Mediterranean Neogene, Athens, 1979.

(1989). Die Kleinsäugerfauna des Untermiozäns (Eggenburgien) von Maigen, Niederösterreich. *Annales Naturhistorisches Museum Wien*, **90A**, 49–58.

Miall, A. D. (1997). *The Geology of Stratigraphic Sequences*: Springer-Verlag, Berlin.

(2004). Empiricism and model building in stratigraphy: the historical roots of present-day practices. *Stratigraphy*, **1**, 3–25.

Miall, A. D. & Miall, C. E. (2001). Sequence stratigraphy as a scientific enterprise: the evolution and persistence of conflicting paradigms. *Earth Science Reviews*, **54**, 321–348.

(2004). Empiricism and model building in stratigraphy: around the hermeneutic circle in the pursuit of stratigraphic correlation. *Stratigraphy*, **1**, 27–46.

Miall, C. E. & Miall, A. D. (2002). The Exxon factor: the roles of corporate and academic science in the emergence and legitimation of a new global model of sequence stratigraphy. *The Sociological Quarterly*, **43**, 307–334.

Miller, A. I. (1997). Coordinated stasis or coincident relative stability? *Paleobiology*, **23**, 155–164.

Miller, F. X. (1977). The graphic correlation method in biostratigraphy. In *Concepts and Methods of Biostratigraphy*, ed. E. G. Kauffman & J. E. Hazel, pp. 165–186. Stroudsburg: Dowden, Hutchinson & Ross.

Miller, K. G. (1994) The rise and fall of sea level studies: are we at a stillstand? *Paleoceanography*, **9**, 183–184.

Miller, K. G. & Kent, D. V. (1987). Testing Cenozoic eustatic changes: The critical role of stratigaphic resolution. *Cushman Foundation for Foraminiferal Research, Special Publication*, **24**, 51–55.

Miller, K. G., Fairbanks, R. G. & Mountain, G. S. (1987). Tertiary oxygen isotope synthesis, sea-level history, and continental margin erosion. *Paleoceanography*, **1**, 1–19.

Miller, K. G., Mountain, G. S., Browning & J. V., Kominz, M., Sugarman, P. J., Christie-Blick, N., Katz, M. E., & Wright, J. D. (1998). Cenozoic global sea level, sequences, and the New Jersey transect: results from coastal plain and continental slope drilling. *Reviews of Geophysics*, **36**, 569–601.

Miller, K. G., Sugarman, P. J., Browning, J. V., Kominz, M. A., Hernández, J. C., Olsson, R. K., Wright, J. D., Feigenson, M. D. & Van Sickel, W. V. (2003). Late Cretaceous chronology of large, rapid sea-level changes: glacioeustasy during the greenhouse world. *Geology*, **31**, 585–588.

Miller, K. G., Thompson, P. R. & Kent, D. V. (1993). Integrated late Eocene–Oligocene stratigraphy of the Alabama coastal plain: correlation of hiatuses and stratal surfaces to glacioeustatic lowerings. *Paleoceanography*, **8**, 313–331.

Miller, K. G., Wright, J. D. & Fairbanks, R. G. (1991) Unlocking the ice house: Oligocene-Miocene oxygen isotopes, eustasy, and margin erosion. *Journal of Geophysical Research*, **96**(B4), 6829–6848.

Miller, W. III (1986). Paleoecology of benthic community replacement. *Lethaia*, **19**, 225–231.

(1990). Hierarchy, individuality and paleoecosystems. *Paleontological Society Special Publication*, **5**, 31–47.

(1993). Models of recurrent fossil assemblages. *Lethaia*, **26**, 182–183.

Mishler, B. & Brandon, R. (1987). Indivuality, pluralism, and the biological species concept. *Biology and Philosophy*, **2**, 397–414.

Montanari, A., Odin, G. S. & Coccioni, R., Editors (1997). *Miocene Stratigraphy: an Integrated Approach*. New York: Elsevier.

Monty, C. L. V. (1968). D'Orbigny's concepts of stage and zone. *Journal of Paleontology*, **42**, 689–701.

Moore, R. C. (1941). Stratigraphy. In, *Geology*, 188–938, 50th Anniversary volume, Geological Society of America, 178–220.

(1948). Stratigraphical paleontology. *Geological Society of America, Bulletin*, **59**, 301–326

Moore, T. C. & Romine, K. (1981). In search of biostratigraphic resolution. In *The Deep Sea Drilling Program: a Decade of Progress*, ed. J. E. Warme, R. G. Douglas & J. E.

Winterer. Tulsa: Society of Economic Paleontologists and Mineralogists Special Publication **32**, 317–334.

Morgans, H. E. G., Scott, G. H., Beu, A. G., Graham, I. J., Mumme, T. C., George, W. St. & Strong, C. P. (1996). New Zealand Cenozoic time scale (version 11-96). Institute of Geological & Nuclear Sciences, Science Report 96/38, 12 pp.

Morris, P. J., Ivany, L. C., Schopf, K. M. & Brett, C. E. (1995). The challenge of paleo-ecological stasis: reassessing sources of evolutionary stability. *Proceedings of the National Academy of Sciences USA*, **92**, 1126–1273.

Moss, G. D. & McGowran, B. (2003). Oligocene neritic foraminifera in Southern Australia: spatiotemporal biotic patterns reflect sequence–stratigraphic environmental patterns. In *Paleobiological, Geochemical, and Other Proxies of Sea Level Change*, ed. H. C. Olson H. & R. M. Leckie M., SEPM (Society of Sedimentary Geology) Special Volume **75**, 117–138.

Murphy, M. A. (1977). On time-stratigraphic units. *Journal of Paleontology*, **51**, 213–219.

Murray, G. E. (1961). *Geology of the Atlantic and Gulf Coastal Province of North America*. New York: Harper & Row.

Murray, J. (1912). *The Ocean: A General Account of the Science of the Sea*. London: Williams & Norgate.

Nagappa, Y. (1959). Foraminiferal biostratigraphy of the Cretaceous–Eocene succession in the India–Pakistan–Burma region. *Micropaleontology*, **5**, 145–192.

Nagel, E. (1952). The logic of historical analysis. *Science Monthly*, **74**, 162–169.

Neal, J. E., Stein, J. A. & Gamber, J. A. (1998). Nested stratigraphic cycles and depositional systems of the Paleogene central North Sea. In *Mesozoic and Cenozoic Sequence Stratigraphy of European Basins*, ed. P.-C. de Graciansky, J. Hardenbol, T. Jacquin & P. R. Vail, SEPM (Society of Sedimentary Geology) Special Publication No. **60**, 261–288.

Norell, M. A. (1992). Taxic origin and temporal diversity: the effect of phylogeny. In *Extinction and phylogeny*, ed. M. J. Novacek & Q. D. Wheeler, pp. 89–118. New York: Columbia University Press.

Norell, M. A. & Novacek, M. J. (1992). The fossil record and evolution: comparing cladistic and paleontological evidence for vertebrate history. *Science*, **255**, 1690–1693.

Norris, R. D. (1991a). Biased extinctions and evolutionary trends. *Paleobiology*, **17**, 388–399.

(1991b). Parallel evolution in the keel structure of planktonic foraminifera. *Journal of Foraminiferal Research*, **21**, 319–331.

(1992). Extinction, selectivity and ecology in planktonic foraminifera. *Palaeogeography, Palaeoecology, Palaeoclimatology*, **95**, 1–17.

(1996). Symbiosis as an evolutionary innovation in the radiation of Paleocene planktonic foraminifera. *Paleobiology*, **22**, 461–480.

(1999). Hydrographic and tectonic control of plankton distribution and evolution. In *Reconstructing Ocean History: A Window into the Future*, ed. F. Abrantes & A. Mix, pp. 173–193. London: Plenum.

(2000). Pelagic species diversity, biogeography, and evolution. In *Deep time: Paleobiology's perspective*, ed. D. H Erwin & S. L. Wing, *Paleobiology*, Supplement to vol. **26**(4), 236–258.

Norris, R. D. & de Vargas, C. (2000). Evolution all at sea. *Nature*, **405**, 23–24.

Norris, R. D. & Nishi, H. (2001). Evolutionary trends in coiling of tropical Paleogene planktic foraminifera. *Paleobiology*, **27**, 327–347.

Norris, R. D. & Röhl, U. (1999). Carbon cycling and chronology of climate warming during the Paleocene/Eocene transition. *Nature*, **401**, 775–778.

Norris, R. D. & Wilson, P. A. (1998). Low-latitude sea-surface temperatures for the mid-Cretaceous and the evolution of planktic foraminifera. *Geology*, **26**, 823–826.

Norris, R. D., Corfield, R. M. & Cartlidge, J. E. (1994). Evolutionary ecology of *Globorotalia* (*Globoconella*) (planktic foraminifera). *Marine Micropaleontology*, **23**, 121–45.

(1996). What is gradualism? Cryptic species in globorotaliid planktonic foraminifera. *Paleobiology*, **22**, 386–405.

Odin, G. S. & Curry, D. (1985). Palaeogene time scale: radiometric dating versus magnetostratigraphic approach. *Journal Geological Society of London*, **142**, 1179–1188.

Odin, G. S., & Curry, D., Gale, N. H. & Kennedy, W. J. (1982). The Phanerozoic time scale in 1981. In *Numerical Dating in Stratigraphy*, ed. G. S. Odin, pp. 957–960. New York: John Wiley & Sons.

Okada, H. & Bukry, D. (1980). Supplementary modification and introduction of code numbers to the low latitude coccolith biostratigraphic zonation. *Marine Micropaleontology*, **5**, 321–325.

Oldroyd, D. (1979). Historicism and the rise of historical geology, *History of Science*, **17**, 191–213; 227–257.

(1996). *Thinking about the Earth: A History of Ideas in Geology*. London: Athlone, 410 p.

Olson, E. C. (1952). The evolution of a Permian vertebrate chrono-fauna. *Evolution*, **6**, 181–196.

Olsson, R. K. (1988). Foraminiferal modeling of sea-level change in the late Cretaceous of New Jersey. In *Sea-level Changes, An Integrated Approach*, ed. C. K. Wilgus, B. S. Hastings, C. G. S. C. Kendall, H. W. Posamentier, C. A. Ross & J. C. van Wagoner, 289–297. Society of Sedimentary Geology, Memoir, **42**,

(1991). Cretaceous to Eocene sea level fluctuations on the New Jersey margin. *Sedimentary Geology*, **70**, 195–208.

Olsson, R. K. & Wise, W. W. (1987). Upper Paleocene to middle Eocene depositional sequences and hiatuses in the New Jersey Atlantic margin. In *Timing and Depositionial History of Eustatic Sequences: Constraints on Seismic Statrigraphy*, ed. C. Ross and D. Harnan. Cushman Foundation for Foraminiferal Research, Special Publication, 24, 85–97.

Olsson, R. K., Berggren, W. A., Hemleber, C. H. & Huber, B. T. (1999). *Atlas of Paleocene Planktonic foraminifera*. Washington, DC: Smithsorian Institution Press.

Olsson, R. K., Miller, K. G., Browning, J. V., Wright, J. D. & Cramer, B. S. (2002). Sequence stratigraphy and sea level change across the Cretaceous-Tertiary boundary on the New Jersey passive margin. In *Catastropic events and mass extinctions: impacts and beyond*. Geological Society of America Special Paper, **56**, 97–108.

Orue-Etxebarria, X., Pujalte, V., Bernaola, G., Apellaniz, E., Baceta, J. I., Payros, A., Nuñez-Betelu, Serra-Kiel, J. & Tosquella, J. (2001). Did the Late Paleocene thermal maximum affect the evolution of large foraminifers? Evidence from the Campo section (Pyrenees, Spain). *Marine Micropaleontology*, **41**, 45–71.

Osborn, H. F. (1910). *The Age of Mammals in Europe, Asia and North America*. New York: Macmillan, 635 pp.

(1934). Aristogenesis, the creative principle in the origin of species. *American Naturalist*, **68**, 193–235.

Oslick, J. S., Miller, K. G., Feigenson, M. D. & Wright, J. D. (1994). Oligocene–Miocene strontium isotopes: stratigraphic revisions and correlations to an inferred glacioeustatic record. *Paleoceanography*, **9**, 427–444.

Padian, K., Lindberg, D. R. & Polly, P. D. (1994). Cladistics and the fossil record: the uses of history. *Annual Review of Earth and Planetary Sciences*, **22**, 63–69.

Paley, W. (1827). *The Works of William Paley, D.D., Archdeacon of Carlisle*. Edinburgh: Peter Brown and Thomas Nelson.

Pälike, H. & Shackleton, N. J. (2000). Constraints on astronomical parameters from the geological record for the last 25 Myr. *Earth and Planetary Science Letters*, 182, 1–14.

Pandolfi, J. M. (1996). Limited membership in Pleistocene reef coral assemblages from the Huon Peninsula, Papua New Guinea: constancy during global change. *Paleobiology*, **22**, 152–176.

Papp, A. & Schmid, M. E. (1985). Die Fossilen Foraminiferen des Tertiären Beckens von Wien: Revision der Monographie von Alcide d'Orbigny (1846). *Abhandlungen der Geologischen Bundesanstalt*, Band 37.

Papp, A., Grill, R., Janoschek, R., Kapounek, J, Kollmann, K. Turnovsky, K. (1968). Zur Nomenklatur des Neogens in Österreich. *Sonderabdruck aus den Verhandlungen der Geologischen Bundesanstalt 1968*, Heft 1/2, 9–18 [English text by E. J. Tynan, p. 19–27].

Park, J., D'Hondt, S. L., King, J. W. & Gibson, C. (1993). Late Cretaceous precessional cycles in double time: a warm-earth Milankovitch response. *Science*, **261**, 1431–1434.

Pawlowski, J. (2000). Introduction to the molecular systematics of foraminifera. In *Advances in the Biology of Foraminifera*, ed. J. J. Lee & P. Hallock, p. 1–12. *Micropaleontology*, **46**, Supplement 1.

Pearson, P. N. (1998a). Stable isotopes and the study of evolution in planktonic foraminifera. *Paleontological Society Papers*, **4**, 138–178.

(1998b). Evolutionary concepts in biostratigraphy. In *Unlocking the Stratigraphical Record*, ed. P. Doyle & M. R. Bennett, pp. 123–144. John Wiley & Sons, Chichester.

(1998c). Speciation and extinction asymmetries in paleontological phylogenies: evidence for evolutionary progress? *Paleobiology*, **24**: 305–335.

Pearson, P. N., Olsson, R. K., Huber, B. T., Hemleben, C. & Berggren, W. A. (eds.) (in press) *Atlas of Eocene Planktonic Foraminifera*, Cushman Foundation for Foraminiferal Research, Special Publication.

Pearson, P. N., Shackleton, N. J. & Hall, M. A. (1997). Stable isotope evidence for the sympatric divergence of *Globigerinoides trilobus* and *Orbulina universa* (planktonic foraminifera). *Journal of the Geological Society*, **154**, 295–302.

Pessagno, E. A. (1967). Upper Cretaceous planktonic foraminifera from the western Gulf Coastal Plain. *Paleontographica Americana*, **5**, 259–444.

Phillips, J. (1840). Palaeozoic Series. In *The Penny Cyclopedia* (cited in Berry, 1968).
(1861). *Life on the Earth: its Origin and Succession*. Cambridge & London.
(1829). Organic remains of the eastern part of Yorkshire. In Conkin & Conkin (1984), pp. 86–93.

Pimm, A. C., McGowran, B. & Gartner, S. (1974). Early sinking history of Ninetyeast Ridge, north-eastern Indian Ocean. *Bulletin Geological Society America*, **85**, 1219–1224.

Poag, C. W. (1977). Biostratigraphy in Gulf Coast petroleum exploration. In *Concepts and Methods of Biostratigraphy*, ed. E. G. Kauffman & J. E. Hazel, pp. 213–234. Stroudsberg, PA: Dowden, Hutchison & Ross.

Pokorny, V. (1963). *Principles of Zoological Micropalaeontology*. Translated by K. A. Allen. Oxford: Pergamon.

Pomerol, Ch. & Premoli Silva, I., Editors (1986). *Terminal Eocene Events*. New York: Elsevier Science.

Popper, K. R. (1957). *The Poverty of Historicism*. London: Routledge & Kegan Paul.

Prell, W. L., Imbrie, J., Morley, J. J., Pisias, N. G., Shackleton, N. J. & Streeter, H. F. (1986). Graphic correlation of oxygen isotope stratigraphy: applications to the late Quaternary. *Paleoceanography*, **1**, 137–162.

Premoli Silva, I., Coccioni, R. & Montanari, A., Editors (1988). *The Eocene–Oligocene boundary in the Marche–Umbria basin (Italy)*. International Subcommission on Paleogene Stratigraphy, International Union of Geological Sciences, Eocene–Oligocene meeting, Ancona (Italy), Special Publication.

Press, F. & Siever, R. (1978). *Earth*, 2nd Edition. San Francisco: W. H. Freeman.

Prothero, D. R. (1992). Punctuated equilibrium at twenty: a paleontological perspective. *Skeptic*, **1**, 38–47.
(1994a). The late Eocene–Oligocene extinctions. *Annual Reviews of Earth & Planetary Science*, **22**, 145–165.
(1994b). *The Eocene–Oligocene Transition: Paradise Lost*. New York: Columbia University Press, 291 pp.
(1995). Geochronology and magnetostratigraphy of Paleogene North American land mammal 'ages': an update. In *Geochronology Time Scales and Global Stratigraphic Correlation*, ed. W. A. Berggren, D. V., Kent, M.-P., Aubry & J. Hardenbol. Tulsa: SEPM (Society of Sedimentary Geology) Special Publication, **54**, 305–316.
(1998). The chronological, climatic, and paleogeographic background to North American mammalian evolution. In *Evolution of the Tertiary Mammals of North America*, Volume 1, ed. C. M Janis, K. M. Scott & L. L. Jacobs, pp. 9–36. Cambridge: Cambridge University Press.
(1999). Does climatic change drive mammalian evolution? *GSA Today*, **9**, 1–7.

Prothero, D. R. & Berggren, W. A., Editors (1992). *Eocene–Oligocene Climatic and Biotic Evolution*. Princeton: Princeton University Press, 568 pp.

Prothero, D. R. & Heaton, T. H. (1996). Faunal stability during the early Oligocene climatic crash. *Palaeogeography, Palaeoecology, Palaeoclimatology*, **127**, 257–283.

Quillévéré, F., Norris, R. D., Moussa, I. & Berggren, W. A. (2001). Role of photosymbiosis and biogeography in the diversification of early Paleogene acarininids (planktonic foraminifera). *Paleobiology* **27**, 311–326.

Rabeder, G. (1986). Herkunft und frühe Evolution der Gattung *Microtus* (Arvicolidae, Rodentia). *Sonderdruck aus Z. f. Säugetierkunde*, Bd.**51**, H.6, 350–367.

Raffi, I. (1999). Precision and accuracy of nannofossil biostratigraphic correlation. *Philosophical Transactions of the Royal Society of London A*, **357**: 1975–1994.

Rama Rao, L. (1968). The problem of the Cretacous–Tertiary boundary. In *Cretaceous–Tertiary Formations of South India*, ed. L Rama Rao. Bangalore: Geological Society of India, Memoir **2**, 1–9.

Raup, D. M. (1991). *Extinction: Bad Genes or Bad Luck?* New York: W. W. Norton.

Raup, D. M. & Sepkoski, J. J. Jr (1984). Periodicity of extinctions in the geologic past. *Proceeding of the National Academy of Sciences USA*, **81**, 801–805.

Raup, D. M. & Stanley, S. M. (1978). *Principles of Paleontology*. New York, San Francisco: W. H. Freeman, 481 pp.

Ravelo, A. C. & Fairbanks, R. G. (1992). Oxygen isotopic composition of multiple species of planktonic foraminifera: recorders of the modern photic zone temperature gradient. *Paleoceanography*, **7**, 815–831.

Raymo, M. (1994) The initiation of northern hemisphere glaciation. *Annual Reviews of Earth and Planetary Science*, **22**, 353–384.

Rea, D. K., Zachos, J. C., Owen, R. M. & Gingerich, P. D. (1990). Global change at the Paleocene/Eocene boundary: climatic and evolutionary consequences of tectonic events. *Palaeogeography, Palaeoclimatology, Palaeoecology*, **79**, 117–128.

Reeckmann, A. (1994). Geology of the onshore Torquay Sub-basin: a sequence-stratigraphic approach. In *Otway Basin Symposium, Extended Abstracts*, compiled D. M. Finlayson. Melbourne: Australian Geological Survey Organisation, Record 1994/13, pp. 3–6.

Reif, W.-E. (1993). Afterword. In Schindewolf (1993), pp. 435–453.

Reiss, Z. (1968). Planktonic foraminiferids, stratotypes, and a reappraisal of Neogene chronostratigraphy in Israel. *Israel Journal of Earth Sciences*, **17**, 153–169.

Remane, J. (1997). Chronostratigraphic standards: how are they defined and when should they be changed? *Quaternary International*, **40**, 3–4.

Remane, J., Bassett, M. G., Cowie, J. W., Gohrbandt, K. H., Lane, H. R., Michelsen, O. & Naiwen, W. (1996). Revised guidelines for the establishment of global stratigraphic standards by the International Commission of Stratigraphy (ICS). *Episodes*, **18**, 77–81.

Rensch, B. (1983). The abandonment of Lamarckian explanations: the case of climatic parallelism of animal characteristics. In *Dimensions of Darwinism*, ed. M. Grene, 31–42. Cambridge: Cambridge University Press.

Ridley, M. (1986). *Evolution and Classification: The Reformation of Cladism*. London, New York: Longman, 201 pp.

Riedel, W. R. (1973). Cenozoic planktonic micropaleontology and biostratigraphy. *Annual Review of Earth and Planetary Sciences*, **1**, 241–268.

Rieppel, O. (1988). *Fundamentals of Comparative Biology*. Basel, Boston: Birkhauser Verlag, 202 pp.

Rio, D., Sprovieri, R. & Thunell, R. (1991). Pliocene–Pleistocene chronostratigraphy: a reevaluation of Mediterranean type sections. *Geological Society of America Bulletin*, **103**, 1049–1058.

ROCC Research on Cretaceous Cycles Group (1986). Rhythmic bedding in Upper Cretaceous pelagic carbonate sequences: varying sedimentary response to climatic forcing. *Geology*, **14**, 153–156.

Rodgers, J. (1959). The meaning of correlation. *American Journal of Science*, **257**, 684–691.

Rögl, F. (1985). Late Oligocene and Miocene planktic foraminifera of the central Paratethys. In *Plankton Stratigraphy*, ed. H. M. Bolli, J. B. Saunders & K. Perch-Nielsen, pp. 315–328. Cambridge: Cambridge University Press.

(1998). Palaeogeographic considerations for Mediterranean and Paratethys seaways (Oligocene to Miocene). *Annalen des Naturhistorischen Museums in Wien*, **99A**, 279–310.

(1999). Mediterranean and Paratethys. Facts and hypotheses of an Oligocene to Miocene paleogeography (short overview). *Geologica Carpathica*, **50**, 339–349.

Rögl, F. & Steininger, F. F. (1983). Vom Zerfall der Tethys zu Mediterran und Paratethys. *Die neogene Paläogeographie und Palinspastik des zirkummediterranen Raumes. Ann. Nat. Hist. Mus. Wien* **85A**, 135–163.

(1984). Neogene Paratethys, Mediterranean and Indo-Pacific seaways. Implications for the palaeobiogeography of marine and terrestrial biotas. In *Fossils and Climate*, ed. P. Brenchley, pp. 171–200. New York: Wiley.

Rohling, E. J. & Cooke, S. (1999). Stable oxygen and carbon isotopes in foraminiferal carbonate shells. In *Modern Foraminifera*, ed. B. K Sen Gupta, pp. 239–259. Dordrecht, Kluwer Academic Publishers.

Rosenberg, A. (1985). *The Structure of Biological Science*. New York: Cambridge University Press.

Ross, C. A. & Ross, J. P. (1985). Late Paleozoic depositional sequences are synchronous and worldwide. *Geology*, **13**, 194–197.

Ruddiman, W. F., Raymo, M. & McIntyre, A. (1986). Matuyana 41,000 year cycles: North Atlantic Ocean and northern hemisphere ice sheets. *Earth and Planetary Science Letters*, **80**, 117–29.

Rudwick, M. J. S. (1972). *The Meaning of Fossils: Episodes in the History of Palaeontology*. London, New York: Macdonald and Co., American Elsevier (2nd Edition, 1985).

(1982a). Cognitive styles in geology. In *Essays in the Sociology of Perception*, ed. M. Douglas, pp. 210–242. London: Routledge, Kegan, Paul.

(1982b). Charles Lyell's dream of a statistical palaeontology. *Palaeontology*, **21**, 225–244.

(1990). Introduction. *Introduction to Facsimile of C. Lyell, Principles of Geology*, First Edition, London: J. Murray, 1830–1833. pp. vii–viii Chicago: University of Chicago Press.

(1998). Lyell and the Principles of Geology. In *Lyell: The Past is the Key to the Present*, ed. D. J. Blundell & A. C. Scott. Geological Society Special Publication No. **143**, 3–15.

Ruse, M. (1986). *Taking Darwin Seriously: A Naturalistic Approach to Philosophy*. New York: Blackwell.

(1998). All my love is towards individuals. *Evolution*, **52**, 283–288.

Sadler, P. M. (2004). Quantitative biostratigraphy – achieving finer resolution in global correlation. *Annual Reviews of Earth and Planetary Sciences*, **32**, 187–213.

Sadler, P. M. & Cooper, R. A. (2003). Best-fit intervals and consensus sequences: comparison of the resolving power of traditional biostratigraphy and computer-assisted correlation. In *High Resolution Approaches to Stratigraphic Paleontology*, ed. P. Harries, pp. 49–94. Dordrecht: Kluwer Academic Publishers.

Saito, T. (1977). Late Cenozoic planktonic foraminiferal datum levels: the present state of knowledge toward accomplishing Pan-Pacific correlations. In *First International Conference on Pacific Neogene Stratigraphy*, Tokyo, 1976, Proceedings, pp. 61–80.

  (1984). Planktonic foraminiferal datum planes for biostratigraphic correlation of Pacific Neogene sequences – 1982 status report. In *Pacific Neogene Datum Planes: Contributions to Biostratigraphy and Chronology*, ed. N. Ikebe & R. Tsuchi, pp. 47–67. Tokyo: University of Tokyo Press.

Salthe, S. N. (1985). *Evolving Hierarchical Systems: Their Structure and Representation*. New York: Columbia University Press, 343 pp.

Salvador, A., Editor (1994). *International Stratigraphic Guide: A Guide to Stratigraphic Classification, Terminology, and Procedure*, Second Edition. International Union of Geological Sciences & Geological Society of America, 214 pp.

Savage, D. E. (1977). Aspects of vertebrate paleontological stratigraphy and geochronology. In *Concepts and Methods of Biostratigraphy*, ed. E. G. Kauffman & J. E. Hazel, pp. 427–442. Stroudsberg, PA: Dowden, Hutchison & Ross.

Savage, D. E. & Russell, D. E. (1983). *Mammalian Paleofaunas of the World*. Reading, MA: Addison-Wesley.

Savin, S. M., Douglas, R. G. & Stehli, F. G. (1975). Tertiary marine paleotemperatures. *Geological Society of America, Bulletin*, **86**, 1499–1510.

Schaeffer, B., Hecht, M. K. & Eldredge, N. (1972). Phylogeny and paleontology. In *Evolutionary Biology*, ed. Th. Dobzhansky *et al.*, pp. 31–46. New York: Appleton-Century-Crofts.

Schenck, H. G. & Muller, S. W. (1941). Stratigraphic terminology. *Geological Society of America, Bulletin*, **52**, 1419–1426.

Schindewolf, O. H. (1993). *Basic Questions in Paleontology: Geologic Time, Organic Evolution, and Biological Systematics*. Translated by Judith Schaefer; edited by Wolf-Ernst Reif. Chicago: University of Chicago Press. Originally published (1950) as *Grundfragen der Paläontologie*. Stuttgart: E. Schweizerbart'sche Verlagsbuchhandlung.

Schneider, C. & Kennett, J. P. (1996). Isotopic evidence for interspecies habitat differences during evolution of the Neogene planktonic foraminiferal clade *Globoconella*. *Paleobiology*, **22**, 282–303.

Schoch, R. M. (1989). *Stratigraphy: Principles and Methods*. New York: Van Nostrand Reinhold, 375 pp.

Schwan, W. (1980). Geodynamic peaks in alpinotype orogenies and changes in ocean-floor spreading during Late Jurassic–Late Tertiary time. *American Association of Petroleum Geologists Bulletin*, **64**, 359–373.

Schwarzacher, W. & Fischer, A. G. (1982). Limestone–shale bedding and perturbations of the Earth's orbit. In *Cyclic and Event Stratification*, ed. G. Einsele & A. Seilacher, pp. 72–95. Berlin/Heidelberg/New York: Springer-Verlag.

Scott, G. H. (1960). The type locality concept in time-stratigraphy. *New Zealand Journal of Geology & Geophysics*, **3**, 580–584.

(1985). Homotaxy and biostratigraphical theory. *Palaeontology*, **28**, 777–782.

Seibold, E. & Berger, W. H. (1993). *The Sea Floor: an Introduction to Marine Geology* (2nd ed.). Berlin: Springer Verlag.

Sepkoski, J. J., (1978). A kinetic model of Phanerozoic taxonomic diversity. I. Analysis of marine orders. *Paleobiology*, **5**, 223–251.

Sepkoski, J. J., Jr. (1986). Global bioevents and the question of periodicity. In *Global Bioevents*, ed. O. H. Walliser, pp. 47–61. Berlin, Springer-Verlag.

Sereno, P. C. (1997). The origin and evolution of dinosaurs. *Annual Reviews of Earth and Planetary Sciences*, **25**, 435–490.

(1999). The evolution of dinosaurs. *Science*, **284**, 2137–2147.

Shackleton, N. J. (1985). Oceanic carbon isotope constraints on oxygen and carbon dioxide in the atmosphere. In *The Carbon Cycle and Atmospheric CO_2: Natural Variations Archaean to Present*, ed. E. T. Sundquist W. S. Broecker, pp. 412–417. Washington, D.C., American Geophysical Union, Geophysical Monograph 32.

(1986). Paleogene stable isotope events. *Palaeogeography, Palaeoclimatology, Palaeoecology*, **57**, 91–102.

Shackleton, N. J., Crowhurst, S. J., Weedon, G. P. & Laskar, J. (1999). Astronomical calibration of Oligocene–Miocene time, *Philosophical Transactions of the Royal Society of London A*, **357**: 1907–1929.

Shackleton, N. J. & Kennett, J. P. (1975). Paleotemperature history of the Cenozoic and the initiation of Antarctic glaciation: oxygen and carbon isotope analyses in DSDP sites 277, 279, and 281. In *Initial Reports of the Deep Sea Drilling Project*, J. P Kennett, R. E Houtz, *et al.* Washingto, D.C.: US Government Printing Office.

Shackleton, N. J. & Opdyke, N. D. (1973). Oxygen isotope and paleomagnetic stratigraphy of equatorial Pacific core V28-38: oxygene isotope temperatures and ice volumes on a 10^5 and 10^6 year scale. *Quaternary Research*, **3**, 39–55.

Shackleton, N. J., Hall, M. A., Raffi, I., Tauxe, L. & Zachos, J. (2000). Astronomical calibration age for the Oligocene–Miocene boundary. *Geology*, **28**: 447–450.

Shaw, A. B. (1964). *Time in Stratigraphy*. New York: McGraw-Hill, 365 pp.

(1969). Adam and Eve, paleontology, and the non-objective arts. *Journal of Paleontology*, **43**, 1085–1093.

Sheehan, P. M., (1996). A new look at ecological evolutionary units (EEUs). *Palaeogeography, Palaeoecology, Palaeoclimatology*, **127**, 21–32.

Silver, L. T. & Schultz, P. H. (1982). Geological implications of impacts of large asteroids and comets on the earth. *Geological Society of America, Special Paper* **190**.

Simmons, M. D. (1998). Biostratigraphy – surviving extinction. *Palaios*, **13**(3) Online, 2 pp.

Simmons, M. D., Berggren, W. A., O'Neill, B. J., Scott, R. W., Steininger, F. F. & Ziegler, W. (2000). Biostratigraphy and geochronology. In *Fossils and the Future: Paleontology in the 21st Century*, ed. R. H. Lane, F. F Steininger, R. L. Kaesler, W. Ziegler & J. H. Lipps. Senckenberg–Buch Nr. 74, 119–132.

Simmons, M. D. & Williams, C. L. (1992). Sequence stratigraphy and eustatic sea-level change: the role of micropalaeontology. *Journal of Micropalaeontology*, **11**, 112.

Simons, A. M., (2002). The continuity of macroevolution and microevolution. *Journal of Evolutionary Biology*, **15**, 688–701.

Simpson, G. G. (1944). *Tempo and Mode in Evolution*. New York: Columbia University Press.

  (1949; revised edition, 1967). *The Meaning of Evolution*. New Haven: Yale University Press.

  (1951). The species concept. *Evolution*, **5**, 285–298.

  (1952). Periodicity in vertebrate evolution. *Journal of Paleontology*, **26**, 359–370.

  (1953). *The Major Features of Evolution*. New York: Columbia University Press, 434 pp.

  (1961). *Principles of Animal Taxonomy*. New York: Columbia University Press.

  (1964). *This View of Life*. New York: Harcourt, Brace & World.

  (1965). *The Geography of Evolution*. New York: Capricorn.

Simpson, G. G. (1970). Uniformitarianism: an inquiry into principle, theory, and method in geohistory and biohistory. In *Essays in Evolution and Genetics*, ed. M. K. Hecht & W. C. Steere, pp. 43–96. New York: Appleton-Century-Crofts.

Singleton, O. P. (1968). Otway region. In *A Regional Guide to Victorian Geology*. Ed. J. McAndrew & M. A. H. Marsden, pp. 117–131. Melbourne: Geology Department, University of Melbourne.

Sloss, L. L. (1963). Sequences in the cratonic interior of North America. *Geological Society of America Bulletin*, **100**, 1661–1665.

Sloss, L. L., Krumbein, W. C. & Dapples, E. C. (1949). Integrated facies analysis. *Geological Society of America Memoir*, **39**, 91–124.

Smith, A. B. (1994). *Systemmatics and the Fossil Record: Documenting Evolutionary Patterns*. Oxford: Blackwell Sctientific Publication.

Spencer-Cervato, C., Lazarus, D. B., Beckmann, J.-P., von Salis Perch-Nielsen, K. & Biolzi, M. (1993). New calibration of Neogene radiolarian events in the North Pacific. *Marine Micropaleontology*, **21**, 261–293.

Spencer-Cervato, C., Thierstein, H. R., Lazarus, D. B. & Beckmann, J.-P. (1994). How synchronous are Neogene marine plankton events? *Paleoceanography*, **9**, 739–763.

Spero, H. J. (1992). Do planktic foraminifera accurately recordshifts in the carbon isotopic composition of sea water ΣCO_2? *Marine Micropaleontology*, **19**, 275–285.

Spero, H. J. & DeNiro, M. J. (1987). The influence of symbiont photosynthesis on the $\delta^{18}O$ and $\delta^{13}C$ values of planktonic foraminiferal shell calcite. *Symbiosis*, **4**, 213–228.

Srinivasan, M. S. & Kennett, J. P. (1976). Evolution and phenotypic variation in the Late Cenozoic *Neogloboquadrina dutertrei* plexus. In *Progress in Micropaleontology, Selected Papers in Honour of Prof. Kiyoshi Asano*, ed. Y. Takayanagi and T. Saito, 329–354. New York: Micropaleontology Press.

  (1981a). A review of Neogene planktonic foraminiferal biostratigraphy: applications in the equatorial and south Pacific. In *The Deep Sea Drilling Program: a Decade of Progress*, ed. J. E. Warme, R. G. Douglas & J. E. Winterer. Tulsa: Society of Economic Paleontologists and Mineralogists Special Publication **32**, 395–432.

(1981b). Neogene planktonic foraminiferal biostratigraphy and evolution: equatorial to Subantarctic South Pacific. *Marine Micropaleontology*, **6**, 499–533.

Srinivasan, M. S. & Sinha, D. K. (1991). Improved correlation of the Late Neogene planktonic foraminiferal datums in the equatorial to cool subtropical DSDP sites, southwest Pacific: application of the graphic correlation method. In *The World of Martin F. Glaessner*, ed. B. P. Radhakrishna. Geological Society of India, Memoir **20**, 55–94.

Stainforth, R. M., Lamb., J. M., Luterbacher, H., Beard, J. H. & Jeffords, R. M. (1975). *Cenozoic Planktonic Foraminifera Zonation and Characteristics of Index Forms*. Lawrence, University of Kansas Paleontological Contributions **62**, 1–245 (in two parts).

Stanley, S. M. (1979). *Macroevolution, Pattern and Process*. Baltimore: Johns Hopkins University Press, 332 pp.

Stanley, S. M., Wetmore, K. L. & Kennett, J. P. (1988). Macroevolutionary differences between the two major clades of Neogene planktonic foraminifera. *Paleobiology* **14**, 235–249.

Steininger, F. F. (1977). Integrated assemblage-zone biostratigraphy at marine–non-marine boundaries: examples from the Neogene of central Europe. In *Concepts and Methods of Biostratigraphy*, ed. E. G. Kauffman & J. E. Hazel, pp. 235–256. Stroudsburg, PA: Dowden, Hutchinson & Ross, Inc.

Steininger, F. F. & Papp, A. (1979). Current biostratigraphic and radiometric correlations of Late Miocene Central Paratethys stages (Sarmatian s.str., Pannonian s.str., and Pontian) and Mediterranean stages (Tortonian and Messinian) and the Messinian event in the Paratethys. *Newsletters in Stratigraphy,* **8**, 1000–110.

Steininger, F. F. & Proponents of Working Group (1994). Proposal for the Global Stratotype Section and Point (GSSP) for the base of the Neogene (the Paleogene–Neogene boundary). *IUGS/ICS/SNS/Working Group for the Paleogene–Neogene Boundary*. Vienna: Institute of Paleontology, University of Vienna, pp. 1–41.

Steininger, F. F., Aubry, M.-P., Berggren, W. A., *et al.* (1997). The global stratotype and point (GSSP) for the base of the Neogene. *Episodes*, **2**, 23–28.

Steininger, F. F., Berggren, W. A., Kent, D. V., Bernor, R. C., Sen, S., and Agusti, J. (1995). Circum-Mediterranean (Miocene and Pliocene) marine–continental chronologic correlations of European mammal units and zones. In *The Evolution of Western Eurasian Neogene Mammal Faunas*, ed. R. L. Bernor, V. Fahlbusch, V. Mittmann & S. Rietschel, pp. 23–46, New York: Columbia University Press.

Steininger, F. F., Rögl, F., Hochuli, P. & Müller, C. (1989). Lignite deposition and marine cycles: the Austrian Tertiary lignite deposits – a case history. *Öst. Akad. Wiss., Math.-nat. Kl.*, Abt.I, Bd 5:309–332.

Steininger, F. F., Rögl, F. & Martini, E. (1976). Current Oligocene/Miocene biostratigraphic concept of the Central Paratethys (middle Europe). *Newsletters in Stratigraphy*, **4**, 174–202.

Steininger, F. F., Senes, I., Kleemann, K. Rögl, F. (1985). *Neogene of the Mediterranean, Tethys and Paratethys*, 2 Vols, Wien: Universität Wien, Paläontologisches Institut.

Stott L. D. & Kennett J. P. (1990). Antarctic Paleogene planktonic foraminifer biostratigraphy: ODP Leg 113, Sites 689 and 690. In *Proceedings of the Ocean Drilling Program*, Scientific Results, 113, P. F. Barker, J. P. Kennett *et al.*, pp. 549–569, College Station, Texas.

Strasser, A., Hillgärtner, H., Hug, W. & Pittet, B. (2000). Third-order depositional sequences reflecting Milankovitch cyclicity. *Terra Nova*, **12**, 303–311.

Strasser, A., Pittet, B., Hillgärtner, H., Hug, W. & Pasquier, J.-B. (1999). Depositional sequences in shallow carbonate-dominated sedimentary systems: concepts for a high-resolution analysis. *Sedimentary Geology*, **128**, 201–221.

Subbotina, N. N. (1953). [Globigerinidae, Hantkeninidae, and Globorotaliidae. Fossil foraminifera of the U.S.S.R.] *Vses. Neft. Nauchno–Issled. Geol.-Razved. Inst. (VNIGRI), Trudy*, n.s. **6**, 9, 1–296, 41 pls.

Sylvester–Bradley, P. C., (1951). The subspecies in palaeontology. *Geological Magazine*, **88**, 88–102.

Tan Sin Hok (1932). On the genus *Cycloclypeus* Carpenter. I. *Wetensch. Meded.* 19. Dienst van der Mijnbouw.

(1936). Zur Kenntnis der Miogypsiniden. *De Ingenieur in Nederlandsch-Indië*, No. **3**(IV), 45–61, 84–98, 109–123.

(1937). Weitere Untersuchungen über die Miogypsiniden. *De Ingenieur in Nederlandsch-Indië*, No. **4**(IV), 35–45, 87–111.

(1939a). On *Polylepidina, Orbitocyclina* and *Lepidorbitoides. De Ingenieur in Nederlandsch-Indië*, No. **6**(5), 53–83.

(1939b). The results of phylomorphogenetic studies of some larger Foraminifera (a review). *Mijnbouw en Geologie*, **6**, 93–97.

(1939c). Remarks on the 'letter classification' of the East Indian Tertiary. *De Ingenieur in Nederlandsch-Indië*, **6** (7), **98**–101.

Tang, C. M. & Bottjer, D. J. (1996). Long-term faunal stasis without evolutionary coordination: Jurassic marine benthic paleocommunities, Western Interior, United states. *Geology*, **24**, 815–818.

Tappan, H. & Lipps, J. H. (1966). Wall structures, classification, and evolution in planktonic foraminifera. *American Association of Petroleum Geologists, Bulletin*, 50, 637.

Tedford, R. H. (1970). Principles and practices of mammalian geochronology in North America. In, *North American Paleontological Convention (Chicago, 1969), Proc., Part F (Correlation by Fossils)*, 666–703.

Teggart, F. J. (1925). *Theory of History*. New Haven: Yale University Press.

Teichert, C., (1958). Some biostratigraphical concepts. *Geological Society of America Bulletin*, **69**, 99–120.

Templeton, A. (1989). The meaning of species and speciation: a genetic perspective. In Otte & Endler.

Ten Kate, W. G. H. & Sprenger, A. (1993). Orbital cyclicities above and below the cretaceous–Paleogene boundary at Zumaya (N. Spain), Agost and Relleu (SE Spain). *Sedimentary Geology*, **87**, 69–101.

Thaler, L. (1965). Une échelle de zones biochronologiques pour les mammifères du Tertiaire d'Europe. *C. R. Somm. S.G.P.F.*, 1965, 118.

(1966). Les rongeurs fossiles du Bas Languedoc dans leur rapport avec l'histoire des faunes et la stratigraphie du Tertiaire d'Europe. *Mém. Mus. Nat. Hist. Natur., Nouvelle Séries., Séries C,* **17**, 1–295.

Thalmann, H. E. (1934). Die regional–stratigraphische Verbreitung der oberkretazischen Foraminiferen–Gattung Globotruncana Cushman 1927. *Ecologae Geologicae Helvetiae,* **27**, 413–428.

Thierstein, H. R., Geitzenauer, K., Molfino, B. & Shackleton, N. J. (1977). Global synchronicity of Late Quaternary coccolith datums: validation by oxygen isotopes. *Geology,* **5**, 400–404.

Thomas, E. (1992). Middle Eocene-late Oligocene bathyal benthic foraminifera (Weddell Sea): faunal changes and implications for ocean circulation. In *Eocene–Oligocene Climatic and Biotic Evolution,* ed. D. R. Prothero & W. A. Berggren, pp. 245–271. Princeton: Princeton University Press.

(1999). Introduction to 'Biotic responses to major paleoceanographic changes'. In *Reconstructing Ocean History: A Window into the Future,* ed. F. Abrantes & A. Mix, pp. 163–171. London: Plenum.

Thomas, E., Zachos, J. C. & Bralower, T. J. 2000. Deep-sea environments on a warm earth: latest Paleocene-early Eocene. In *Warm Climates in Earth History,* ed. B. T. Huber, K. G. MacLeod & S. L. Wing, 132–160. Cambridge University Press, Cambridge.

Tian, J., Wang, P., Cheng, X. & Li, Q. (2002). Astronomically tuned Plio-Pleistocene benthic $\delta^{18}O$ record from South China Sea and Atlantic–Pacific comparison. *Earth and Planetary Science Letters,* **203**, 1015–1029.

Tjalsma, R. C. & Lohmann, G. P. (1983). *Paleocene–Eocene Bathyal and Abyssal Benthic Foraminifera from the Atlantic Ocean.* New York: Micropaleontology, Special Paper No. 4.

Toulmin, L. D. (1977). Stratigraphic distribution of Paleocene and Eocene fossils in the eastern Gulf Coast regions. *Alabama Geological Survey Monograph 13,* vol. **1**, 602 pp.

Toumarkine, M. & Luterbacher H.-P. (1985). Paleocene and Eocene planktic foraminifera. In *Plankton Stratigraphy,* ed. H. M. Bolli, J. B. Saunders & K. Perch-Nielsen, pp. 87–154. Cambridge, Cambridge University Press, 2 volumes.

Van Couvering, J. A. & Berggren, W. A. (1977). Biostratigraphical basis of the Neogene time scale. In *Concepts and Methods of Biostratigraphy,* ed. E. G. Kauffman & J. E. Hazel, 283–306. Stroudsburg: Dowden, Hutchinson & Ross, Inc.

Trümpy, R. (1973). The timing of orogenic events in the central Alps. In *Gravity and tectonics,* ed. K. A. de Jong & R. Scholten, pp. 229–251. New York: John Wiley.

Vail, P. R., Audemard, F., Bowman, S. A., Eisner, P. N. & Perez-Cruz, C. (1991). The stratigraphic signatures of tectonics, eustasy and sedimentology – an overview. In *Cycles and Events in Stratigraphy,* ed. G. Einsele, W. Ricken & A. Seilacher, pp. 617–628. Berlin: Springer-Verlag.

Vail, P. R., Mitchum Jr, R. M., Todd, R. G., Widmier, J. M., Thompson III, S., Sangree, J. B., Bubb, J. N. & Hatlelid, W. G. (1977). Seismic stratigraphy and global changes of sea-level. In *Seismic Stratigraphy: Applications to Hydrocarbon Exploration,* ed. C. E Payton. American Association of Petroleum Geologists Memoir, **26**, 49–212.

Vakarcs, G., Hardenbol, J., Abreu, V. A., Vail, P. R., Várnai, P. & Tari, G. (1998). Oligocene–Middle Miocene depositional sequences of the Central Paratethys and their correlation with regiona stages. In *Mesozoic and Cenozoic Sequence Stratigraphy of European Basins*, ed. P.-C. de Graciansky, J. Hardenbol, T. Jacquin & P. R. Vail, SEPM (Society of Sedimentary Geology) Special Publication No. **60**, 209–232.

Valentine, J. W. & May, C. L. (1996). Hierarchies in biology and paleontology. *Paleobiology*, **22**, 23–33.

Van Bemmelen, R. W. (1949). *The Geology of Indonesia*. The Hague: Government Printing Office, 3 Volumes.

Van Couvering, J. A., Editor (1997). *The Pleistocene Boundary and the Beginning of the Quaternary*. Cambridge: Cambridge University Press.

Van Couvering, J. A., Castradori, D., Cita, M. B., Hilgen, F. J. & Rio, D. (2000). The base of the Zanclean Stage and of the Pliocene Series. *Episodes*, **23**, 179–187.

Van Couvering, J. A.; Hedberg, H. D. (1977). Review and response to review of: Hedberg, H. D., Editor, 1976, *International Stratigraphic Guide: A Guide to Stratigraphic Classification, Terminology and Procedure*, New York: John Wiley & Sons. *Micropaleontology*, **23**, 227–232.

Van der Vlerk, I. M. (1955). Correlation of the Tertiary of the Far East and Europe. *Micropaleontology*, **1**, 72–75.

Van der Vlerk, I. M. (1959). Problems and principles of Tertiary and Quaternary stratigraphy. *Quarterly Journal Geological Society of London*, **115**, 49–63.

Van der Vlerk, I. M. & Umbgrove, J. H. F. (1927). Tertiaire gidsforaminiferenvan Nederlandsch Oost–Indië. Dutch East Indies, *Dienst. Mijnb., Wetensch. Meded.*, No. 9.

Van Harten, D. (1988). Chronoecology, a non-taxonomic application of ostracods. In *Ostracoda in the Earth Sciences*, ed. P. DeDeckker, J.-P. Colin & J.-P. Peypouquet, p. 47–54. Amsterdam: Elsevier.

Van Harten, D. & van Hinte, J. E, 1984. Ostracod range charts as a chronoecologic tool. *Marine Micropaleontology*, **8**, 425–433.

van Hinte., J. E. (1969). The nature of biostratigraphic zones. In *Proceedings of the First International Conference on Planktonic Microfossils*, ed. P. Brönnimann & H. H. Renz. Leiden: E. J. Brill, vol. **2**, pp. 267–272.

Vaughan, T. W. (1924). American and European Tertiary larger foraminifera. *Bulletin Geological Society of America*, **35**, 785–822.

Vella, P. (1965). Sedimentary cycles, correlation, and stratigraphic classification. *Royal Society of New Zealand, Geology, Transactions*, **3**, 1–9.

Vénec-Peyré, M.-T. (2004). Beyond frontiers and time: the scientific and cultural heritage of Alcide d'Orbigny (1802–1857). *Marine Micropaleontology*, **50**, 149–159.

Vincent, E. & Berger, W. H. (1985). Carbon dioxide and polar cooling in the Miocene: the Monterey hypothesis. In *The Carbon Cycle and Atmospheric CO_2: Natural Variations Archaean to Present*, ed. E. T. Sundquist & W. S. Broecker W. S., pp. 455–468. Washington, D.C., American Geophysical Union, Geophysical Monograph 32.

Visser, W. A. & Hermes, J. J., Compilers (1962). *Geological Results of the Exploration for Oil in Netherlands New Guinea*. The Hague: Geologische serie, Nederlands Geologisch Mijnbouwkundig Genootschap, Special nummer, 265 pp.

Von Engelhardt, W. (1982). Neptunismus und Plutonismus. *Fortschritte der Mineralogie*, **60**, 21–43.

Vrba, E. S. (1980). Evolution, species and fossils: how does life evolve? *South African Journal of Science*, **76**, 61–84.

(1985). Environment and evolution: alternative causes of the temporal distribution of evolutionary events. *South African Journal of Science*, **81**, 229–236.

(1995). The fossil record of African antelopes (Mammalia, Bovidae) in relation to human evolution and paleoclimate. In *Paleoclimate and Evolution, With Emphasis on Human Origins*, ed. E. S. Vrba, G. H. Denton, T. C. Partridge & L. H. Burckle, pp. 385–424. New Haven: Yale University Press.

Vrba, E., Denton, G. H., Partridge, T. H. & Burckle, L. H., Editors (1995). *Environmental Change and Evolution*. Yale University Press, New Haven

Wade, M. (1964). Application of lineage concept to biostratigraphic zoning based on planktonic foraminifera. *Micropaleontology*, **10**, 273–290.

(1966). Lineages of planktonic foraminifera in Australia. In *Committee on Mediterranean Stratigraphy*, ed. C. W. Drooger, Z. Reiss, R. F. Rutsch & P. Marks, pp. 30–39. Proceedings of the third session in Berne, 8–13 June 1964. International Union of Geological Sciences Commission on Stratigraphy, Leiden: E. J. Brill.

Wagner, P. J. (1995). Stratigraphic tests of cladistic hypotheses. *Paleobiology*, **21**, 153–178.

(2000). Phylogenetic analyses and the fossil record: tests and inferences, hypotheses and models. In *Deep Time: Paleobiologys Perspective*, ed. D. H. Erwin & S. L. Wing. *Paleobiology*, Supplement to vol. **26(4)**, 341–371.

Wagner, P. J. & Erwin, D. H. (1995). Phylogenetic patterns as tests of speciation models. In *New Approaches to Speciation in the Fossil Record*, ed. D. H. Erwin & R. L. Anstey. pp. 87–122, Columbia University Press: New York.

Wakefield, M. I. & Monteil, E. (20002). Biosequence stratigraphical and palaeoenvironmental findings from the Cretaceous through tertiary succession, Central Indus Basin, Pakistan. *Journal of Micropalaeontology*, **21**, 115–130.

Walsh, S. L. (1998). Fossil datum and paleobiological event terms, paleontostratigraphy, chronostratigraphy, and the definition of land mammal 'age' boundaries. *Journal of Vertebrate Paleontology*, **18**, 150–179.

Webb, S. D. (1984). On two kinds of rapid faunal turnover. In *Catastrophes and Earth History*, ed. W. A. Berggren & J. A. van Couvering, pp. 417–436. Princeton: Princeton University Press.

Webb, S. D. & Opdyke, N. D. (1995). Global climatic influence on Cenozoic land mammal faunas. In *Effects of Past Global Change on Life*, ed. J. P. Kennett & S. M. Stanley, pp. 184–208. Washington, D.C.: National Academy of Sciences, Studies in Geophysics.

Weedon, G. P. (1993). The recognition and implications of orbital forcing of climatic and sedimentary cycles. In *Sedimentology Review*, ed. V. P. Wright, pp. 31–50. oxford: Blackwell.

Weedon, G. P. Shackleton, N. J. Pearson, P. N. (1997). The Oligocene time scale and cyclostratigraphy on the Ceara Rise, western equatorial Atlantic. *Proceedings of the Ocean Drilling Program, Scientific Results*, **154**, 101–116. College Station, Texas.

Wei K.-Y. (1994). Allometric heterochrony in the Pliocene–Pleistocene planktonic foraminiferal clade *Globoconella. Paleobiology*, **20**, 66–84.

Wei K.-Y. & Kennett, J. P. (1983). Nonconstant extinction rates of Neogene planktonic foraminifera. *Nature*, **305**, 218–220.

  (1988). Phyletic gradualism and punctuated equilibrium in the late Neogene planktonic foraminiferal clade *Globoconella. Paleobiology*, **14**, 345–363.

Wheeler, H. E. (1958). Time-stratigraphy. *American Association of Petroleum Geologists Bulletin*, **42**, 1047–1063.

Wiley, E. O. (1981). *Phylogenetics: The Theory and Practice of Phylogenetic Systematics.* New York: Wiley, 439 pp.

Wilson, D. (1993). Confirmation of the astronomical calibration of the magnetic polarity timescale from seafloor spreading rates. *Nature*, **364**, 788–790.

Wilson, E. O. (1998). *Consilience: The Unity of Knowledge.* New York: Knopf.

Wilson, R. C. L. (1998). Sequence stratigraphy: a revolution without a cause. In *Lyell: The Past is the Key to the Present*, ed. D. J. Blundell & A. C. Scott. Geological Society Special Publication No. **143**, 303–314.

Winchester, S. (2001). *The Map that Changed the World: The Tale of William Smith and the Birth of a Science.* London: Viking.

Wing, S. L. & Tiffney, B. H. (1987). The reciprocal interaction of angiosperm evolution and tetrapod herbivory. *Review of Palaeobotany and Palynology*, **50**, 179–210.

Wing, S. L., Sues, H.-D. *et al.* (1992). Mesozoic and early Cenozoic terrestrial ecosystems. In *Terrestrial Ecosystems Through Time: Evolutionary Paleoecology of Terrestrial Plants and Animals*, ed. A. K. Behrensmeyer, J. D. Damuth, W. A. DiMichele, R. Potts, H.-D. Sues & S. L. Wing, The Evolution of Terrestrial Ecosystems Consortium, pp. 327–416. Chicago & London: The University of Chicago Press.

Wolfe, J. A. (1978). A paleobotanical interpretation of Tertiary climates in the Northern Hemisphere. *American Scientist*, **66**, 694–703.

Wood, H. E. II, Chaney, R. W. Jr, Clark, J., Colbert, E. H., Jepsen, G. W., Reeside, J. B., Stock, C. (1941). Nomenclature and correlation of the North American continental Tertiary. *Geological Society of America Bulletin*, **52**, 1–48.

Woodburne, M. O. (1977). Definition and characterization of mammalian chronostratigraphy. *Journal of Paleontology*, **51**, 220–234.

  ed. (1987). *Cenozoic Mammals of North America: Geochronology and Biostratigraphy.* Berkeley: University of California Press.

Woodburne, M. & Swisher, C. (1995). Mammal high–resolution geochronology, intercontinental overland dispersals, sea level, climate and vicariance. In *Geochronology Time Scales and Global Stratigraphic Correlation*, ed. W. A. Berggren, D. V. Kent, M. -P. Aubry J. Hardenbol. Tulsa, SEPM (Society of Sedimentary Geology) Special Publication **54**, 335–364.

Woodburne, M. O., Tedford, R. H., Archer, M., Turnbull, W., Plane, M. & Lundelius, E. L. (1985). Biochronology of the continental mammal record of Australia and New Guinea. In *Stratigraphy, Palaeontology, Malacology, Papers in Honour of Dr Nell Ludbrook*, ed. J. M. Lindsay, pp. 347–365. Adelaide Department of Mines and Energy, Special Publication 5, D. J. Woolman, Government Printer.

Woodruff, F. & Savin, S. M. (1991). Mid-Miocene isotope stratigraphy in the deep sea: high-resolution correlations, paleoclimatic cycles, and sedimentary preservation. *Paleoceanography*, **6**, 755–801.

Worsley, T. R. & Jorgens, M. L. (1977). Automated biostratigraphy. In *Oceanic Micropalaeontology*, ed. A. T. S. Ramsay, p. 1201–1229. London: Academic Press.

Wright, J. D & Miller, K. G. (1993). Southern Ocean influences on late Eocene to Miocene deepwater circulation. In *The Antarctic Paleonviroment: a Perspective on Global Change*, ed. J. P. Kennett & D. A. Warnke, pp. 125–144. Washington, D.C.: Antarctic Research Series, Part Two, v. 60, American Geophysical Union.

Wright, J. D., Miller, K. G. & Fairbanks, R. G. (1992). Early and Middle Miocene stable isotopes: implications for deepwater circulation and climate. *Paleoceanography*, **7**, 357–389.

Young, K. (1960). Biostratigraphy and the new paleontology. *Journal of Paleontology*, **34**, 347–358.

Zachos, J., Arthur, M. A. & Dean, W. E. (1989). Geochemical evidence for the suppression of pelagic marine productivity at the Cretaceous–Tertiary boundary. *Nature*, **337**, 61–64.

Zachos, J. C., Pagani, M., Sloan, L., Thomas, E. & Billups, K. (2001b). Trends, rythyms, and aberrations in global climate 65 Ma to Present. *Science*, **292**, 686.

Zachos, J. C., Quinn, T. C. & Salamy, K. C. (1996). High-resolution (10^4 years) deep-sea stable isotope records of the Eocene–Oligocene transition. *Paleoceanography*, **11**, 251–266.

Zachos, J. C., Shackleton, N. J., Revenaugh, J. S., Pälike, H. & Flower, B. P. (2001a). Periodic and non–periodic climate response to orbital forcing across the Oligocene–Miocene boundary. *Science*, **292**, 274–277.

Zalasiewicz, J., Smith, A., Brenchley, P., Evans, J., Knox, R., Riley, N., Gale, A., Gregory, F. J., Rushton, A., Gibbard, P., Hesselbo, S., Marshall, J., Oates, M., Rawson, P. & Trewin, N. (2004). Simplifying the stratigraphy of time. *Geology*, **32**, 1–4.

Zinsmeister, W. J. & Feldmann, R. M. (1984). Cenozoic high latitude heterochroneity of southern hemisphere marine faunas. *Science*, **224**, 281–283.

Index